Advanced Information Processing

Springer
Berlin
Heidelberg
New York
Hong Kong
London
Milan
Paris
Tokyo

S. M. Deen (Ed.)

Agent-Based Manufacturing

Advances in the Holonic Approach

With 152 Figures and 17 Tables

Springer

S. M. Deen
Department of Computer Science
University of Keele
Keele
Staffs. ST5 5BG
UK
deen@cs.keele.ac.uk

Cataloging-in-Publication Data applied for
A catalog record for this book is available from the Library of Congress.
Bibliographic information published by Die Deutsche Bibliothek.
Die Deutsche Bibliothek lists this publication in the Deutsche Nationalbibliografie; detailed bibliographic data is available in the Internet at http://dnb.ddb.de.

ACM Subject Classification (1998): I.2, J.2, J.6

ISBN 3-540-44069-0 Springer-Verlag Berlin Heidelberg New York

Springer-Verlag Berlin Heidelberg New York
a member of BertelsmannSpringer Science+Business Media GmbH

Printed in Germany

Cover Design: KünkelLopka, Heidelberg
Typesetting: Computer to film by author´s data
Printed on acid-free paper 45/3142YL 5 4 3 2 1 0

Foreword

In writing this Foreword, I recall the start of the IMS (Intelligent Manufacturing Systems) programme, founded on *international cooperation* in research and development in manufacturing technology. I first made the proposal for such an international programme in informal factory automation meetings in Japan between 1988 and 1989. It called for a change of attitude from a purely competitive stance to one that demanded both *disclosure* of manufacturing know-how within companies and *cooperation* to develop the future technologies on a sound basis.

There were of course much uncertainty. The demand of disclosure and cooperation goes against the very ethos of a free economy, it weakens the incentives for technological progress, and above all it requires the corporate managers to commit an apparent heresy – to disclose manufacturing know-how to the competitors. After much debate and deliberation, a decision was taken in 1989 to draft the IMS programme, as a ten-year 150 billion yen venture in the form of an international consortium for research and development in manufacturing. The United States and the European countries were invited to participate. At first international discussions were conducted at the inter-company level, but later the Governments of those countries took the view that international joint research in manufacturing technology should be conducted under close Government scrutiny. This change of perspective led to many redraftings of the proposal and also to the establishment of many structures to control and guide the programme. Australia, Canada and the EFTA countries also joined in.

The programme was launched with a number of projects as a feasibility study in 1993, which was subsequently followed by the full-scale programme. I personally remained in touch with its activities as a member (and rotating chairman) of its International Steering Committee.

Holonic Manufacturing Systems (HMS), with the objective of low-volume high-variety manufacturing in an agile environment, was one of the original IMS projects in the feasibility study and has remained an important project ever since. The HMS project has been a successful consortium, and I gained some first hand knowledge of its activities when I attended one of its international meetings at Kitakiyushu (Japan) in the year 2000. I was very impressed by the work the members had done and the international cooperation (which was unthinkable a few years ago) they had achieved. Of the various research and development topics covered internationally in the HMS project, distributed manufacturing provided the general foundation.

This book focuses on the aspects of distributed manufacturing where holons, as autonomous units of production, cooperate effectively to manufacture products in a robust, scalable, stable and dynamically reconfigurable environment – with agility, adaptability and flexibility. This is undoubtedly an area of significant new developments in manufacturing science in the 21st Century. I understand that this area is of particular interest also to the researchers and developers in multi-agent systems, where manufacturing applications offer a formidable challenge. The articles in the book address both practical and theoretical issues with many invigorating ideas and approaches, which should provide new inspiration to explore not only the individual agent behaviour, but also the emergent behaviour of multi-agent systems.

As implied above, the book presents real-world problems and suggests possible solutions, which should stimulate all those researchers and developers interested in multi-agent systems and future manufacturing. I recommend the book to all such persons in industry, universities and research centres.

Tokyo, Japan
April 2003

H. Yoshikawa
Professor
President of the Science Council of Japan,
and also President of the National Institute of
Advanced Industrial Science and Technology, Japan

Preface

This book is a collection of advanced articles covering in a comprehensive fashion various aspects of agent-based manufacturing, drawing from the research work carried out in the international project called Holonic Manufacturing Systems (HMS) from 1993 onward. This HMS project is one of several initiatives undertaken in an international programme called IMS (Intelligent Manufacturing Systems (IMS), as explained in the Foreword by Professor Yoshikawa.

While the book is intended to report the major findings of the HMS project on agent-based manufacturing, it is not an official HMS/IMS publication. Its contents reflect the choice the editor and the contributors have made. It should also be stated that the HMS project included also other excellent work on various aspects of manufacturing, which in the opinion of the editor fell outside the scope of agent-based manufacturing, and hence does not appear in this volume. The editor is wholly responsible for this choice.

Today multi-agent systems (MAS) are an important research area in Computer Science and Distributed Artificial Intelligence (DAI). The field of agent-based manufacturing systems offers a major challenge for the MAS researchers, as it has to deal with both logical and physical objects. For example, in the event of a malfunction, a logical object can be logically discarded from the operational environment and the software itself restarted. But this is not feasible with physical objects – they may have to be inspected by human beings for damage and physically removed from the operational environment. In addition the offending machine may have to be replaced and the objects rerouted to another physical location. A malfunction in one part of a production line can cause partially completed objects to tumble over each other, thus creating a major operational catastrophe. Therefore the issues are far more complex in systems that are involved in the processing of physical objects.

The MAS/DAI researchers often like to employ human mental categories in the form of Belief, Desire and Intention (BDI) as the principal problem-solving concepts, usually within a deliberative approach involving symbolic logic processing. Some MAS researchers, such as Rodney Brooks, prefer the alternative reactive approach, while some other multi-agent researchers have constructed hybrid approaches, mixing the deliberative with the reactive. Much of manufacturing seems to be better suited to the reactive treatment – at any rate, the papers presented in this volume follow largely the reactive path. It should be of interest to the multi-agent researchers to see how the top researchers in agent-based manufacturing are examining and providing solutions to real-world problems. They would find much in this book to grapple with.

The 21st Century manufacturing is meant to be offering choices to the customers, which in turn requires (i) low-volume high-variety manufacturing to handle those choices, (ii) dynamic shopfloor reconfiguration to meet new requirement flexibility and (iii) an agile environment to respond to changes and new demands quickly. These facilities, it is generally believed, can be best offered by agent-based manufacturing – the central justification for undertaking the HMS project. Observe that HMS was (is) an industry-led project to produce ideas that can lead to product developments. Therefore this book should appeal to the researchers, advanced thinkers and innovators in industry, academia and research centres in manufacturing – in addition to those interested in the multi-agent approach as a research topic.

Acknowledgements

First I would like to thank my colleagues in the HMS projects who have been able to contribute to this book, despite their busy commitments elsewhere during that period. I am also indebted to the members of my research group, particularly to James Cole, Rashid Jayousi, Thomas Neligwa, Kapila Ponnamperuma and Ryad Soobhany for helping me with much of the administrative and editorial tasks. In particular, Rashid Jayousi spent an enormous amount of time in editing the chapters to ensure a uniform standard. Finally and very importantly, I am especially thankful to Prof Yoshikawa for the Foreword endorsing this book.

University of Keele, England — *S. M. Deen*
April 2003 — *Editor*

Table of Contents

Part IV: Application Environments and Issues

Part I

Starters

1. Introduction

S. M. Deen

Department of Computer Science
University of Keele
England
Email: deen@cs.keele.ac.uk

1.1 The Idea

As explained in the Foreword the idea of an international programme on Manufacturing Science was originally initiated in 1989 by Professor H. Yoshikawa of Japan, the then Professor of Precision Engineering at the University of Tokyo. His 1989 paper states

> The intelligent manufacturing system takes the intelligent activities in manufacturing and uses them to better harmonise human beings and intelligent machines ... integrating the entire corporation from marketing through the design, production and distribution, in a flexible manner, which improves productivity.

He elaborated his ideas further in a paper in 1993 [1.1], in which he described the technical knowledge production process as a four-stage process shown diagrammatically as in Fig. 1.1.

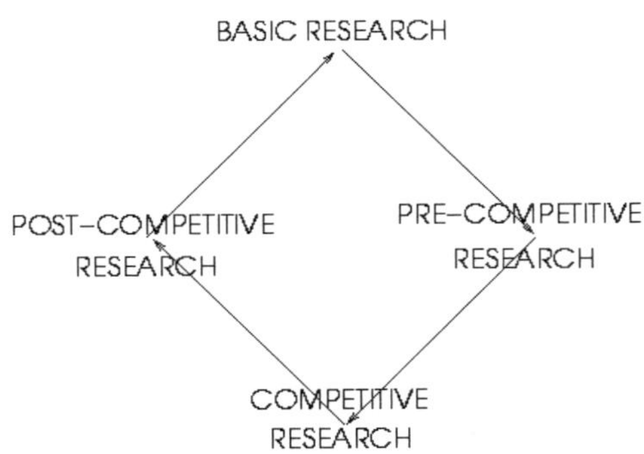

Fig. 1.1. Technical Knowledge Production Process

In this view, basic research when considered for applications leads to pre-competitive research, which in turn leads to competitive (private) research for product development. This is followed by demand for new products that requires post-competitive research, which is public research. When post-competitive research loses its effectiveness in innovation, the market looks for new ideas from new basic research. The cycle is thus completed. This scenario of the knowledge production cycle has inspired the launch of the IMS initiative, the largest international collaborative programme in manufacturing science for pre-competitive research.

In the rest of this chapter, we shall provide a background to the HMS project and an overview of the structure and content of the book in order to give the reader a clearer and comprehensive picture of the themes in and the inter-relationships of the topics covered in the book.

1.2 The HMS Project

The HMS (Holonic Manufacturing Systems) project for low-volume high-variety production with agility was one of the five IMS projects that began in 1993 with the feasibility study phase. Low-volume high-variety products would include things, such as personalised cars, jet engines, or lifts (elevators in the USA) in hotels, and agility implies an ability to respond to changes quickly. The feasibility phase for the HMS project proved to be very successful, demonstrating that international collaboration on such a large scale was not only feasible, but it could also flourish. (*Aside*: it was interesting to observe how Australians, Japanese, Europeans, Canadians and Americans could all deliberate together on research issues without regional bias, but with real passion and on first-name terms!)

Following the successful completion of the feasibility study, the full-scale programme was launched in 1996. Phase I of this full-scale programme was completed successfully in 2001, with about 30 partners, including the following major industries:

- Australia [BHP, …]
- Japan [Toshiba, Hitachi, Fanuc, Yaskawa-Electric, ...]
- Europe [Atos, DaimlerChrysler, ...]
- USA [Rockwell Automation, …]

In addition to these giants, there were many small and medium scale industries, research centres and universities from Australia, Japan, Europe, Canada and the USA. This book is based on the research results of this Phase I – which was the principal research phase of the HMS project.

Holons and Agents

The term *holon* was originally coined by Arthur Koestler [1.2] as a concept to embody the characteristics of hierarchies, which for him were relativity and ambiguity as in "part" and "whole" relationships when applied to subassemblies. For this, he combined the Greek word *holos* meaning "whole", with *on* to imply part[icle] as in neutr*on*, prot*on* and mes*on*. He viewed a holon as an autonomous, cooperative and self-reliant entity with fixed rules and flexible strategies, applicable to biological, social, organisational and other hierarchies. Human beings are the lowest-level holons. Given this socio-biological concept of a holon, one can associate with it some additional interesting properties desirable in manufacturing, such as efficiency in resource-utilisation, flexibility in applications, adaptability to changes, agility to challenges and stability against disruptions. The term *holon* was suggested by Suda [1.3] to characterise what became the HMS project, since many of the properties described above are directly applicable to manufacturing environments.

As the project progressed, a holon was redefined as an autonomous and cooperative unit of production, with a compulsory software component and an optional hardware component, but without any human element in it. The term "intelligence" used in the earlier definition was dropped after some discussions, as this property cannot be specified precisely. The notion of a holon being made up of other holons was retained, but such compound holons were not studied to any significant degree. In fact a holon can be regarded as a special kind of large-grain reactive agent, familiar in multi-agent systems. A holonic manufacturing system is thus a kind of agent-based manufacturing system. In some of the chapters in this book the terms holon and agent have been used interchangeably. Another term *holonics* used in this book means *the study of holons*. Some other terms, such as *holonomy*, enthusiastically defined earlier in the project, seem to have fallen out of use.

Workpackages and the Book

The work in the HMS project was divided into two types by workpackages: Generic and Application-oriented. The generic work included systems architecture and systems operation as the generic facilities. The application-oriented work focused on well-defined application systems to solve problems in those systems. It was quite obvious from the beginning of the project that the time and resource constraints would not permit the deployment of the results from generic workpackages into those application systems, and therefore each application developed its own approach as suited to its needs. All these different approaches were intended to be (and in fact were) evaluated for their suitability at the end of the project.

Some of the work in the project (e.g. machining) did not deal with the distributed aspects of manufacturing, and hence are not included in this book on agent-based manufacturing. The works presented here are individual (or group) contributions, and *not* an official project view. Researchers (individually or in groups) presented their work and views as they thought appropriate, without being constrained follow a sanitised project view. Therefore this book has preserved their insights and the diversity of their approaches with the full richness of the contributions. This will permit the reader to get a glimpse of the real issues, problems, concepts and solutions as perceived by the top researchers themselves, without any artificial imposition.

1.3 Book Overview

For convenience of presentation, the book is divided into four parts relating to the following topics:

- Starters
- Systems Architecture
- Systems Operations
- Application Environments and Issues

Each part has a number of chapters, also referred to as papers below.

Part I: Starters

Apart from this Introduction, we have included two other chapters in this start-up part, which examines the advanced manufacturing context within which the holonic approach falls:

- Chapter 2: From Fractals and Bionics to Holonics [Dr A. Tharumarajah]
- Chapter 3: From FMS to HMS [Dr R.W. Brennan and Prof D.H. Norrie]

Chapter 2 explores some new concepts for future manufacturing, which include fractals and bionics, and discusses the holonic concepts within that perspective. As stated in the Introduction of that chapter: "Future manufacturing companies to stay competitive have to have the ability to respond quickly, responsibly and cost-effectively to the unmistakable trend towards high product variety ...". A fractal can produce self-similar patterns recursively with more details. The concept of a fractal factory exploits the abstract property of self-similarity in the product and production hierarchy. The fractals are considered to be able to adapt dynamically to a turbulent environment without any formal organisational structure – a clear advantage in manufacturing. Bionic manufacturing draws its inspiration from biological systems, and attempts to exploit features of biological systems, such as autonomy, spontaneity and social harmony within hierarchically ordered relationships, using the cell as the basic unit. The instructions can perhaps be coded in an equivalent of a DNA structure. These features of fractals and bionics are intended to be captured in holonic manufacturing in a practical way. The chapter closes with some discussions on issues and aspirations.

Chapter 3 examines recent trends in manufacturing, particularly tracing the development from Flexible Manufacturing Systems (FMS) to Holonic Manufacturing Systems (HMS), by the experts in the field. It starts with car production. I suppose we all have heard about the slogan of Ford Model T cars: "You can have it in any colour as long as it is black". This perhaps epitomised the philosophy of mass-production and production efficiency over consumer choice in the early 20th Century. The chapter explains why mass production had to give way to lean production in the seventies. Along with the concept of "continual improvement", CAD/CAM developments, more automation, the need for "intelligent" adaptation, and then FMS for more desirable small batch sizes. The chapter goes on to explore some of the current concepts in agent-based manufacturing, placing HMS in context. Do you ever wonder how Japan came to dominate the automotive industry? It's all there.

Part II: Systems Architecture

This part focuses on the architectural issues and provides three different perspectives.

- Chapter 4: HMS/FB Architecture and Its Implementation
  [Dr J. H. Christensen]
- Chapter 5: FIPA Standards and Holonic Manufacturing
  [Prof V. Mařík, Dr M. Pěchouček, Dr P. Vrba and V. Hrdonka]
- Chapter 6: Towards a Formalised HMS Model [Dr C. A. Johnson]

The IEC standard 61499 (in which Rockwell Automation has a major stake) has defined a distributed controller architecture for manufacturing, using a concept of Function Blocks (FBs), which are special-purpose procedures. This standard was then adapted to develop an HMS control architecture referred to here as the HMS/FB architecture. This architecture has two levels: (a) the lower-level, which is fully defined and deals with the intra- and inter-holon control of physical machine operations and (b) the upper-level, which is only partly defined and deals with things such as inter-holon cooperation and negotiation on task executions. The chapter concentrates principally on the lower-level and describes an open standard-based architecture for HMS that is capable of fulfilling the economic and technical requirements for global adoption and deployment.

As is well-known in the multi-agent community, FIPA (Foundation for Intelligent Physical Agents) is interested to develop standard facilities for multi-agent applications, among which manufacturing is deemed as a key domain. FIPA has introduced a systematic approach to development and maintenance of specifications and standards for multi-agent application domains including interoperability. Prof V. Marik and his group are developing a FIPA version for the HMS architecture, focusing on the upper-level of control including inter-holon interoperability. The paper explores how the HMS architecture can gainfully employ the ideas, procedures and standards developed within FIPA for its implementation. This study leads the authors to advocate the deployment of the FIPA abstract architecture as a basis for holonic software solutions.

Although Chapter 4 and 5 complement each other to some extent, they are geared towards specific approaches. What is also required is an overall abstract model for HMS that can provide not only a sound theoretical foundation from which other ideas can be sprouted, but also a formal basis from which different but equally valid implementation versions can be derived by different approaches. Additionally such a model can be used to study the logical correctness and consistency of the various concepts, procedures, processes and interactions in the failure-prone distributed environment. If a specification is found to be logically incorrect or inconsistent, then it must be changed. Dr Chris Johnson has presented such a model in Chapter 6 in his usual lucid and precise style. The HMS/FB model can be viewed as a partial implementation version of this theoretical foundation. Anyone interested in the theoretical meat of the HMS architecture will find plenty here to munch on.

Part III: Systems Operation

The chapters in this part describe some of the operational features of distributed manufacturing as outlined below.

- Chapter 7: An Investigation into a Computational Model for HMS [Prof S. M. Deen]
- Chapter 8: An HMS Operational Model [T. Neligwa and Dr M Fletcher]
- Chapter 9: Holonic Diagnostics for an Automotive Assembly Line [Dr D.H. Jarvis and J.H. Jarvis]

Since the HMS environment is complex and highly distributed, the potential for something going wrong during operations is naturally high. A computational model, presented in Chapter 7, specifies the correct systems behaviour through all its operational stages (particularly failures) during execution The system must guarantee that an execution carried out in the normal error-prone environment is equivalent to that carried out in an error-free environment, in spite of system malfunctions. The execution must terminate in a finite time at a recoverable point, and must guarantee stability (*cf.* the neural network in the human brain). A system must have structures and rules that ensure these behaviours. Chapter 7 examines the issues and suggests structures and rules to ensure correctness, termination and stability.

The second paper in this part presents an HMS operational model and examines its systems behaviour and effectiveness. This operational model is constructed by blending ideas from both the HMS/FB model in Chapter 4 and the formalised HMS model in Chapter 6. The paper examines the operational requirements along with constraints and systems dynamics, and develops some important concepts. The paper also studies the operational dynamics with respect to performance, robustness and systems effectiveness, using two case-examples: (i) a case study of motor assembly, demonstrating the performance advantages of the basic HMS approach, and (ii) a more general case study of a production shopfloor, demonstrating systems robustness (including fault-tolerance) and systems effectiveness (in terms of throughput) with a more elaborate HMS approach. For the second approach, the authors develop and employ what is called *a temperature model* and show how the operational environment with inter-linked holons can maintain both systems robustness and effectiveness. If you are interested to find out how the HMS model might work on a shop-floor and the operational advantages it is meant to offer, then this is the chapter to read.

The third paper addresses the diagnostics issues in HMS. Diagnostics is an essential function of a holonic manufacturing system, since without it we cannot provide any degree of flexibility, adaptability or stability. A diagnostics approach can be built by incorporating the desirable holonic behaviours (such as those listed above) into operational industrial systems. In the absence of such an operational system (as is the case now), one can develop the ideas on a conventional system as is done in this paper. The paper presents a diagnostic model with devices, sensors and their states, which was verified with four subsystems (holons). While concluding in favour of a model based approach as being most appropriate for most applications, the paper also warns against applying it to all applications, since some applications can benefit more from other approaches (such as conventional rule-based or neural networks).

Part IV: Application Environments and Issues

This part of the book describes some important developments in the application of agent-based approaches in the HMS project. The HMS project never claimed to have the whole truth. As it was a large project, it very wisely encouraged many ideas to flourish, so that future researchers and developers can have choices. Partly because of this philosophy and also because of the time constraints, the applications (as presented below) did not and could not use the results (models) described in Parts II and III. Instead, each used an agent-based approach that seemed most appropriate to the needs of the application concerned. The contributions in this part are:

- Chapter 10: HMS Development and Implementation Environments [Dr S. Tamura, Dr T. Seki and T. Hasegawa]
- Chapter 11: Transport Agents [A. Ritter, A. Braatz, M. Höpf, [C. Schaeffer and Prof E. Westkämper]
- Chapter 12: A Holonic Shot-Blasting System [Dr T. Heikkilä, L. Rannanjärvi, M. Sallinen and M. Rintala]
- Chapter 13: Holonic Manufacturing Control: Rationales, Developments and Open Issues [Dr D. C. McFarlane and S. Bussmann]

The first paper describes a major application development carried out in Toshiba, with two main elements: an HMS Shell and JDPS. The HMS Shell is a GUI-based design tool for the cooperation mechanism that facilitates inter-holon decision-making in a transparent way, which retains the autonomy of the application programs (holons). It also permits easy dynamic systems reconfiguration and incremental developments. JDPS is meant to provide the HMS implementation and execution environment, based on a reliable broadcast messaging. Holons can enter or leave the system during execution and can adapt to different roles, so that they can contribute to the varying objectives in the most effective way. The need to support these nondeterministic behaviours in a rigorous way gave rise to some new issues which required development and implementation of new strategies, as discussed in the paper.

The second paper addresses mobile holons for transportation, particularly automated Guided Vehicles (AGVs) developed by IPA/FhG, a leading European research centre in engineering. Mobile agents have been visualised, not only as those working in a manufacturing shopfloor, but also as those which can work in hostile remote environments, such as examining hot nuclear reactors and detecting landmines. Communication, control, coordination, reliability, robustness – are all major issues requiring investigation. The paper also considers architectures, agent models and prototyping tools.

The Holonic Shot-Blasting System described in the next chapter was another major multi-agent application development carried out jointly by a number of researchers from academia, research centres and industry. Shot blasting is dangerous and dusty work for human beings, and hence the increasing usage of robots in it. The blasting process is dynamic, with dynamically changing parameters. The recycling of the blasting material causes particle size to change and tools to suffer from gradual wear. Therefore it is a challenging application domain for holon-based robots. The shot-blasting system described is a flexible cooperative multi-robot application for physical surface treatments (e.g. to polish or paint a physical surface), based on a multi-layered holon and control structure, permitting dynamic changes in cooperation and coordination. This is a major real-world application prototype on a multi-agent theme which includes also strategic human inputs.

The final chapter of this part and of the book reviews the nature of holonic control in terms of rationality and development, and it invites the readers to examine some of the open issues. The authors speak from real experience: the first author worked in the HMS project earlier as a researcher at BHP, Australia, while the industrial credentials of the second author need no elaboration from me. Even though, a part of their observations is based on a specific architecture, this is only a small part. The substance of their observations has a general validity transcending all architectures and models, and therefore these observations should be taken into account in all future research and development in this area.

References

[1.1] H. Yoshikawa: "Intelligent Manufacturing Systems ...", in *Information Infrastructure Systems for Manufacturing* (IFIP transactions B-14), eds: Yoshikawa and Goossenaerts, Elsevier, London (1993), p 19.
[1.2] A. Koestler: *The Ghost in the Machine*, PAN Books, London (1967).
[1.3] H. Suda: "Future Factory Systems Formulated in Japan", Techno Japan **23**(3), (1990).

2. From Fractals and Bionics to Holonics

A. Tharumarajah

CSIRO Manufacturing Science & Technology
Preston, Victoria, Australia
Email: Rajah.Tharumarajah@csiro.au

Abstract. Future manufacturing, to stay competitive, should possess the capabilities to permit much faster response to changes in the business and technology environment. Companies should be able to transform their operations, organisation and technology at much shorter notice than what is prevalent today. A promising structure, in this respect, would be organic in nature and would resemble a conglomerate of distributed and autonomous units. The units would have the ability to self-determine their actions while communicating and cooperating with others to steer the manufacturing system towards meeting the challenges placed upon the organisation. This paper examines the fractal, bionic and holonic concepts that advocate such organic structures. The concepts are first explained and the application in manufacturing noted. Comparison of these concepts with respect to their design, operation and self-organising abilities is undertaken. This comparison reveals a rich insight into the similarities and differences among them that can be fruitfully exploited to fill any shortcomings in one or another concept.

2.1 Introduction

Future manufacturing companies, to stay competitive, have to have the ability to respond quickly, responsibly and cost-effectively to the unmistakable trend towards high product variety, shortening technology and product life cycles, smaller numbers of repeat orders and smaller batch sizes. In production, this would translate as the ability of a manufacturing system to change its capability at will, create additional capability if required, adapt its internal organisation to support and respond to the changes, migrate from old to new capability, and ensure a performance that continuously supersedes its competition in product choice, price, quality and delivery.

The type of manufacturing system that would satisfy these requirements would have to be agile, in that it would possess typical characteristics of being flexible, adaptive and responsive to changes in the environment a company operates in. Ideally, an agile manufacturing system would construct and reconstruct itself in response to changes in product demands and technology offerings while creating new market opportunities. To appreciate what these changes mean to an operating enterprise, the concept of agile manufacturing can be compared with that of lean manufacturing. Lean manufacturing is concerned with reducing waste or fat in an organisation, including non-value-added activities and unnecessary inventory and resources, through the application of just-in-time, concurrent engineering, total quality management and improved customer-supplier relationships. As a result, the operation becomes more efficient, increasing yield and reducing costs and lead times. However, the improvements achieved through becoming lean or more efficient do not by themselves result in an enterprise becoming more responsive or gaining an ability to change its internal organisation or technology at will, as is envisaged by agile-manufacturing concepts.

However, while there is a general agreement that a manufacturing system has to be agile, that is, it should be flexible or changeable in what it does and how it does it, there is still a search for the most suitable structure. A promising structure, in this respect, would be organic in nature very similar to a conglomerate of distributed and autonomous units. These units, while self-determining their actions, would communicate and cooperate with others to carry out the expected actions and pursue goals both in isolation and in a group. Through these concerted actions, the assemblage would function very similarly to a self-organising whole, where the challenges of a variable environment would be met by internal regulation and modification of functions, and, if needed, of the content and form of the organisation.

In this paper we examine the emerging fractal, bionic, and holonic concepts that advocate such organic manufacturing structures, and compare the concepts with respect to their design, operation and self-organising abilities. The next three sections review the fractal, bionic, and holonic concepts of manufacturing. Following this, the concepts are compared with respect to the three aspects alluded to above. Finally, conclusions are drawn. Since the proposed concepts are quite general in their scope and are currently being developed further to realise their practical potential, the discussion and comparison are undertaken in general terms.

2.2 Fractal Factories

The term "fractal" comes from fractal geometry for describing and analysing objects in multi-dimensional spaces, and is specially focused on the fractional dimension where Euclidean geometry is not suitable. The main characteristic of fractals is self-similarity. That is to say, the overall pattern is repeated in the details of the pattern. The fractal pattern in Mandelbrot's sets [2.4] display self-similarity, because they not only produce detail at finer scales, but also produce details with certain constant proportions or ratios, though they are not identical. The self-similarity can also be interpreted as self-referencing, i.e. a fractal exists only in relation to itself, meaning the underlying structure is a fractal itself and so on. Self-similarity is then a property of the object and not of the steps taken to build the object [2.24]. This can also be viewed as a property where a fractal is made of parts similar to the whole.

The concept of fractal factories (FFs, [2.41]) draws on such features of fractals. Consider a general description of a unit in manufacturing as providing service (i.e. production or some output) to other units. The structures become similar, as shown in Fig. 2.1. However, each unit may arrange its internal structure differently to focus on one or more criteria, such as product, processes or materials. For example, in structuring a production shop, concentration on processes and materials brings about familiar job shop production, while focus on products leads to concentration on product-centred manufacture. However, both structures are similar in their production function in the performance of service and pursuit of goals.

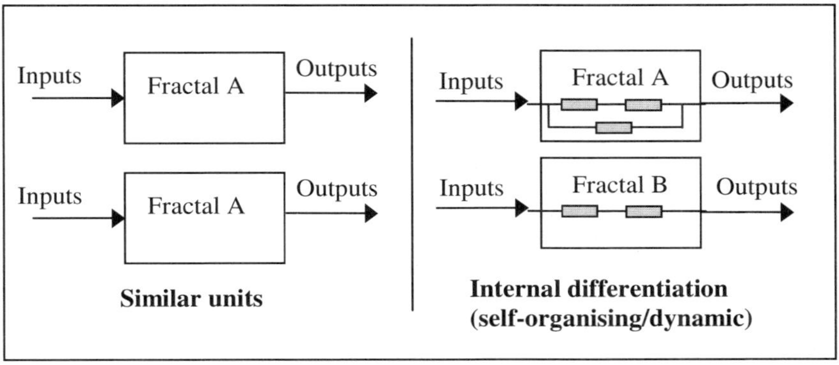

Fig. 2.1. Similar properties of production units

Based on the self-similar features of fractals, the concept of fractal factories proposes a manufacturing company to be composed of small units, or *fractal entities*. But what will be the features of such fractal entities? Strauss and Hummel [2.29] offer three basic features: self-organisation, dynamics and self-similarity. Self-organisation is the freedom that the entities have in organising, executing and regulating tasks and, in some cases in creating the necessary freedom and responsibility for results. The entities may choose their own methods of problem-solving and self-optimisation, such as for improving processes. The second feature, of dynamics, points to the ability of fractals to adapt to turbulent influences from the environment without any formal hindrance of organisational structure. Finally, self-similarity among the fractals conveys a similarity of goals [2.26] and hence the need for the entities to be aligned with a common target system [2.42]. It does not necessarily imply similarity in the internal organisation of the fractals.

In realising the above features, emphasis is given to factory fractals acting independently on the one hand, and as a coherent whole on the other. However, for them to function as a coherent whole, goal consistency should be maintained. The goal formation and revision process, shown in Fig. 2.2, is envisaged as a process of par-

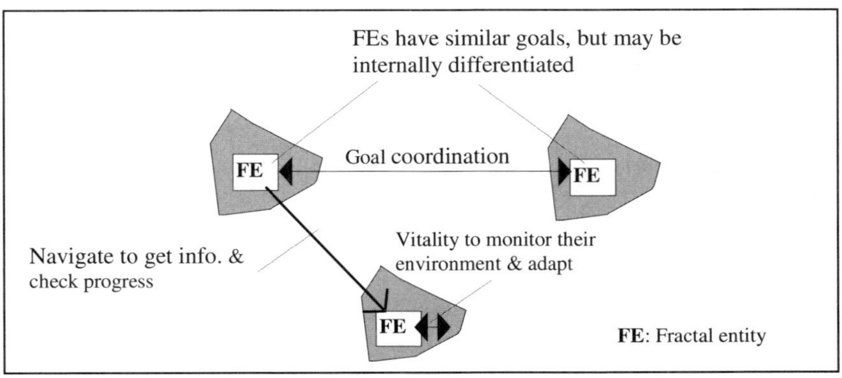

Fig. 2.2. Goal formation process among fractals

ticipation and coordination among the fractals, supported by an inheritance mechanism to ensure consistency of the goal system. In fact, according to Strauss and Hummel [2.29], fractals are always structured bottom-up, building fractals of a higher order. Units at a higher level always assume only those responsibilities in the process that cannot be fulfilled in the lower-order fractals. This principle guarantees teamwork among the fractals and also forces distribution of power and ability.

In operation, the characteristic of *vitality* is proposed as a means for the fractals to adapt and react to the influences of their respective environments. As Sihn [2.25] points out, using this characteristic, a fractal is able to record and evaluate changes in six dimensions of a system's environment (i.e. six sub-environments): cultural, strategic, socio-informal, financial, informational and technological. Response to changes in these dimensions includes dynamic (re)structuring of groupings in the interest of serving the whole company. A navigation system that constantly checks target areas is proposed as supporting the fractals to monitor and act on changes.

While the above-discussed features of fractal entities provide a generic description of the desired capabilities, its application requires some specialisation of these entities to represent such items as machines, transport, orders and so on. In fact, such specialisation can be made easier by defining a basic fractal entity [2.34] having the generic features and using it as a base class to define specialised fractal entities. As illustrated in [2.27], a multi-level architecture can be defined that has specialised entities occupying the strategic, tactical and operational layers. Machines, transport system, and storage can occupy the operational level, shop floor and logistics can control the tactical level, and customer relations and other specialist management functions can control the strategic level. An interesting aspect of this work is the set-up of an outline model that defines the flows, interfaces and statistics of lower-level fractals. Other studies [2.25] indicate designing fractals and combining typical shop floor technologies such as kanban and scheduling to self-manage production. Implementing variable layout, together with flexible use of personnel, enhances adaptability of production to changing market conditions.

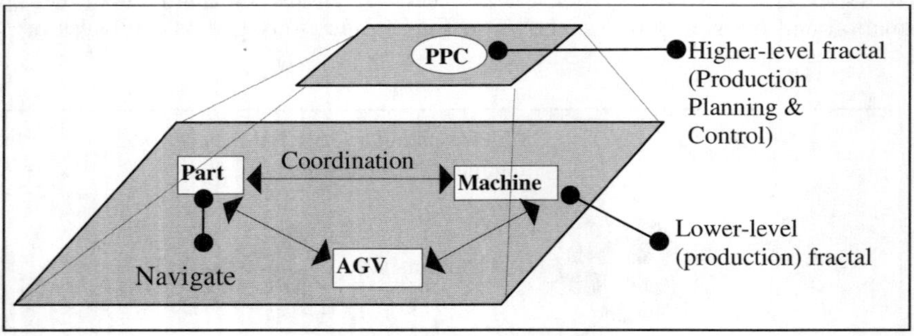

Fig. 2.3. An example of fractal factory

In order to illustrate the concept, a simple example is given in Fig. 2.3. Here, the part, the transport and the machine are fractals with capabilities to coordinate and form their own goals, navigate and assess their respective environments and act independently so as to optimise their operation. Using the concept of self-similarity, a higher-level fractal (that references the lower-level fractals) is defined. This fractal captures the abstracted operation of the lower level fractals, such as the ability to control the flow of parts (shown as "PPC" in the figure).

It could be said that the concept of a fractal factory foresees a set of highly dynamic and adaptive fractal entities that pursue a self-similar goal system, which can result in organisational structures that are able to change quickly in response to changing situations. Needless to say, many isolated elements of the fractal factory can be found in daily practice, including ([2.41], p. 215): business units, an orientation towards business processes, introduction of manufacturing segmentation, and cellular production structures. Nevertheless, these do not contain the full capability envisaged by the fractal factory.

2.3 Bionic Manufacturing

The bionic manufacturing system (BMS) [2.17–2.20, 2.35–2.37, 2.39] draws parallels with biological systems and proposes concepts for realising some essential properties of future manufacturing systems.

A biological system exhibits many features, including autonomous and spontaneous behaviour, and social harmony within hierarchically ordered relationships. Structurally, the cell is the basic unit from which all other parts of a biological system are derived, and the cells group together to form the hierarchical layers that make up an organism (listed in Fig. 2.4). For instance, cells with similar functions and shape form tissues (e.g. muscle tissue), and different tissues can group together to form organs (e.g. heart).

Cell	Basic unit having specialised function(s)
Tissue	Group of similar cells
Organ	Group of tissues with one or more functions
Body System	Group of organs
Organism	Group of body systems

Fig. 2.4. Hierarchical layers in biological systems

In order to preserve the stability of its internal chemical environment, a living organism must possess a means of regulating the rate of its metabolic reactions. This is done at two levels. At the cell level, the enzymes act as catalysts or inhibitors to control the reactions. In fact, the cell regulates the amount of, and activity of, enzymes that are produced and engages in self-control. The second level of control is through hormones. Hormones are secreted by cells, tissues or organs and transported in body fluids to other cells, tissues or organs, where they exert a specific physiological action. For instance, the colour change in some animals in response to visual cues occurs through appropriate hormone secretion that affects the functions of certain cells. Also, hormones may change the properties of the membranes of the cells (osmoregulation) and, thus, alter the rate and direction of cellular activities, leading to a widespread modification of physiological processes in the organism.

Apart from the control exerted by enzymes and hormones, there is the central nervous system. This system helps the organism to react rapidly to changes in the external environment: for instance, the almost immediate contraction of muscles and shivering when the body is exposed to unbearably cold weather. In this respect, the hormonal activity can be seen as control in the long term for gradual and sustained response.

The above characteristics of biological systems resemble closely those of manufacturing systems with autonomously operating units, as shown in Fig. 2.5. These units, similar to cells, obtain the needed inputs from the factory floor environment and perform operations. Outputs of these operations flow back to the environment. The operations require information from both the internal and the external environment. Regulation of the system can be effected at the level of the unit via coordination and appropriate control of material flow. Also, regulatory mechanisms similar to that of hormones (such as operating policies and strategies for longer-term control and coordination) and central nervous system (e.g. centralised production control) may exist.

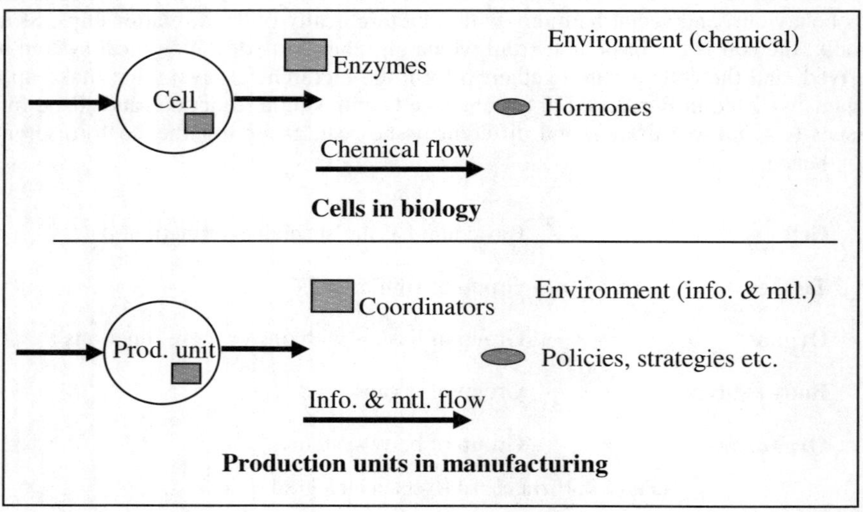

Fig. 2.5. Similarity of biological and manufacturing structures

The manner in which manufacturing systems are structured provides further evidence of the similarity between a biological and manufacturing system. For instance, the manufacturing units can be grouped together to form manufacturing cells, and the cells to form manufacturing shops and so on, with each layer in the hierarchy supporting and being supported by the adjacent layers. As in the biological counterpart, operations to be performed by lower layers are communicated as instructions in a top-down process by higher layers, and the consequent actions of the lower layers accumulate and manifest themselves as operations of the higher layers. Coordination and regulation at the various levels, as discussed before, harmonises the entire operation.

In addition to the operational similarities, the growth process (morphology) of life forms has some potential for application in designing manufacturing systems. The growth process takes place through repeated cell division to create the needed life components and to further develop these to form organs with distinct functionality. This is made possible by DNA-type information that is replicated in the genes of dividing cells. Likewise, functions of manufactured products can be created through provision of information to develop the required functionality of dividing components (such as assemblies, sub-assemblies or components).

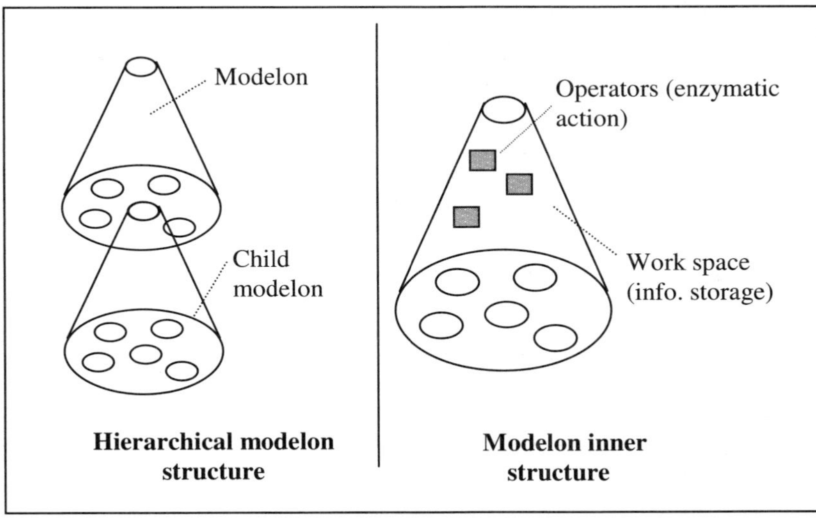

Fig. 2.6. Modelon structure (modified from [2.18])

The above concepts of biology and their parallels in manufacturing are used to propose BMS concepts and their supporting modelling elements. The primary element is a modelon [2.18], as shown in Fig. 2.6. A modelon is composed of lower-level modelons (a hierarchy of child modelons), operators (enzymatic actions) and a common memory or environment (a workspace to temporarily store information exchanged among modelons). The generality of the modelon structure is exploited to realise whole-part relations, self-decision, integration and harmony among autonomous units. Other features of the proposed structure include an environmental-response function to initiate modelons, and communication.

The application of BMS concepts range from modelling factories to design rooms. An example of the former is a bionic factory that will operate with distributed components (machines, automated guided vehicles, etc.) having quasi-life, as foreseen by Okino [2.20]. The operation of such a factory will involve entities, including products and machines that communicate each other of the decisions made by them. In fact,

Okino compares the knowledge of the processes possessed by the entities to genetic information. Thus, typical information about how a product should be made or assembled resides in the product, and the product communicates with the machines and decides on the schedule. The representation of genetic information in products can also be extended to manufacturing equipment so that it has knowledge of the types of processes, functions and even material flow paths. Armed with such genetic information, modelons representing products, equipment and other resources cooperate and engage in producing the required physical products.

A bionic design room [2.19, 2.20], in contrast to a bionic factory, uses the biological growth process of repeated cell division to form organs with differentiated functions. Adopting a similar process, a typical design activity would be carried out by developing specifications for a product (a parent modelon), followed by creating the required sub-assemblies or parts (sub-modelons) to fulfil lower-level functions. In this process, the design specifications are stored and passed, similarly to DNA-type information, by higher-level modelons to lower levels. An example of this process, given by Okino [2.20], is provided by the design of a winch. First, the design specification of the winch is developed and a parent modelon is created. This is followed by the winch modelon firing in sequence the sub-modelons representing the components, viz. rope, drum, motor, reduction gear equipment, clutch and brake. This process, which is self-organising, creates the components (i.e. the required functions).

Recent research in BMSs has also explored the application of evolution, that is, the process of acquiring new functions and features over successive generations (note that DNA-type information reflects an organism's current state of evolution). References [2.37, 2.39] study such evolution by simulating the material flows among a set of production units. The units, essentially through a self-organising process, learn those material flow patterns that are reinforced over a period. Interestingly, such an evolution could be thought of for a product where the requirements and hence the design (i.e. DNA-type information) evolves over time owing to a product's experience of being used by consumers.

In essence, the concept of bionic manufacturing can be extended to accommodate both production- and product-oriented views of manufacturing [2.39]. In the production view, the manufacturing resources – machines, tools and manufacturing cells – act similarly to biological cells or the derived higher levels that respond to stimuli from the production environment to jointly engage in production activities. In the product-oriented view, the product is seen as similar to an organism, with well-defined sub-assemblies and components that follow a growth process using DNA-type information.

2.4 Holonic Manufacturing

Koestler [2.10], in his attempt to explain the general principles of open hierarchical systems, proposes the concept of the holon, an entity that is capable of functioning as a self-contained *whole* while simultaneously acting as a *part* of a *whole* in a hierarchically[1] ordered system. The *wholes* (called holarchies [2.23]), as distinct from *parts*, exist only when viewed at some level of a system; this distinction soon disappears when one moves to a higher level, where the holarchy exhibits characteristics similar to a holon. In this sense, a holarchy can be seen as an *operational closure* that encapsulates the interactions among its member holons and thus acts similarly to a unit working together to achieve the set of goals of the collective. This way, holons, in a hierarchically ordered system, would possess *hierarchic awareness* ([2.10], p. 247), where every holon on every level is conscious of its goals as a whole and its goals as a part.

A holon has both self-assertive and integrative tendencies. The self-assertive tendency is the dynamic expression of a holon's wholeness, and the integrative tendency is the dynamic tendency of its partness. These two opposing tendencies manifest themselves as autonomous and cooperative attributes, which are essential to all holons, as illustrated in Fig. 2.7. The holons counterbalance these two attributes to achieve dynamic equilibrium in their operation. Suppose that a change in the environment causes a holon to become unstable; the holon would increase its autonomy (and thereby decrease its cooperation with others) or do the reverse to maintain stability. Such adjustments may lead to changes to the structure of the organisation, or to its behaviour or to both.

In order to counterbalance the twin attributes of autonomy and cooperation, Koestler suggests that holons operate using a fixed set of rules and flexible strategies. The rules, termed the system's canon,[2] determine a holon's invariant structural con-

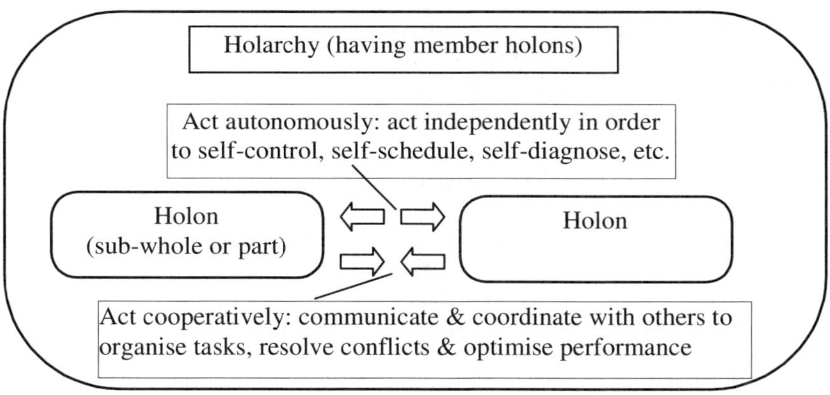

Fig. 2.7. A basic holonic system

[1] Hierarchies can be broadly classified into structural (the spatial aspect – anatomy topology) and functional (a process in time). Though they are complimentary aspects of an indivisible spatio-temporal process, it is often convenient to focus on one aspect or the other ([2.10], p. 59). Social hierarchies can also exist, though this term has been largely used with reference to humans.

[2] This can be explained as follows [2.21]: from the point of view of the entity, the behavioural impulses would be creative, while the necessary configurations of the outer world winnow these impulses through exigencies of the actual situation, thereby diminishing their numbers to a realistic few.

figuration (i.e. whole-part relations) and/or functional pattern, while the strategies define the permissible steps in the holon's activity. Using flexible strategies, a holon self-regulates and reacts to changes in the environment, and the changes are fed back to the centre (which can also be interpreted as the holarchy that it is a part of) controlling it. In response, the centre continually adjusts its course of action.

Koestler's work provides useful background and is a source of inspiration. The Holonic Manufacturing Systems (HMS) Consortium, under the Intelligent Manufacturing Systems Program, has adopted it. HMS has developed working definitions (see [2.23]) of the holonic concepts, and numerous models of their application in manufacturing have been proposed. These range from purely conceptual to practical models and explore both architectural and operational aspects.

The architectural aspects specify the structure of holonic systems and can be divided into models having a single- or a multi-level structure. Models with a single-level holonic structure will have a set of atomic-level holons (i.e. holons that are not decomposed any further) without any hierarchical ordering, similar to that of the holons in the machining cell of Fig. 2.8. Approaches to specifying the holons can take a functional or a manufacturing view. A functional view would model the holons to represent entities capable of generic functions such as scheduling, planning, execution and monitoring [2.1, 2.8, 2.9, 2.13]. A manufacturing view, on the other hand, models entities such as machines and parts that have embedded capabilities to plan and schedule [2.5, 2.6]. Coordination in such systems can be an in-built function [2.33] of the holons or can be a separate function embedded in a specialised coordinator holon [2.5, 2.6, 2.16].

A multi-level holonic structure would have a number of hierarchically ordered levels. In Fig. 2.8, the manufacturing shop is a holarchy that has member holons (such as the machining-cell holarchy). These holons may, in turn, contain member holons and so on. When there are holarchies representing diverse characteristics of a system, as

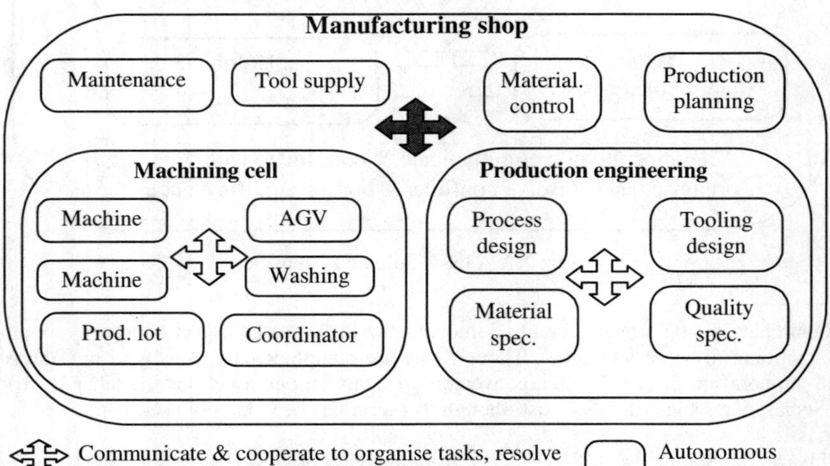

Fig. 2.8. A holonic manufacturing shop

the example in the figure illustrates, generalising and extending single-level models can be difficult. The HMS global architecture of [2.3] addresses this by proposing a set of generic holon types and cooperation blocks to provide the mechanism for constructing holarchies (or compound holons as the authors of [2.3] call it). Similar general-purpose multi-level models, having virtual clusters for collaborative planning and action, have also been attempted [2.38]. In this architecture, mediator agents (holons) act to form clusters (groupings of resources) and create the needed control components.

A manufacturing-specific multi-level model has also been proposed [2.40]. This model, called the PROSA architecture, is built on four basic holon types: a resource holon, to represent one or more physical entities (such as a machine); a product holon, to represent a product; an order holon, to represent a customer or make-to-stock order; and a staff holon, to represent specialist knowledge (such as process planning). The holon types possess procedural knowledge and functions to make decisions (such as allocation) and organise cooperation with other holons. Clustering holons of a type together creates multi-level structures.

At a more practical level, operational aspects of the concept have been investigated, with the majority focus being on strategies for cooperation and contracting for scheduling tasks among competing manufacturing holons [2.6, 2.15, 2.16, 2.28, 2.32, 2.33]. A cross-section of other applications includes: the development of a holonic multi-cell control system [2.22]; development of wrapper and emulation drivers for incorporating physical devices and logical-level holons into a machining and assembly cell [2.7]; visual interfaces for the human visualisation of holons [2.12]; and automotive vehicle data management covering the part of the life cycle from factory to recycling [2.2]. Also, a comparison of hierarchical and holonic control has been attempted, though less rigorously [2.11]. Further studies of purely conceptual interest include those of Suda [2.30, 2.31] and Mathews [2.14].

In summary, the application of holonic concepts in manufacturing would result in highly distributed and flexible HMSs having a conglomeration of specialised manufacturing holons. The holons would counterbalance their desire to act autonomously with the need for cooperation to integrate their actions and to bring about performance and stability. Through the formation of holarchies, HMSs would contain complexities of interactions among holons while permitting one to build hierarchically aware and responsive multi-level manufacturing architectures.

2.5 Comparison of Concepts

The above concepts describe, in quite general terms, the underlying principles and features of the process of, designing manufacturing systems that are highly flexible in their structure and operation. The next three subsections discuss the similarities and differences in the design, operation and self-organising features of the systems as envisaged by the concepts. This discussion is pursued for some critical parameters pertinent to highly modular and responsive systems in general.

2.5.1 Design

The concepts described above, in particular the architectural aspects, stress recursive part-whole relations. However, the approaches differ in the manner in which these aspects are used to describe and design manufacturing systems. These differences are highlighted for some critical system parameters in Table 2.1.

The definition of a unit in a BMS is a modelon that may be capable of multiple operations. Such units may be defined through repeated cell division (recursion) using DNA-type growth process. The definition of holons is similar. Their definition, however, may be carried out top-down by defining the holarchy first and the constituent holons subsequently. Designing fractals requires consideration of the six dimensions of the environment, as alluded to before. The design of fractals is bottom-up, with the operations being assigned to lower-level fractals according to their ability before higher-order fractals are formed to subsume the already assigned functions.

Table 2.1. Comparison of design features

System parameter	Bionic manufacturing	Fractal factory	Holonic manufacturing
Definition of unit	Cells: multi-operation and defined flexibly through genesis	Fractal: corporate service entity, multi-dimensional including technical, human and cultural	Functional and predefined
Definition of group	Organs: cell division to support needed functionality, more dynamic	Predefined and recursive as similar service fractals, also dynamic re-grouping	Predefined set of holons in holarchy to support specific functions
Autonomy of unit	High, cells able to define operations in response to changes in operating environment	High, setting of individual goals and adaptability through vitality	High, independence in negotiation and co-operation to set goals and tasks, limited by canons
Autonomy of group	Predefined functions of organs through genesis and operational autonomy	Inheritance of self-similar fractals and autonomy of goals; also dynamic re-structuring	Flexible strategies subject to fixed canons with stable intermediate forms

The design of the autonomous properties of units in a BMS comes from their ability to respond to changes in the environment, largely through performing multiple operations. Also, a unit may have self-regulation through enzymatic actions. The regulation within the group may be contained in objects similar to enzymes or hormones (this aspect is discussed in the next subsection). In the case of holons and holarchy, the autonomy is represented by the ability to manage the interactions. The holarchy acts as an operational closure, where member holons share the interactions and set basic rules of cooperation (i.e. canons) and strategies to deal with contingencies. The fractals, by the setting of goals along with vitality measures that monitor and act on the environment, may have a high level of autonomy. The autonomy of a group is in the negotiation of goals that are essentially similar. The fractals may contemplate regrouping in response to changes in the vitality measures.

In summary, the design of a fractal and its dynamic evolution can be considered to be an all-encompassing design task that includes technical, human and cultural dimensions and can lead to self-governing units. Early attempts in designing holonic manufacturing systems have taken a more technically oriented approach. Only elements that can be precisely designed are included. All the knowledge is assumed to be explicit and to be incorporated in the system. However, holonic manufacturing admits human components of holons – such an element is not precisely defined. Admittedly no testbed has explored socio-technical holons such as work cells. Design of bionic manufacturing is realised through genesis. This has the flexibility to evolve different forms according to the needs of the system (the design requirements of life). Also, factory entities modelled as quasi-life elements permit self-regulation, though such elements are confined to single-level architectures.

2.5.2 Operation

Operational features of manufacturing systems exhibit functional interdependence among units, both hierarchical and heterarchical. This is captured by various flows and efforts in coordination to ensure harmony in operation in particular, to minimise the possibility of chaotic behaviour as the autonomous units pursue their own actions.

Table 2.2. Short- vs. long-term regulation

Characteristic	Short-term	Long-term
Method of transmission of regulation	Messages	Messages, policies or other indirect means that change the operating environment
Speed of transmission	Rapid	Slow
Speed of response to regulation	Immediate	Usually slow
Duration of response	Short-lived	Long-lasting
Area of response	Activity of a unit	Goals or objectives of one or more units

In order to ensure harmony, regulatory mechanisms for control and coordination are called for. Regulatory mechanisms are mechanisms whereby the various activities of the component parts of a system are modified so that they contribute to the coherent functioning of the totality. The regulation provided by these mechanisms may vary depending on whether it is for short- or long-term coordination. Table 2.2 provides some key characteristics of short- and long-term regulation. For instance, a typical short-term regulation such as a change to production quantity may be communicated and acted upon immediately. In contrast, a long-term regulation such as lowering of inventory levels, reduction in production cycle times or even improving employee satisfaction may involve analysis accompanied by modification of practices. This type of regulation may need wider consultation and may not be immediate in its effect.

The mechanism to implement regulation can be global, filtered through hierarchical whole-part relations (i.e. regulation between adjacent layers) and/or local (among units of a single-level hierarchy). Table 2.3 shows the type of mechanism promoted by each of the concepts. The mechanisms given in italics have been attempted in the literature reviewed.

Table 2.3. Comparison of regulatory mechanisms[a]

Concept	Type of regulation	Type of mechanism		
		Global	Hierarchical	Local (entity-level)
BMS	Short-term	CNS	Enzymatic action	Enzymatic action (coordination)
	Long-term	Hormones	Hormones	Hormones, BN-type learning
FF	Short-term	No	Fractal navigation, JIT, kanban	JIT, kanban
	Long-term	No	Goal coordination	Vitality measures of environment
HMS	Short-term	No	Plan coordination	CN, learning coordination through conflicts
	Long-term	No	Hierarchic awareness	Balance A&C through canons and strategies

[a]CNS: central nervous system, BN: brain neuron, CN: contract-net, A&C: autonomy & cooperation, JIT: just-in-time

Referring to Table 2.3, we see that BMS supports all types of mechanisms for both short- and long-term regulation. However, a direct correspondence with the biological features has yet to be developed. FF provides for both hierarchical and local mechanisms with less emphasis on explicit global control, except as global goals that are taken up by the fractals (at the local level). HMS is similar to FF, but the mechanisms implemented are quite generic compared with FF, which implements manufacturing-specific control strategies (such as JIT and kanban). This comparison brings up another interesting aspect of regulation. A regulatory mechanism can be predefined, such as contract-net, or can be left to the learning process of an entity. Learning can be for both short- and long-term regulation. An example of the former is the learning of scheduling behaviours among a set of holons [2.33], and an example of the latter is the BN-type learning of material flow patterns in BMS [2.37].

In essence, the regulatory mechanisms to harmonise operations are an important part not only of achieving the desired short-term performance but also of recognising and amending entity inter-relationships, functional divisions and the organisational arrangement to avoid the long-term dysfunctional effects of an organisation. Thus, there is a need to identify and define both short- and long-term mechanisms as part of the design for an efficient operation.

2.5.3 Self-Organisation

Self-organisation can be described as the ability of a system (or a single entity) to adapt itself, without an external agency, to the prevailing conditions of its environment. This quality can be manifested in different ways [2.43]: a self-regulating system would have the capability to actively control the course of its internal transformations, typically with respect to one or more parameters and a self-configuring system would actively determine the arrangement of its constituent parts. Self-organisation can be brought about through modification of individual behaviours and/or organisational structures. Behaviour modification can be achieved through a variety of means, such as global, hierarchical or local regulatory mechanisms. Structural modifications can take place through regrouping or reconfiguration of the functional make-up of the entities. These characteristics, discussed in the preceding two subsections, are advocated, though somewhat differently, by the various concepts.

From an overall point of view, the extent to which self-organisation is taken into account by the concepts can be captured through the framework shown in Fig. 2.9. This framework is based on the one proposed by Hackman for Autonomous Work Groups (AWGs) (cited in [2.29]). It shows four levels of progressively increasing functional responsibility of a unit, and the type of unit that results. In a manager-led unit, the level of authority is only over the task being executed. At the other extreme, a self-governing unit assumes responsibility for all major functions. As the unit moves towards this end, its sphere of influence extends beyond the immediate control of the processes to other aspects, including human and business aspects, of the organisation.

Considered within these four types of units, the focus of the concepts varies somewhat with respect to the types of units envisaged. Holons (and holarchies) are basically formed from a functional decomposition of a system, i.e. the tasks to be done. In this way, the concept may focus on self-managing units with limited capabilities for self-design or self-governing. Similarly, BMS considers the functions of modelons (cells) with multiple operations, where the focus is on self-management. On the other hand, the features of a fractal transcend the immediate microscopic division of functions and cover the symbiosis of a fractal with its business and operational environ-

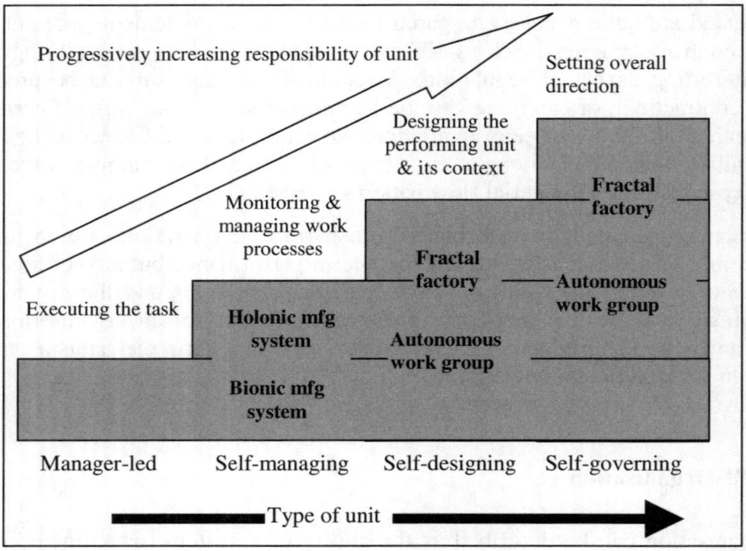

Fig. 2.9. Level of autonomy of units

ment (the six dimensions of a system's environment [2.25]). Consequently, fractals may be seen as dynamic, with the ability to reconfigure themselves in response to the environment. Thus, a fractal unit may be designed to be self-governing, more like a business unit.

However, this seemingly broader perspective of autonomy of fractals may not necessarily translate to the development of such units at the technology end. Consider the six dimensions of a system's environment, transposed to a continuum in Fig. 2.10. The technology end of the spectrum addressed by the fractal factory is oriented more towards applying principles of flexibility in layout and application of technologies such as kanban and JIT. In contrast, currently, both HMS and BMS, in addition to such applications, tend to develop technology that makes the equipment and devices themselves display autonomous behaviour in operation. This provides these physical units with intelligence and the ability to function as quasi-living things. Nevertheless, it may soon be necessary for BMS and HMS to move towards the socio-technical and strategic end to fully realise the potential of these developments.

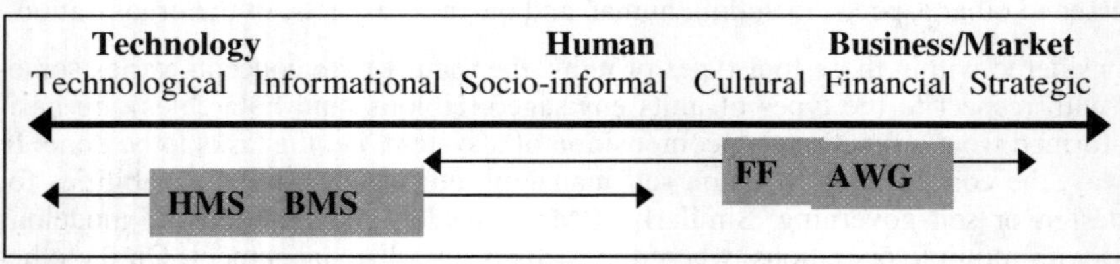

Fig. 2.10. Focus of development of distributed autonomous units

2.6 Conclusion

The concepts of bionic, fractal and holonic manufacturing are unified in that they propose distributed, autonomous, adaptive systems. An important observation that comes from considering these concepts is that to build and operate more responsive manufacturing systems, one should strive to define entities, structures and mechanisms that are operationally and structurally flexible in permitting easy, cost-effective faster evolution of technology, operating practices and, if needed, organisational forms.

In regard to particular aspects, the following can be observed:

- *Applicability and development of basic concepts*: the analogy of biological systems provides BMS with a stronger set of concepts to draw from. For FF and HMS, further effort may be required to develop the design and operational concepts.

- *Difference in development focus*: fractal factory considers the multi-dimensional nature of the manufacturing environment in developing and evolving the factory fractals. In contrast, the others seem to focus on the technology end to provide flexible and intelligent forms of automation of devices and equipment.

- *Difference in level of autonomy*: the autonomy of fractals can be extended include self-governing behaviour including the ability to regroup. HMS and BMS, at present, are more concerned with the operational aspects of autonomy, i.e. execution and control of processes, both at the individual and community levels. Nevertheless, the conceptual richness of HMS could be exploited to develop self-organisation capabilities to the highest levels of an organisation.

- *Design approaches and methodologies*: the approach to design varies, with FF advocating a bottom-up process while others advocate a top-down specification. HMS seems to fit well into some of the existing system description methodologies, namely the object-oriented and structured approaches. So does BMS, but with some difficulty in providing the generality of modelons. Such definitive approaches may not be suited for the FF concept, which requires analysis from multi-dimensional points of view.

- *Regulatory mechanisms*: these can range from simple, fixed regulation to those that permit short- and long-term regulation, and adaptive regulation to deal with changes in the environment. Of these mechanisms, BMS explores short-term regulation at all levels – global, hierarchical and local, whereas other concepts advocate only the latter two. It should be stressed here that longer-term regulation such as hormonal activity is seldom explored in the literature, though it has potential to detect and deal with dysfunctional effects due to gradual deterioration in performance.

- *Applications*: applications of these concepts are mixed with FF leading in reported industrial applications. However, HMS is also advanced in developing concepts and approaches to design that can be put into practice. Reports on industrial applications of BMS have not been widely evident in the literature.

Finally, the types of manufacturing systems envisaged by these concepts promise many advantages. To realise practically these advantages requires further research, experimentation and cross fertilisation of ideas from all three concepts. To this end, this paper aims to provide an insight into similarities and differences between the concepts that can be fruitfully exploited to fill any shortcomings in one or another concept.

References

[2.1] Agre, J. R., Elsley, G., McFarlane, D., Chen, J. and Gunn, B. "Holonic control of a water cooling system for a steel rod mill", Proceedings of the Fourth International Conference on Computer Integrated Manufacturing and Automation Technology (1994), pp. 134–141.

[2.2] Chaxel, F., Bajic, E. and Richard, J. "Automotive vehicle data management based on holon-product paradigm", Proceedings of the IEEE International Conference on Systems, Man and Cybernetics, Vol. 6 (1999), pp. 434–439.

[2.3] Fletcher, M. and Deen, S. M. "Fault-tolerant holonic manufacturing systems", Concurrency and Computation: Practice and Experience, 13(1), pp. 43–70 (2001).

[2.4] Gleick, J. Chaos, Sphere, London (1988).

[2.5] Guo, L., Hasegswa, T., Luh, P. B., Tamura, S. and Oblak, J. M. "Holonic planning and scheduling for a robotic assembly testbed", Proceedings of the Fourth International Conference on Computer Integrated Manufacturing and Automation Technology (1994), pp. 142–149.

[2.6] Guo, L., Luh, P. B. and Kyoya, Y. "Holonic Manufacturing scheduling: architecture, co-operation mechanism and implementation", Computers in Industry, No. 37, pp. 213–231 (1998).

[2.7] Hammerle, A., Bardi, S., Geurts, R. A. and Franz, A. "Internal communication architecture for resource holons", Proceedings of the IEEE International Conference on Systems, Man and Cybernetics, Vol. 3 (2000), pp. 721–726.

[2.8] Heikkila, T., Jarviluoma, M. and Juntunen, T. "Holonic control for manufacturing systems: functional design of a manufacturing robot cell", International Journal of Integrated Computer-Aided Engineering, 4(3), pp. 202–218 (1997).

[2.9] Heikkila, T., Kollingbaum, M., Valckenaers, P. and Bluemink, G. "manAge: an agent architecture for manufacturing control", Proceedings of the Second International Workshop on Intelligent Manufacturing Systems, Leuven, Belgium (1999), pp. 127–136.

[2.10] Koestler, A. The Ghost in the Machine, Arcana, London, (1967).

[2.11] Kotak, D., Bardi, S., Groover, W. A. and Zohrevand, K. "Comparison of hierarchical and holonic shop floor control using virtual manufacturing environment", Proceedings of the IEEE International Conference on Systems, Man and Cybernetics, Vol. 3 (2000), pp. 1667–1672.

[2.12] Kremer, R. and Norrie, D. "Architecture and design of a holonic visual interface", Proceedings of the IEEE International Conference on Systems, Man and Cybernetics, Vol. 3 (2000), pp. 1715–1720.

[2.13] McFarlane, D., Marett, B., Elsley, G. and Jarvis, D. "Application of holonic methodologies to problem diagnosis in a steel rod mill", IEEE International Conference on Systems, Man and Cybernetic, International Systems for 21st Century, Vancouver, Vol. 1 (1995), pp. 940–945.

[2.14] Mathews, J. "Organisational foundations of intelligent manufacturing systems-the holonic viewpoint", Computer Integrated Manufacturing Systems, 8(4), pp. 237–243 (1995).

[2.15] Moriwaki, T., Sugimura, N., Matawirya Yatna, Y. and S. H. Wirjomartono Hadjoko, "Production scheduling in autonomous distributed manufacturing system", Quality Assurance Through Integration of Manufacturing Processes and Systems, PED-Vol. 56, ASME (1993), pp. 175–186.

[2.16] Ng, A. H. C., Yeung, R. W. H. and Cheung, E. H. M. "HSCS-the design of a holonic shopfloor control system", Proceedings of the 1996 IEEE Conference on Emerging Technologies and Factory Automation, ETFA '96, 18–21 Nov., Kauai, HI, Vol. 1 (1996), pp. 179–185.

[2.17] Okino, N. "Bionical manufacturing systems", Organisation of Engineering Knowledge for Product Modelling in Computer Integrated Manufacture, ed. T. Sata, Elsevier, (1989), pp. 65–81.

[2.18] Okino, N. "Bionic manufacturing systems-modelon based approach" Proceedings of the CAM-I 18th Annual International Conference, New Orleans (1989), pp. 485–492.

[2.19] Okino, N. "A prototyping of bionic manufacturing system", Proceedings of the International Conference on Object-oriented Manufacturing Systems, Calgary (1992), pp. 297–302.

[2.20] Okino, N. "Bionic manufacturing system", CIRP, Flexible Manufacturing Systems Past-Present-Future, ed. J. Peklenik, Ljubljana (1993), pp.73–95.

[2.21] Salthe, S. N. "Evolving hierarchical systems", Columbia University Press, (1985).

[2.22] Schnell, J., Anderson, S., Langer, G. and Sorensen, C. "Development of a robot holon using an open modular controller", Proceedings of the IEEE International Conference on Systems, Man and Cybernetics, Vol. 2 (1999), pp. 1642–1647.

[2.23] Seidel, D. and Mey, M. "IMS-Holonic Manufacturing Systems: glossary of terms", IMS – Holonic Manufacturing Systems: Strategies Vol. 1, eds. D. Seidel and M. Mey, IFW, Hannover (1994).

[2.24] Shoder Education Foundation Inc. Self-similarity discussion, http://www.shodor.org/interactivate/discussions/selfsim.html (2001).

[2.25] Sihn, W. "Re-engineering through fractal structures", Proceedings of IFIP WG5.7 working conference Re-engineering the Enterprise, Galway, Ireland (1995), pp. 21–30.

[2.26] Sihn, W. and von Briel, R. "Process cost calculation in a fractal company", International Journal of Technology Management, 23(1), pp. 68–77, (1997).

[2.27] Sihn, W., Lickefett, M. and Pirron, J. "Simulation-based planning and control of production fractals", IEEE Conference on Control Applications – Proceedings, Vol. 2 (1999), pp. 1632–1635.

[2.28] Sousa, P. and Ramos, C. "A dynamic scheduling holon for manufacturing orders", Journal of Intelligent Manufacturing, 9, pp. 107–112, (1998).

[2.29] Strauss, R. E. and Hummel, T. "The new industrial engineering revisited – information technology, business process reengineering, and lean management in the self-organizing fractal company", Proceedings of the 1995 IEEE Annual International Engineering Management Conference, Theme "Global Engineering Management: Emerging Trends in the Asia Pacific", ed. F. S. Wei (1995), pp. 287–292.

[2.30] Suda, H. "Future factory system formulated in Japan (1)", Techno Japan, 22(10), pp. 15–25 (1989).

[2.31] Suda, H. "Future factory system formulated in Japan (2)", Techno Japan, 23(3), pp. 51–61, (1990).

[2.32] Sugimura, N., Hiroi, M., Moriwaki, T. and Hozumi, K. "Study on holonic scheduling for manufacturing system of composite parts", Proceedings of the Japan/USA Symposium on Flexible Automation, Vol. 2 (1996), pp. 1407–1410.

[2.33] Tharumarajah, A. and Wells, A. J. "A Behavior-based approach to scheduling in distributed manufacturing systems", Integrated Computer-Aided Engineering, 4(4), pp. 235–249, (1997).

[2.34] Tirpak, T. M., Daniel, S. M., LaLonde, J. D. and Davis, W. J. "A note on a fractal architecture for modelling and controlling flexible manufacturing systems", IEEEE Transactions on Systems, Man and Cybernetics, 22(3), pp. 564–567, (1992).

[2.35] Ueda, N. "A concept for bionic manufacturing systems based on DNA-type information, Proceedings of the 8th International PROLAMAT Conference, Man in CIM, Tokyo, (1992), pp. 853–863.

[2.36] Ueda, N. "A genetic approach toward future manufacturing systems" CIRP Flexible Manufacturing Systems Past-Present-Future, ed. J. Peklenik, Ljubljana (1993), pp. 211–228.

[2.37] Ueda, K., Hatono, I., Fujii, N. and Vaario, J. "Reinforcement learning approaches to biological manufacturing systems", Annals of CIRP, 49(1), pp. 343–346, (2000).

[2.38] Ulieru, M., Stefanoiu, D. and Norrie, D. "Holonic self-organization of multi-agent systems by fuzzy modelling with application to intelligent manufacturing", Proceedings of the IEEE International Conference on Systems, Man and Cybernetics, Vol. 3 (2000), pp. 1661–1666,

[2.39] Vaario, J. and Ueda K. "Self-organisation in manufacturing systems", Japan/USA Symposium on Flexible Automation, Vol. 2, ASME (1996), pp. 1481–1484.

[2.40] Van Brussel, H., Wyns, J., Valkenaers, P., Bongaerts, L. and Peeters, P. "Reference architecture for holonic manufacturing systems: PROSA", Computers in Industry, 37, pp. 255–274(1998).

[2.41] Warnecke, H. J. The Fractal Company, Springer Berlin, Heidelberg (1993).

[2.42] Westkamper, E., Huser, M. and Briel, R. von. "Transition processes: a survey of German industry", Journal of Materials Processing Technology, 106, pp. 141–151 (2000).

[2.43] Whitaker, R. Encyclopaedia Autopoietica,
www.informatik.umu.se/~rwhit/ EAIntro.html1998 (1998)

3. From FMS to HMS

Robert W. Brennan and Douglas H. Norrie

Department of Mechanical and Manufacturing Engineering
University of Calgary
Calgary, Alberta, Canada
Email: {brennan | norrie}@enme.ucalgary.ca

Abstract. For most of the 20th century, a central theme of industrial and academic research in manufacturing has been the development of flexible systems. In this chapter we discuss the notion of manufacturing flexibility and, in particular, the efforts toward achieving flexibility through advanced automation. After years of research into these flexible manufacturing systems it has become evident that the existing technologies alone are not the solution; effective system integration and coordination are also necessary. The second part of this chapter looks at recent efforts to achieve system integration and coordination through emerging technologies such as multi-agent systems and holonic manufacturing systems.

3.1 Introduction

The 20th century saw a series of fundamental changes in the manufacturing environment as industry moved from traditional labour-intensive manufacturing to more automated systems. In particular, the first steps towards mass production were made possible early in the 20th century through the development of precision jigs and repetitive flow techniques. Most notably, the stationary assembly line first made its appearance at Oldsmobile Motor in 1903. This was shortly followed by Cadillac's introduction of part interchangeability in 1908 and Ford's introduction of the moving assembly line with interchangeable parts in 1913 [3.52].

An important concept during the early years of mass production was the rationalisation of production. In other words, manufacturers started to see the wisdom of positioning specialised machines according to process flow requirements (to achieve progressive production) rather than using the conventional process or functional layout where machines are grouped by type and material flow is not considered (e.g. a job shop). Rationalisation led to a product layout where machines were subsequently arranged such that the product always follows the same routing. Rather than scheduling being the primary concern, set-up and balancing tasks now became critical. This approach was well suited to high-volume, standardised products and, as a result, allowed manufacturers to realise important economies of scale.

In the post-WWII United States, the virtual lack of competition internationally, combined with a domestic population eager for new products, allowed manufacturers to focus myopically on production efficiency rather than on the consumer [3.52]. In this environment, customers could be taken for granted (i.e. they were more than pleased to buy what was offered). As a result, mass production of high-quality, standardised goods was the norm and efficiency of production was all-important.

The latter part of the 20th century, however, saw more sophisticated consumers and an increase in foreign competition that quickly made this approach obsolete. Customers became more demanding and started to seek more variety, lower costs and superb quality; as well, they no longer had any qualms about where their products were manufactured. As a result, rather than efficiency of production alone, flexibility and responsiveness became key benchmarks for world-class manufacturing.

Among the first to address this issue in detail was the Massachusetts Institute of Technology's team to study "the future of the automobile", which was led by James Womak [3.63]. This group studied how Japan was able to dominate the automotive industry in the 1970s and 1980s through its move from mass production to lean production.

Lean production differs from mass production in a number of ways. For example, mass production sets very limited "good enough" goals (e.g. "acceptable number of defects" and "acceptable low product variety") since continuous improvement was never an important consideration. In contrast, lean production is focused on high performance goals, such as zero defects, declining costs, higher flexibility, and more product variety that are in direct alignment with current customer requirements. In their study, Womak et al. [3.63] focused on the importance of eliminating inventory and other forms of waste, increasing flexibility in production scheduling, reducing lead times, and enhancing levels of quality (i.e. both product and customer service). As a result, the basic goal for lean production is to use less of everything, i.e. human effort, space, and time-to-market.

To achieve this goal, manufacturers had to look at an entirely new approach to production. Guided by the promise of more flexible systems, manufacturers were at first keen to take advantage of new developments in process automation (e.g. Computer Numerical Controlled (CNC) machine tools and robotics) and computer-aided software (e.g., CAD/CAM) that had previously resulted in significant advantages in productivity for large-batch, repetitive manufacturing. Not surprisingly, it was felt that these advantages could be transferred to the more complicated world of small-batch manufacturing, where a new breed of "flexible manufacturing systems" would cure the inventory, lead-time and quality problems that traditionally plagued this area of manufacturing.

In order to realise the flexibility and productivity that these advanced systems promised, however, it quickly became evident that the existing technologies alone were not the solution; effective system integration and coordination were also necessary. Guided by this promise of more flexible systems, the current trend has been towards "intelligent" systems that are intended to adapt quickly to change while providing extensibility through a modular, distributed design.

In this chapter, we look at how holonic manufacturing systems (HMS) fit into this picture and, in particular, at how HMS can be viewed as the logical next step towards achieving manufacturing flexibility. In the next section, we provide background on flexible manufacturing systems and look at how well these systems addressed the issue of manufacturing flexibility. Next, we provide an overview of the work on distributed manufacturing systems control that ultimately led to the interest in holonic manufacturing systems. Finally, we discuss how HMS fit into the overall manufacturing flexibility picture in Sect. 3.4, and then provide our thoughts on the future of HMS in Sect. 3.5.

3.2 An Advanced Machining Technology Solution

Flexibility became an increasing priority for manufacturers during the 1960s, 1970s and 1980s in response to changes in international competition during this same period (i.e. the competitive advantage of the traditional industrial nations such as the US and UK was being lost to other nations such as Japan [3.11]). Coinciding with this interest in flexible manufacturing were new opportunities to exploit advances in manufacturing technology. For example, there were some very important advances in flexible automation in the 1970s such as Williamson's System 24 [3.62] and the advent of CNC machine tools. However, only large-batch manufacturers with a small, slowly changing range of products enjoyed the process flexibility offered by these new systems. At the time, small-batch manufacturing systems were considered too difficult to coordinate with this new technology. Although these small-batch systems were already capable of producing a wide range of products (and as a result could be considered "flexible"), they were plagued by "high inventory, unpredictable, long lead-times and quality problems" [3.57].

In order to address this problem, a new type of system along the lines of System 24 was introduced in the 1980s. These flexible manufacturing systems (FMS) consisted of several machines connected by a material-handling system and controlled by a central computer (for scheduling, downloading part programs and control). The result of implementing FMS for small- to medium-batch manufacturing was a drastic reduction in inventory (typically in the range of 85%), as well as significant improvements in lead time and quality performance [3.38].

Although many definitions of FMS have been given, it is perhaps best to use the Draper Laboratories FMS Handbook [3.53] definition here, since this definition most accurately reflects the contemporary perception of an FMS during its heyday in the 1980s [3.38]:

> a computer-controlled configuration of semi-independent work stations and a material handling system designed to efficiently manufacture more than one part number at low to medium volumes [3.53].

At the time, manufacturing flexibility was seen as being closely linked to a system's process range and volume, and as such, the real advantage of FMS was its potential to combine "economies of scope *and* scale" [3.57]. In other words, systems that could bridge the gap between dedicated (i.e. hard automation) and custom (i.e. job shop) systems were considered to be flexible. Figure 3.1 is a typical illustration of this notion of flexibility.

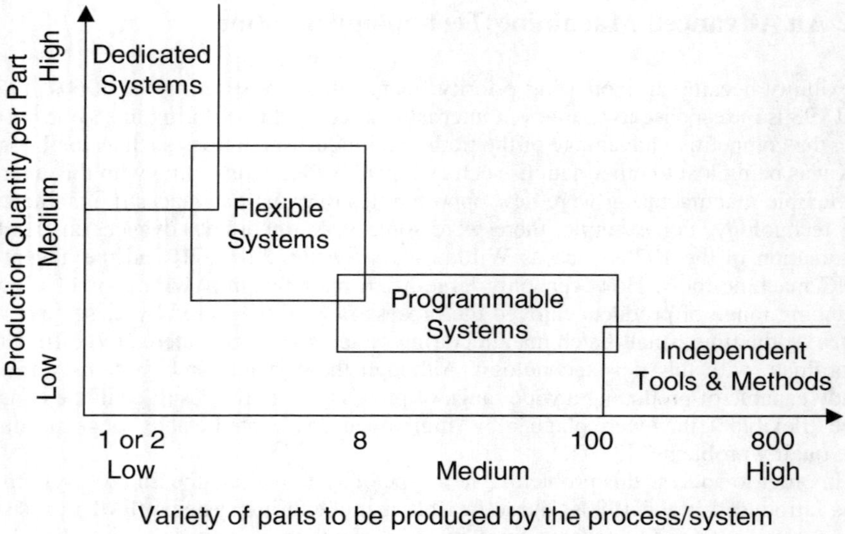

Fig. 3.1. The manufacturing flexibility spectrum (from [3.54])

FMS attempted to achieve this ideal through the use of advanced manufacturing technology. However, as many manufacturers found, blind acceptance of FMS technology as the panacea to achieving flexibility often led to disillusionment. We shall look at this more closely in Sect. 3.2.3, but first it is important to look more closely at the concept of manufacturing flexibility in order to fully understand the motivation for the transition from FMS to HMS.

3.2.1 Manufacturing Flexibility

Manufacturing flexibility is more than just product range. For example, Upton [3.56] notes that flexibility "is about increasing range, increasing mobility [i.e. a plant's ability to change from one product to another], or achieving uniform performance across a specified range". As a result, an important first step to achieving flexibility is to ask what form of flexibility is required.

Swamidass [3.54] proposed three dimensions of manufacturing flexibility, as illustrated in Fig. 3.2: (i) process flexibility, (ii) product design flexibility and (iii) infrastructure flexibility. As can be seen in this figure, process and product design flexibility primarily focus on enabling technologies such as CNC, robotics and FMS (which enable process flexibility) and CAD/CAM (which enables product design flexibility). Infrastructure flexibility, however, is the key to ensuring that these other forms of flexibility result in a flexible system, i.e. infrastructure "is the framework made up of people, procedures, information systems, and decision-making systems" [3.54]. As will be discussed in Sects. 3 and 4, this is the key to many of the fundamental HMS ideals.

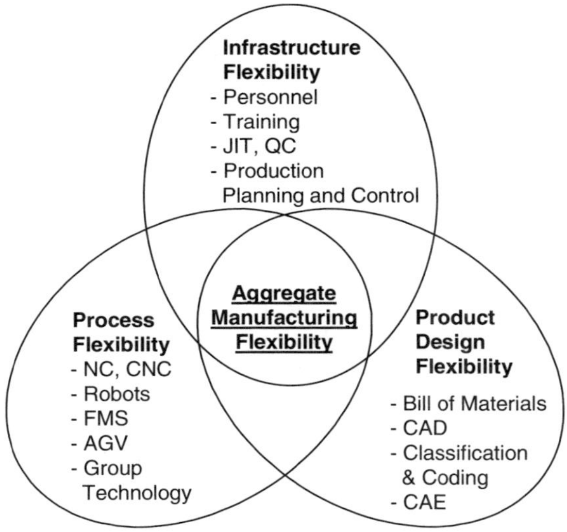

Fig. 3.2. Ingredients of manufacturing flexibility (adapted from [3.54])

More recently, Upton [3.56] provided examples of how flexibility can mean different things to different people. For example, it might be important to one company to define a flexible plant as one that can produce a small number of products that are very different from one another. Alternatively, another company might define a flexible plant as one that has the ability to produce a large number of products concurrently that are only slightly different from each other. Although both are quite different views of flexibility, both are examples of plants with a flexible product range.

Ultimately, it is important to recognise that flexibility does not mean the same thing to all companies. As a result, strictly following one approach to achieving flexibility cannot work in all cases.

3.2.2 Barriers to Flexibility

It should not come as a surprise to hear that, despite its bright beginnings, the success of FMS has been quite limited. In particular, FMS has proven to be useful for a specific class of manufacturing systems where turning, milling and sheet metal work dominate. Even when applied in this specific domain, these systems can quickly become inflexible if required to operate outside of their initial design range. This leads one to the conclusion that FMS alone is not the answer to manufacturing flexibility.

At this point we may ask, why has FMS fallen short of realising the ideals of Williamson [3.62] and others? An important part of the answer to this general question can be found by revisiting the three dimensions of flexibility described above: that is, FMS is only one part of the manufacturing flexibility picture. In fact, it is arguably not even part of the picture in many cases (when we look back at the motivations for FMS, however, it does play an important role in the evolution of flexibility).

As was noted previously, infrastructure is the key to realising both process and product design flexibility. For example, when machines alone are considered, a very technology-oriented view of flexibility can be taken (e.g. processes can be made more flexible through the use of CNC machines, robots, etc.). However, when the entire plant is considered, technology is only one aspect of flexibility. For example, FMS enables process flexibility; other tools such as CAD enable product design flexibility. To make a system truly flexible, however, these "technologies" must be properly integrated through a suitable framework, otherwise the result is a system of disconnected "islands of automation".

It was recognised quite early that the key to integration lies in the system's control software. However, it was also quickly recognised that the main barrier to flexibility lies in the difficulties associated with the development of this software. The software development process is often so time-consuming (and, as a result, expensive) that many companies have been discouraged from implementing a new technology such as FMS. For example, in some cases software development has been shown to account for 20% to 40% of the cost of the FMS installation [3.57].

Traditionally, FMS software has required a considerable amount of expertise to develop and is complex and time-consuming to modify. This expertise is not always available in manufacturing companies. As well, even once the software is fully commissioned, some manufacturers argue that these systems can limit their options when change is required. For example, an experienced operation can often "tweak" a system to obtain the desired results; most existing computer-controlled systems are a "black box" to operators, who may argue that "there isn't a button for that" when a new situation arises.

As a result, the software becomes a liability of the "flexible system" – making it inflexible. A considerable amount of work has been conducted over the last 20 years towards finding a better software solution that takes advantage of the potential of flexible automated machinery. In the next section, we look at the early work in this area that ultimately led to the notion of holonic manufacturing systems.

3.3 A Distributed Control Solution

Recent work in manufacturing systems control has focused on moving away from the centralised form of control popularised by FMS. This centralised strategy typically involves the classical control-theoretic techniques of analysis and design for small-scale systems, which, as Sandell et al. [3.46] note, "rest on the common presupposition of centrality". When large-scale systems such as manufacturing systems are considered, the problem becomes difficult, if not impossible, to solve using classical control-theoretic techniques. The solution to this inadequacy of centralised control of large-scale systems is the decentralised approach, which involves the use of a number of interacting decision-makers in place of a single centralised one.

The inherently difficult problem of shop floor scheduling provides a good example of why a distributed approach is more promising than the traditional centralised approach to computer control. The traditional approach to shop floor scheduling is to use a single, centralised scheduler. As Parunak [3.41] notes, "scheduling and control take place sequentially" and are analogous to a large sequential computer program when this approach is used. If there are no disturbances on the shop floor (e.g., rush orders, missing tools or material, machine processing time variability or failure), this approach is very effective and can generate optimal solutions to the scheduling problem.

Real manufacturing systems are not deterministic, however. Also, they consist of physically distributed materials and resources that behave in a highly concurrent and dynamic fashion. As a result, centralised schedules quickly become obsolete and must be either ignored or frequently updated, although this is not always possible if the schedule generation and dissemination time is great.

An alternative approach is to match the scheduling model more closely with the physical system. In other words, the problem can be decomposed such that the decision-making is performed by many simple, autonomous, cooperative entities. Rather than following a predetermined plan, the schedule emerges from the interaction of these intelligent entities. This approach leads to a scheduling system that is "decentralised rather than centralised, emergent rather than planned, and concurrent rather than sequential" [3.42].

The disadvantages of this approach are that global optima cannot be guaranteed and predictions of the system's behaviour can only be made at the aggregate level [3.42]. However, even with conventional approaches, global optima often cannot be achieved in practice, and the detailed predictions provided by these systems are frequently invalidated. Alternatively, the distributed intelligent control approach provides the advantages of adaptability, ease of upgradeability and maintenance, and emergent behaviour.

3.3.1 New Approaches

The question that follows at this point, however, is, how should one organise these independent decision-makers? This has led industrial and academic researchers to investigate a spectrum of decentralised control architectures ranging from hierarchical decomposition to a completely decentralised approach where individual controllers are assigned to subsystems and may work independently or may share information.

The hierarchical approach to large-scale systems control involves decomposing the overall system into small subsystems that have weak interactions with each other. In order to control a system in this manner, the two most common approaches of hierarchical control are spatial separation (i.e. the multi-level technique described by Mesarovic et al. [3.35] and Singh [3.51]) and temporal separation (i.e. the multi-layer, or frequency decomposition, approach described by Gershwin [3.21] and by Kimemia and Gershwin [3.29]).

In the manufacturing domain, Jones and Saleh [3.28] have combined both the spatial and temporal approaches at the Automated Manufacturing Research Facility (AMRF) in their multi-level/multi-layer hierarchical controller design for manufacturing cells.

In addition to this work on hierarchical control architectures, a considerable amount of work has been conducted in parallel on the basis of the negotiation metaphor that stemmed from the work of Davis and Smith [3.14] on distributed sensor systems. The distributed problem-solving approach of these authors requires the "cooperative solution of problems by a decentralised, loosely coupled collection of problem solvers" [3.14] and differs from distributed computing in that it is concerned with a "single task envisioned for the system"; the goal is to create an "environment for co-operative behaviour".

Parunak [3.40] extended the concept of the contract net from the information domain to the manufacturing domain with the development of the YAMS (Yet Another Manufacturing System) software. This system uses Davis and Smith's negotiation metaphor to achieve coordination of a network of distributed computing nodes. The YAMS system uses an "open system model" that allows the addition and removal of problem solvers and separates knowledge from the control structure through information hiding.

Duffie and Prabhu [3.16] have investigated this distributed approach with a "heterarchical", or non-hierarchical, opportunistic scheduling approach for machine cell control. One characteristic of this approach that sets it apart from traditional hierarchical control structures is that part flow is achieved by part-oriented requests as opposed to machine-oriented requests. This requires the parts to be "intelligent enough to know when they are done at one machine and need another" [3.15]. Similar work in the area of bidding-based control of manufacturing cells has also been performed by other researchers (e.g. Veeramani [3.59]). Work in this area has served as a catalyst for research on extending concepts from the emerging field of multi-agent systems (MAS) to the manufacturing domain. This next phase in the transition from FMS to HMS is discussed next.

3.3.2 Agent-Based Manufacturing

Multi-agent systems can be thought of as a "general software technology that was motivated by fundamental research questions" [3.8] such as that of how groups of software agents work together to solve a common problem [3.37]. Huhns and Singh [3.24] define agents as "active, persistent (software) components that perceive, reason, act and communicate". This definition follows from the notion of a software object, where the focus is on data abstraction, encapsulation, modularity and inheritance [3.4].

A major advantage of agent-based software over conventional software is this autonomous nature of individual agents. As Parunak [3.41] notes, agents "decide locally not only how to act (as subroutines do), and what actions to take (as objects do), but also when to initiate their own activity". This, in combination with the cooperative nature of agents in multi-agent systems, is particularly well suited to distributed intelligent control problems, as noted previously.

As described in the previous section, there has already been considerable utilisation of the agent concept in the manufacturing domain. In particular, autonomous agents or multi-agent systems are an attractive software engineering tool for the development of systems in which "data, control, expertise, or resources are distributed; agents provide a natural metaphor for delivering system functionality; or a number of legacy systems must be made to interwork," [3.64]. Manufacturing applications are characteristically "modular, decentralised, changeable, ill-structured, and complex" [3.44], and as a result supply a host of problems that are well suited to agent technology. Not surprisingly, although manufacturing is traditionally known for its conservatism, it is "one of the oldest and strongest areas of agent research" [3.1].

Research into the application of multi-agent systems and distributed artificial intelligence in the manufacturing domain has been steadily growing over the last ten years and has focused on all areas of the manufacturing enterprise, ranging from product design to real-time control. Examples of this work include work in the areas of product design (e.g. PACT [3.12], ACDS [3.13] and RAPPID [3.43]), enterprise integration and supply chain management (e.g. ISCM [3.20], AIMS [3.39] and CIIMPLEX [3.45]), planning and scheduling (e.g. AARIA [3.2], MASCADA [3.7] and YAMS [3.40]), and real-time control (e.g. [3.3, 3.60, 3.65]). For a comprehensive overview of agent-based systems in manufacturing, see Shen and Norrie [3.48], and for multi-agent systems in concurrent design and manufacturing, Shen et al. [3.49] can be consulted.

In order to provide a more concrete example, however, we conclude this section with a brief description of some of our recent research at the University of Calgary that is focusing on the development of manufacturing (or other distributed) control systems that can automatically adapt to changes in the controlled system. Since these systems will be expected to operate automatically with little or no human intervention, these control systems need to be adaptable in an emergent sense, i.e. they should be intelligent enough to be able to automatically adapt and reconfigure on the basis of the ever-changing requirements of the controlled system. Rather than a fixed architecture, the resulting structure of the new control system will be metamorphic in nature: i.e., to meet the dynamic requirements of agile manufacturing, the control system's architecture will be capable of constant metamorphosis.

An example of earlier work on this concept is the MetaMorph I architecture developed by Maturana and Norrie [3.34]. This architecture takes its name from its ability to change its structure and activities to adapt dynamically to tasks that emerge in the manufacturing system. A key element that enables this behaviour is the use of mediator agents to facilitate the coordination of the heterogeneous resource agents that are associated with the physical manufacturing system.

The mediator agents facilitate coordination by "promoting co-operation among intelligent agents and learning from the agents' behaviour" [3.34]. As well, mediators do not interfere with low-level decision-making in this architecture unless a critical situation occurs which requires their attention. The result of intelligent-agent cooperation is the creation of "dynamic hierarchical decision trees, which occur naturally, without predefined organisational structures" [3.34].

The communication mechanisms used by mediators in MetaMorph I are brokering (where the mediator receives a message from an intelligent agent, and then broadcasts to a selected group of intelligent agents) and recruiting (where intelligent agents are introduced to a group that communicate directly with each other). The organisational knowledge required of the mediator agents (i.e. a list of agent-to-agent relationships) is dynamically acquired through the mediators' learning process noted above.

MetaMorph I supports coordination of manufacturing systems through a process of sub-tasking, coordination cluster creation (by mediator agents), and task execution. The result is that the architecture moves towards global system objectives while enforcing local decision through these coordination clusters, or virtual clusters, that dynamically form or dissolve with the emerging requirements of the manufacturing system.

This approach has been used to implement a distributed concurrent design and manufacturing system in a simulated form using the following high-level mediator agents: Enterprise Mediator, Design Mediator (the CAD mediator), Resource Mediators, and Simulation/Execution Mediator (the AGV mediator). Each of these mediators, shown in Fig. 3.3, provides high-level coordination for the multi-agent systems within its "domain", as well as with the other mediators shown. Thus, each Resource Mediator provides high-level coordination for a manufacturing shop floor of machines, tools and other entities. Similarly, the Design Mediator coordinates one or more design systems. Common enterprise integration protocols are used to facilitate agent interoperability.

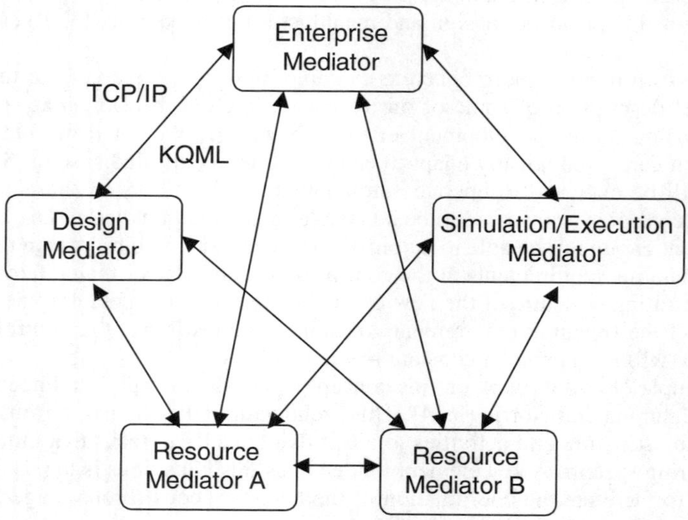

Fig. 3.3. The MetaMorph I architecture (from [3.47])

3.3.3 Representing the Manufacturing System with Agents

One motivation for the use of the agent model for the representation of manufacturing entities is that, like agents, manufacturing entities typically have well-defined state variables that are distinct from those of their environment [3.44]. As a result, a physical decomposition of the manufacturing system is the most obvious: agents are used to represent entities in the physical world, such as workers, machines, tools, fixtures, products, parts, features and operations. Evidence from natural systems suggests that it may be more effective to assign agents to physical entities in the system [3.61]. In the manufacturing domain, critical information is usually organised by physical entities, and the agents that represent these entities are the natural locus for maintaining the information. This approach naturally defines distinct sets of state variables that can be managed efficiently by individual agents with limited interactions; however, it requires a large number of resource-related agents. This leads to other problems, such as communication overheads and complex agent management. A common example of this approach is the use of part and machine agents for manufacturing planning and scheduling (e.g. [3.16, 3.32, 3.40]).

An alternative approach to problem decomposition that has been used in the manufacturing domain is the functional decomposition approach. In this approach, agents are used to encapsulate functionality such as order acquisition, planning, scheduling, material handling, transportation management and product distribution. and have no explicit relationship with physical entities. This approach tends to share many state variables across different functions, and as a result leads to problems of consistency and unintended interactions (i.e. high coupling [3.4]). Examples of this approach include the use of agents to encapsulate special functionality (e.g. facilitator agents [3.12], broker agents [3.45] and mediator agents [3.34]) and the use of agents to integrate existing systems to resolve legacy problems (e.g. ARCHON [3.10], EXPORT [3.36] and CIIMPLEX [3.45]).

As noted previously, multi-agent systems can be thought of as a general software technology; manufacturing, however, is fundamentally concerned with "ironware" (i.e. physical equipment and products). As a result, when these two worlds merge, it becomes useful to start thinking about a different type of agent from those that populate the classical agent-based systems. In particular, it becomes useful to think about "physical agents" that possess both classical software agent capabilities and the capability to interface with physical equipment on the shop floor. In the next section, we provide a brief overview of the international collaborative effort to develop systems of these unique manufacturing-specific agents.

3.4 The Holonic Approach

The shortcomings of the centralised approach to control originally proposed for FMS can be succinctly summarised as follows: these centralised systems are "inflexible, fragile, and difficult to maintain" [3.8]. As will be discussed in this section, work on the development of holonic manufacturing systems was motivated by the need to address these shortcomings, as well as by new pressures faced by manufactures in the 1990s such as increasingly stringent customer requirements (e.g. for high-quality, customisable, low-cost products that can be delivered quickly) and increasing manufacturing system complexity (i.e. these systems are, by their nature, distributed, concurrent and stochastic).

The research in distributed control as described in the previous section led many to the realisation that an "autonomous, distributed and co-operative" [3.58] approach was required to address these issues. This new control software and hardware approach appeared to hold the most promise of realising manufacturing systems that were both flexible (e.g. capable of reconfiguration) and responsive (e.g. capable of recovering from disturbances). However, past experiences with "green field" approaches such as FMS left a bad taste in the mouths of many manufacturers, so clearly, if new approaches to flexible automation were to be sold to industry, an incremental approach would be desired.

In the remainder of this section we shall look at this new dynamic and decentralised approach to manufacturing systems control and provide a brief overview of the HMS project.

3.4.1 Holonic Concepts

Holonic Manufacturing Systems is one of the Intelligent Manufacturing Systems (IMS) programme's six major projects resulting from a two-year feasibility study started in February 1992 [3.9]. In very general terms, holonics can be thought of as a marriage between a general philosophy (of living organisms and social organisations) and an emerging software approach (distributed artificial intelligence) that is intended to lead ultimately to a better understanding of how next-generation manufacturing systems should be designed. The objective of the work of the HMS consortium is to "attain in manufacturing the benefits that holonic organisation provides to living organisms and societies, e.g., stability in the face of disturbances, adaptability and flexibility in the face of change, and efficient use of available resources" [3.23].

The term "holon" was coined by Arthur Koestler [3.30]. He observed a dichotomy of wholeness and partness in living organisms and social organisations, and summarised this with the statement "wholes and parts in the absolute sense do not exist anywhere". To understand these systems, Koestler used the "Janus effect" as a metaphor for this dichotomy of wholeness and partness observed in many such systems: i.e., "like the Roman god Janus, members of a hierarchy have two faces looking in opposite directions". In other words, these members can be thought of as self-contained wholes looking towards the subordinate level and/or as dependent parts looking upward. To explain this concept, Koestler suggested a new term to describe the members of these systems: "holon" (from the Greek *holos*, meaning "whole", and the suffix "on", implying "part" as in "proton" or "neutron").

In order to achieve the benefits of Koestler's holonic organisations, an HMS "consists of autonomous, self-reliant manufacturing units, called holons" [3.8] that cooperate to achieve the overall manufacturing system objectives. A holon is defined by the HMS Consortium [3.23] as "an autonomous and co-operative building block of a manufacturing system for transforming, transporting, storing and/or validating information and physical objects".

As was noted previously, another other important characteristic of holons that is particularly relevant to the manufacturing domain is that they consist of an information-processing part and, often, a physical processing part. This distinction really sets HMS apart from the mainstream multi-agent systems community, though another organisation formed in 1996, the Foundation for Intelligent Physical Agents [3.19], also recognised this use of agent technology quite early. FIPA, however, did not pursue the notion of *physical agents* closely until recently [3.55].

Another important holonic concept that is also relevant to the manufacturing domain is the notion of functional decomposition. Simon's [3.50] observation that "complex systems evolve from simple systems much more rapidly if there are stable intermediate forms than if there are not" is at the heart of this. In other words, the complexity of manufacturing systems can be dealt with by decomposing the system into smaller parts. A consequence of this is the idea that holons can contain other holons (i.e. they are recursive). As well, problem solving is achieved by holarchies, or groups of autonomous and cooperative basic holons and/or recursive holons that are themselves holarchies.

3.4.2 Holonic Manufacturing Systems

As noted previously, Holonic Manufacturing Systems began in 1992 as a feasibility study. At the time, it was felt that an international effort would be required if manufacturing systems based on the concepts described above were to be developed in a reasonable period of time. It was not clear, however, whether a collaborative effort among competitive industries and across various cultures would work in practice [3.58]. Despite these early fears, the results of this feasibility study were very promising [3.31], leading to the launch of the full-scale programme in 1995.

During this first phase of the HMS project (1995–2000), work was conducted around three main areas: generic technologies, benchmarking and organisation (e.g. project management, technology transfer, dissemination and exploitation). In order to provide a clear focus for the project, this work was organised into the seven workpackages summarised below [3.22]:

- *Systems Architecture and Engineering (WP1)*: the generation of reference models, information models, standards and support tools for holons, holonic systems, and the holonic systems engineering process.
- *Systems Operation (WP2)*: generic operational aspects of HMS such as communication, negotiation and coordination, planning, resource allocation, fault management, and reconfiguration.
- *Holonic Resource Management (WP3)*: the development of strategies, techniques and architectures for resource management in HMS.
- *Holonic Manufacturing Unit (WP4)*: machining, finishing, inspection and tooling.
- *Holonic Fixturing (WP5)*: fixturing in the machining and assembly process.
- *Holonic Handling Systems (WP6)*: robots, end-effectors, fixtures, feeders, mobile handling systems, and sensors.
- *Holomobiles (WP7)*: mobile systems for transport, maintenance, supervision, etc.

As can be seen from these workpackage descriptions, this first phase of the HMS project was focused on the development of a generic holonic systems technology (i.e. WP1–WP3) as well as the demonstration of this approach in specific application areas (i.e. WP4–WP7).

During this phase of the project, significant progress was made towards the definition of standards for holonic systems that were primarily based on the International Electrotechnical Commission (IEC) 61499 function block standard [3.25]. For example Fletcher et al. [3.18] describe a suitable architecture for holonic systems that uses a layered framework of IEC 61499 function blocks called a holonic kernel. In order to enable holonic cooperation and autonomy, this kernel consists of four specialised services (each implemented by an IEC 61499 service interface function block): a Function Block Manager, a Coordination Manager, a Cooperation Domain Interface and a Data/Knowledge Base Interface.

The primary motivation for the use of the function block model was its suitability for event-driven control. As Marik and Pechoucek [3.33] note, the traditional scan-based approach to control (e.g. that used by traditional programmable logic controllers) “does not enable the required flexibility in scheduling of events, especially in the case of hard real-time control". Alternatively, by explicitly representing events, the function block model focuses on process abstraction and synchronisation; this makes this approach particularly suitable for control of an environment that is concurrent, asynchronous and distributed.

However, the work during this phase of the HMS consortium showed that this model does have some limitations in its current form. For example, as Christensen [3.9] pointed out, the model must support mechanisms for dynamic reconfiguration if the agility requirements of FMS are to be met. As well, although the function block is particularly suited for low-level, real-time process/machine control and lends itself well to various hardware implementations (e.g. [3.17]), it is not well suited for higher-level reasoning [3.5].

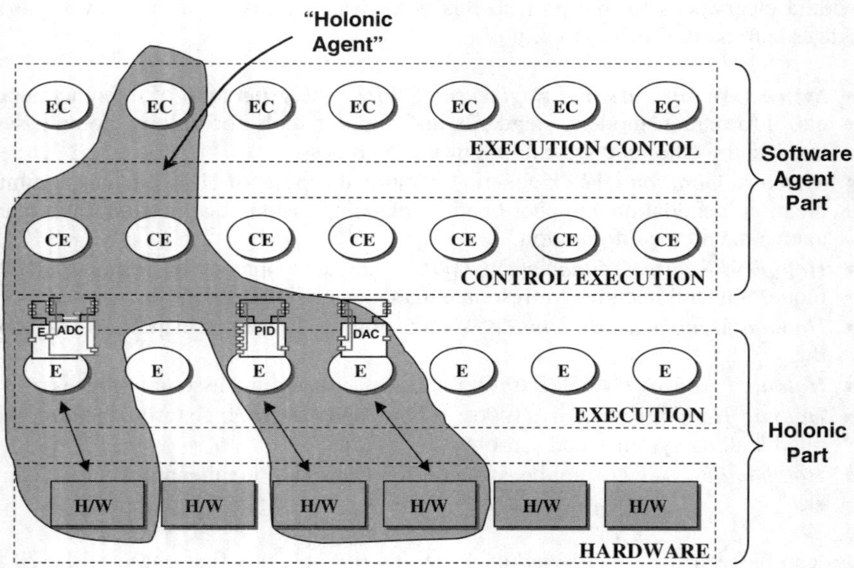

Fig. 3.4. Holonic agents

As a result, it was recognised that the best approach is to encapsulate the function block solution into higher-level software when and where it is required to enable more sophisticated reasoning and a richer knowledge representation than function blocks alone. These resulting components have been referred to as "holonic agents" [3.33] because of their integration of a holonic part (for hard real-time control) and a software agent part (responsible for higher-level, soft real-time or non-real-time intelligent decision-making). This concept is illustrated in Fig. 3.4, which shows a multi-layer architecture consisting of four temporally decomposed layers of agents (the ovals in Fig. 3.4) and devices: execution control (EC), control execution (EC), execution (E), and hardware [3.6]. As one moves down the layers, timescales become shorter and real-time constraints change from soft to hard real-time; as well, the degree of agency decreases (i.e. higher-level agents are more sophisticated but slower, while lower-level agents are fast and light weight).

Having explored holonic systems technologies since 1995, the HMS consortium entered its second phase in 2001, focusing on the integration and application of these technologies. In order to provide for the entire scope of the manufacturing enterprise, the consortium took the conventional view of manufacturing planning and control systems as a hierarchy of decision levels that range from long-term, strategic decisions at the highest level to short-term, detailed execution and control at the lowest level. As a result, the focus of HMS is now on three fundamental levels of the manufacturing enterprise [3.22]:

- *Holonic Production Execution Systems*: the factory and supply chain level.
- *Holonic Production Sites and Physical Equipment*: the work cell level (which consists of a number of holonic control devices).
- *Holonic Control Devices*: the physical, real-time control level.

This decomposition can be thought of as a simplification of the widely accepted National Bureau of Standards (NBS) [3.27] five-level model and the International Organization for Standardization (ISO) [3.26] six-level model. As well, the three HMS levels now have a close parallel with the recently proposed FIPA "holonic enterprise" model [3.55]. Figure 3.5 provides an overview of these models.

Finally, a fourth area of focus of Phase 2 is that of developing a virtual environment for testing holonic systems ("Holonic Man–Machine and Emulation Systems"). The basic idea here is to develop an environment that can be used to demonstrate the technical and economic advantages of holonic systems, support the application and development of these systems, and increase the acceptance of this new approach [3.22].

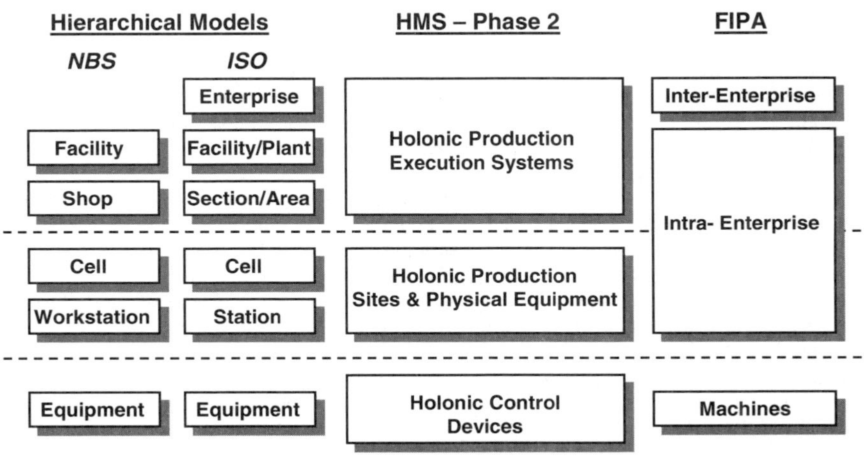

Fig. 3.5. The manufacturing enterprise

3.5 Summary

Flexible manufacturing has come a long way since the days of Williamson's System 24 and the first implementations of FMS, though many of the integration problems still exist. For example, although computer technology has been advancing at a blinding pace (e.g. improvements in processor power, communication speed, software usability and hardware cost), manufacturing control systems still require a significant amount of time to commission and are still difficult to modify.

The first phase of the HMS project demonstrated that the holonic approach has the potential to fully realise manufacturing flexibility. However, if the goals of this project are to be attained, the work from the first phase must be fully integrated and applied at all levels of the manufacturing enterprise. As noted above, this is the main focus of the second phase of the project. As well, efforts are being made to integrate the work of HMS with the standardisation work of FIPA in order to exploit the parallel international effort on software agent systems.

Clearly, we have not yet completed the journey that started with FMS; the work on HMS, however, has certainly got us a lot closer. Some of the key issues that will have to be further addressed by researchers in this area include developing methodologies for dynamic and automatic configuration/reconfiguration, defining ontologies for manufacturing, and developing techniques to allow manufacturers to migrate their systems towards these new systems.

References

[3.1] Aparicio IV, M.: "Point and click manufacturing on the web", IEEE Internet Computing, **3**(5), 38–40 (1999).

[3.2] Baker, A., Parunak, H. and Erol, K.: "Agents and the internet: infrastructure for mass customisation", IEEE Internet Computing, **3**(5), 62–69 (1999).

[3.3] Balasubramanian, S., Brennan, R., and Norrie, D.: "An architecture for metamorphic control of holonic manufacturing systems", Computers in Industry, **46**(1), 13–31 (2001).

[3.4] Booch, G.: *Object-Oriented Analysis and Design with Applications*, 2nd edition, Addison-Wesley (1994).

[3.5] Brennan, R. and Norrie, D.: "Agents, holons and function blocks: distributed intelligent control in manufacturing", Journal of Applied Systems Studies Special Issue on Industrial Applications of Multi-Agent and Holonic Systems, **2**(1), 1–19 (2001).

[3.6] Brennan, R. and Norrie, D.: "Managing fault monitoring and recovery in distributed real-time control systems", *5th IEEE/IFIP International Conference on Information Technology for Balanced Automation Systems in Manufacturing and Services*, (2002), pp. 247–254.

[3.7] Bruckner, S., Wyns, J., Peeters, P. and Kollingbaum, M.: "Designing agents for manufacturing process control", *Proceeding of Artificial Intelligence and Manufacturing Research Planning Workshop*, (1998), pp. 40–46.

[3.8] Bussmann, S.: "An agent-oriented architecture for holonic manufacturing control", *First Open Workshop on Intelligent Manufacturing Systems Europe*, (1998), pp. 1–12.

[3.9] Christensen, J.: "HMS: initial architecture and standards directions", *the 1st European Conference on Holonic Manufacturing Systems*, (1994), pp. 1–20.

[3.10] Cockburn, D. and Jennings, N.: "ARCHON: a DAI system for industrial applications", In *Foundations of Distributed Artificial Intelligence*, eds. G. O'Hare and N. Jennings, Wiley, (1996), 319–344.

[3.11] Corrêa, H.: "Agile manufacturing as the 21st century strategy for improving manufacturing competitiveness", In *Agile Manufacturing: the 21st Century Competitive Strategy*, ed. A. Gunasekaran, Elsevier, (2001), 3–23.

[3.12] Cutkosky, M., Englemor, R., Fikes, R., Gruber, T., Genesereth, M., Mark, W., Tenenbaum, J. and Weber, J.: "PACT: an experiment in integrating concurrent engineering systems", IEEE Computer, **26**(1), 28–37 (1993).

[3.13] Darr, T. and Birmingham, W.: "An attribute-space representation and algorithm for concurrent engineering", AI EDAM, **10**(1), 21–35 (1996).

[3.14] Davis, R. and Smith, R.: "Negotiation as a metaphor for distributed problem solving", Artificial Intelligence, **20**, 63–109 (1983).

[3.15] Duffie, N. and Piper, R.: "Nonhierarchical control of manufacturing systems", Journal of Manufacturing Systems, **5**(2), 137–139 (1986).

[3.16] Duffie, N. and Prabhu, V.: "Real-time distributed scheduling of heterarchical manufacturing systems", Journal of Manufacturing Systems, 13(2), 94–107 (1994).

[3.17] Fletcher, M., Norrie, D. and Christensen, J.: "A foundation for realtime holonic control systems", Journal of Applied Systems Studies, Special Issue on Industrial Applications of Multi-Agent and Holonic Systems, **2**(1), 20–43 (2001).

[3.18] Fletcher, M., Garcia-Herreros, E., Christensen, J., Deen, S. and Mittmann, R.: "An open architecture for holonic cooperation and autonomy" *11th International Workshop on Database and Expert Systems Applications*, (2000), pp. 224–230.

[3.19] Foundation for Intelligent Physical Agents: Web site, http://www.fipa.org/ (2001).

[3.20] Fox, M., Chionglo, J. and Barbuceanu, M.: "The integrated supply chain management system", *Internal Report, Department of Industrial Engineering*, University of Toronto, (1993).

[3.21] Gershwin, S.: "Hierarchical flow control: a framework for scheduling and planning discrete events in manufacturing systems", Proceedings of the IEEE, **77**(1), 195–209 (1989).

[3.22] Gruver, W., Kotak, D., van Leeuwen, E. and Norrie, D.: "Holonic manufacturing systems – phase 2", International IMS Project Forum 2001, Ascona, Switzerland.

[3.23] Holonic Manufacturing Systems Consortium: Holonic manufacturing systems overview, http://hms.ifw.uni-hannover.de/ (2001).

[3.24] Huhns, M., Singh, M. (1998): Readings in Agents. Morgan Kaufmann.

[3.25] International Electrotechnical Commission: Voting Draft – Publicly Available Specification – Function Blocks for Industrial Process-measurement and Control Systems, Part 1 – Architecture, IEC TC65/WG8, Geneva (2000).

[3.26] International Organization for Standardization: *The Ottawa Report on Reference Models for Automated Manufacturing Systems*. ISO, Geneva (1986).

[3.27] Jones, A. and McLean, C.: "A proposed hierarchical control model for automated manufacturing systems", Journal of Manufacturing Systems, **5**(1) (1986).

[3.28] Jones, A. and Saleh, A.: "A multi-level/multi-layer architecture for intelligent shopfloor control" International Journal of Computer Integrated Manufacturing, **3**(1), 60–70 (1990).

[3.29] Kimemia, J. and Gershwin, S.: "An algorithm for the computer control of a flexible manufacturing system", IIE Transactions, **15**(4), 353–362 (1983).

[3.30] Koestler, A.: *The Ghost in the Machine*, Arkana, London (1967).

[3.31] Kriz, D.: *Holonic Manufacturing Systems: Case Study of an IMS Consortium*, Holonic Manufacturing Systems Consortium Web site, http://hms.ifw.uni-hannover.de/ (1995).

[3.32] Lin, G. and Solberg, J.: "Integrated shop floor control using autonomous agents", IIE Transactions: Design and Manufacturing, **24**(3), 57–71 (1992).

[3.33] Marik, V. and Pechoucek, M.: "Holons and agents: recent developments and mutual impacts", *Twelfth International Workshop on Database and Expert Systems Applications*, IEEE Computer Society, (2001), pp. 605–607.
[3.34] Maturana, F. and Norrie, D.: "Multi-agent mediator architecture for distributed manufacturing", Journal of Intelligent Manufacturing, **7**, 257–270 (1996).
[3.35] Mesarovic, M., Macko, D. and Takahara, Y.: *Theory of Hierarchical, Multilevel, Systems*, Academic Press (1970)
[3.36] Monceyron, E. and Barthes, J.: "Architecture for ICAD systems: an example from harbor design", Revue Sciences et Techniques de la Conception, **1**(1), 49–68 (1992).
[3.37] Moulin, B., Chaib-draa, B. (1996): An overview of distributed artificial intelligence. In (G. O'Hare and N. Jennings, ed.) Foundations of Distributed Artificial Intelligence, John Wiley and Sons, 3–56.
[3.38] O'Grady, P. and Menon, U.: "A concise review of flexible manufacturing systems and FMS literature", Computers in Industry, **7**, 155–167 (1996).
[3.39] Park, H., Tenenbaum, J. and Dove, R.: "Agile infrastructure for manufacturing systems: a vision for transforming the US manufacturing base", *Defense Manufacturing Conference* (1993).
[3.40] Parunak, H.: "Manufacturing experience with the contract net", Distributed Artificial Intelligence, **1**, London: Pitman (1987).
[3.41] Parunak, H.: "Autonomous agent architectures: a non-technical introduction", *Industrial Technology Institute Report* (1993).
[3.42] Parunak, H.: "Applications of distributed artificial intelligence in industry", In *Foundations of Distributed Artificial Intelligence*, eds. G. O'Hare and N. Jennings, Wiley, (1996), 139–164.
[3.43] Parunak, H.: "RAPPID project index page", http://www.iti.org/cec/rappid (1997).
[3.44] Parunak, H.: "Industrial and practical applications of DAI", In *Multiagent Systems: A Modern Approach to Distributed Artificial Intelligence*, ed. G. Weiss, The MIT Press, Cambridge, MA, (1999), 377–424.
[3.45] Peng, Y., Finin, T., Labrou, Y., Chu, B., Long, J., Tolone, W. and Boughannam, A.: "A multi-agent system for enterprise integration", *Proceedings of PAAM'98*, (1998), pp. 533–548.
[3.46] Sandell, N., Varaiya, P., Athans, M. and Safonov, M.: "Survey of decentralized control methods for large scale systems", IEEE Transactions on Automatic Control, **AC-23**(2), 108–128 (1978).
[3.47] Shen W., Maturana F. and Norrie, D.: "MetaMorph II: an agent-based architecture for distributed intelligent design and manufacturing", Journal of Intelligent Manufacturing – Special Issue on Distributed Manufacturing Systems. **11**(3), 237–251 (2000).
[3.48] Shen, W. and Norrie, D.: "Agent-based systems for intelligent manufacturing: a state-of-the-art survey", Knowledge and Information Systems, **1**, 129–156 (1999).
[3.49] Shen, W., Norrie, D. and Barthes, J.: *MultiAgent Systems for Concurrent Design and Manufacturing*, Taylor and Francis, London, (2000).
[3.50] Simon, H.: *The Sciences of the Artificial*, MIT Press, Cambridge, MA, (1996).
[3.51] Singh, M. : *Dynamic Hierarchical Control*. North-Holland, Amsterdam, (1980).
[3.52] Sipper, D. and Bulfin, R.: *Production: Planning, Control and Integration*, McGraw-Hill, (1997).
[3.53] Stark, C.: *Flexible Manufacturing System Handbook*, Draper Laboratories, U.S. Department of Commerce NTIS Publication No. AD/A127, (1983).
[3.54] Swamidass, P.: *Manufacturing Flexibility*, Operations Management Association, Monograph No. 2, (1988).
[3.55] Ulieru, M.: "FIPA technology overview: holonic enterprise", FIPA Inform!, **2**(2), 3–4 (2001).
[3.56] Upton, D.: "What really makes factories flexible?" Harvard Business Review, **73**(4), 74–84 (1995).

[3.57] Upton, D.: "A flexible structure for computer-controlled manufacturing systems", Manufacturing Review, **5**(1), 58–74, (1992).
[3.58] Valckenaers, P. and Van Brussel, H.: *IMS TC5: Holonic Manufacturing Systems Technical Overview*, Holonic Manufacturing Systems Consortium Web Site, http://hms.ifw.uni-hannover.de/ (1994).
[3.59] Veeramani, D.: "Task and resource allocation via auctioning", *the Winter Simulation Conference Proceedings*, (1992), pp. 945–954.
[3.60] Wang, L., Brennan, R., Balasubramanian, S. and Norrie, D.: "Realizing holonic control with function blocks", Integrated Computer-Aided Engineering, **8**(1), 81–93 (2001).
[3.61] Weiss, G.: *Multi-Agent Systems*. MIT Press, Cambridge, MA, (1999).
[3.62] Williamson, D.: "System 24 – a new concept of manufacture", *Proceedings of the 8th International Machine Tool and Design Conference*, (1967), pp. 327–376.
[3.63] Womak, J., Jones, D. and Roos, D.: *The Machine that Changed the World: the Story of Lean Production*, Harper Perennial, (1991).
[3.64] Wooldridge, M. and Jennings, N.: "Software engineering with agents: pitfalls and pratfalls", IEEE Internet Computing, **3**(5), 20–27 (1999).
[3.65] Zhang, X., Balasubramanian, S., Brennan, R. and Norrie, D.: "Design and implementation of a real-time holonic control system", Information Science Special Issue on Computational Intelligence for Manufacturing, **127**(1–2), 23–44 (2000).

Part II

Systems Architecture

4. HMS/FB Architecture and its Implementation

James H. Christensen

Rockwell Automation, 1 Allen-Bradley Drive,
Mayfield Heights, Ohio 44121 USA
Email: jhchristensen@ra.rockwell.com

Abstract. Properly architected holonic systems can enhance the ability of each set of players in the industrial automation and control market to deliver added value by encapsulating, reusing and deploying their specialized intellectual property at succeedingly higher levels of integration. Such an architecture expands on the HMS concept of *cooperation domains* to include both low-level control (LLC) and high-level control (HLC) domains. LLC refers to normal, non-holonic control and automation functions, while HLC refers to the integration of these functions into holons through the use of software agent technology. *Function blocks,* as defined in the International Electrotechnical Commission (IEC) 61499 series of standards, can be used for encapsulation, reuse, distribution and integration of both LLC and HLC functions, while HLC functionality can be standardized as defined by the Foundation for Intelligent Physical Agents (FIPA).

4.1 Introduction

This chapter outlines an open, standards-based architecture for holonic manufacturing systems (HMS) which is capable of fulfilling the economic and technical requirements for global adoption, deployment and support. Issues arising in the implementation of this architecture, and means for their resolution, are presented. Throughout this chapter, the following definitions apply:

- *Architecture* The structure and relationship among *functional units* in a *system*, including their mutual *interfaces*. The architecture may also include the system's interfaces with its environment.
- *System* A set of interrelated elements considered in a defined context as a whole and clearly delineated from its environment.
- *Functional unit* An entity of hardware or software, or both, capable of accomplishing a specified purpose, and which may be composed of other *functional units*.
- *Interface* A shared boundary between two *functional units*, defined by functional characteristics, common physical interconnection characteristics, signal characteristics and other characteristics, as appropriate.
- *Holon* A *functional unit* capable of both *autonomy* and *cooperation*.
- *Autonomy* The extent to which an entity can create, control and monitor the execution of its own plans and/or strategies, and can take suitable corrective actions against its own malfunctions.

- *Cooperation* A process whereby a set of entities negotiate and execute mutually acceptable plans and take mutual actions against malfunctions.
- *Holonic system* A *system*, some of whose *functional units* are *holons*.
- *Holonic manufacturing system* A *holonic system* intended for application in the domain of manufacturing.

4.2 Architectural Requirements

This section describes the economic and technical requirements for a practical HMS architecture which is capable of global adoption, deployment and support.

4.2.1 Economic Requirements

Holonic systems will only be succesful in the industrial automation and control market if they enhance the ability of each set of players to deliver added value by encapsulating, reusing and deploying their specialized intellectual property (IP) at succeedingly higher levels of integration, as illustrated in Fig. 4.1.

This market exhibits the characteristics of a *network economy* as described by Shapiro and Varian [4.1]. These characteristics include *large network externalities* (user economies of scale), where the value of a technology increases exponentially with the number of users, and *positive feedback,* where successful application of the technology encourages additional users and vendors to enter the market.

A key ingredient of success in markets of this type is the rapid development of a whole range of *complementary products and services* which facilitate the deployment of the technology. In Fig. 4.1, these are identified as *runtime platforms, engineering methodologies* and *software tools*. Hence a major prerequisite for the success of HMS in this market is that it be *open* and *standards-based* to encourage early entry of suppliers of such products and services. The technological feature of *openness* will be defined in the next section of this chapter.

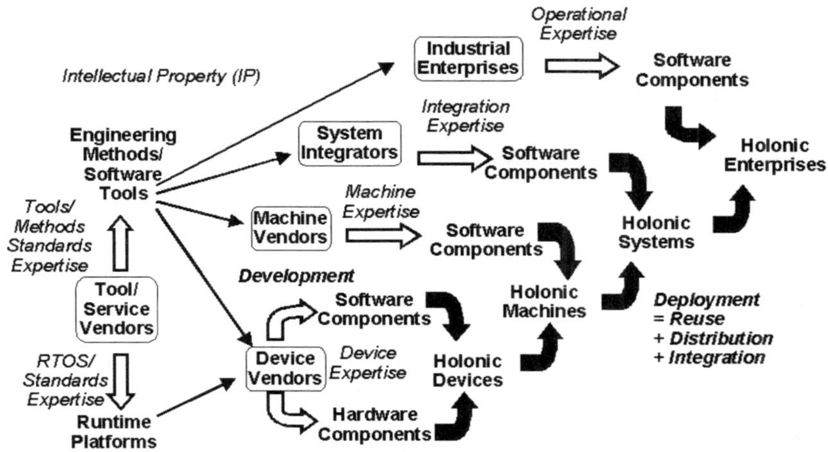

Fig. 4.1. Holonic value-add chain

4.2.2 Technical Requirements

In order to meet the economic requirements expressed above, an HMS architecture must be *component-based* to support the encapsulation and protection of intellectual property (IP). Furthermore, the components developed within this architecture must be *portable* among both software tools and runtime platforms, both in order to increase the value of the encapsulated IP and in order to encourage the emergence of a broad spectrum of software tools for component development and deployment.

HMS applications are strongly differentiated from traditional information technology (IT) applications in that they are directly involved in the control of *physical devices* and *machines*, and are highly *distributed* in nature. Hence, a successful HMS architecture must also be *inherently distributed*, and must provide a straightforward mapping from *distributed applications* to physical devices. This increases the value of the embedded IP for the device vendor, and improves the deployability and hence the value of IP components.

In order to achieve the desired end user economies of scale, the HMS architecture must be *functionally complete*, that is, capable of encapsulation and deployment of IP into components addressing all the technological requirements of holonic systems for industrial process measurement, control and automation, including:

- control and automation components,
- machine and process interface components,
- communication interface components,
- human/machine interface (HMI) components, and
- software agent components.

Industrial automation and control systems have a huge installed base with physical equipment which may have a lifetime of 10 years or longer. Hence, an important requirement for timely adoption of an HMS architecture will be its provisions for *retrofit* of existing systems and their *migration* to the HMS architecture over time.

As discussed in the preceding section, an HMS architecture must be *open*; that is, it must exhibit the following characteristics, as illustrated in Fig. 4.2:

- *Portability*
  Software tools and agents can accept and correctly interpret library elements (software components and system configurations) produced by other software tools.
- *Interoperability*
  Devices can operate together to perform the *autonomous* and/or *cooperative* functions specified by one or more distributed applications.
- *Configurability*
  Devices and their *software components* can be dynamically *configured* (selected, assigned locations, interconnected and parameterized) by multiple software *tools* and/or software *agents*.

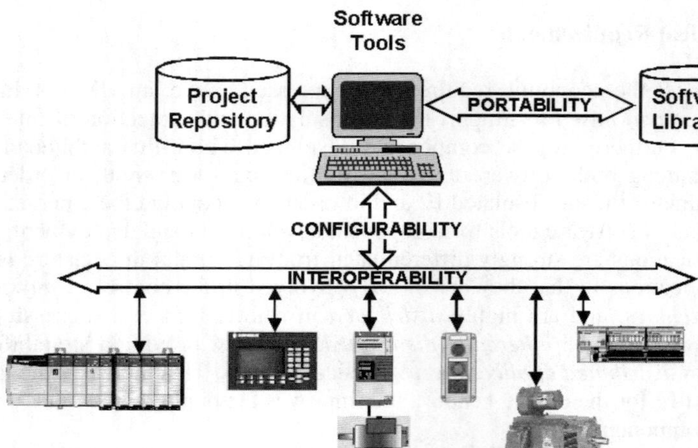

Fig. 4.2. Attributes of open industrial systems

4.3 Architecture Overview

It has been proposed [4.5] that holonic systems be viewed as consisting of one or more *cooperation domains*, that is, logical spaces in which holons may locate, contact, communicate and interact with each other. To meet the requirements expressed in section 4.2, it is proposed that this concept be expanded to include both low-level control (LLC) and high-level control (HLC) domains. LLC refers to normal, non-holonic control and automation functions, while HLC refers to the integration of these functions into holons through the use of software agent technology.

Table 4.1 summarizes the major characteristics of these two domains. To meet the critical requirements for openness, it is proposed that these domains be characterized principally in terms of two sets of standards:

- The IEC (International Electrotechnical Commission) 61499 series of standards [4.2, 4.3, 4.9] for the use of *function blocks* in distributed industrial automation and control systems.[1]
- The FIPA (Foundation for Intelligent Physical Agents) architectural standard for software agent systems [4.4].

The following sections describe in detail the proposed LLC and HLC architectures and their integration.

Table 4.1. LLC (low-level control) and HLC (high-level control = cooperation) domains

Domain	Major functions	Standards	Entities	Ontology	Response times	Message size
LLC	Control/ automation	IEC 61499 [4.2, 4.3]	Function blocks	Physical operations	10 μs to 100 ms	1 bit to 100 bytes
HLC	Negotiation/ coordination	FIPA [4.4]	Agents	Manufacturing tasks	100 ms to 10 sec	100 to 4K bytes

4.4 Low-Level Control Architecture

As shown in Fig. 4.3, the LLC architecture addresses the functions associated with the domain of real-time control, including:

- the control and automation of physical equipment;
- real-time communications among controllers;
- input/output (I/O) between controllers and the controlled machines; and
- interface among the controlled systems and their human operators, designers, installers and maintainers.

The remainder of this section describes the major architectural elements of IEC 61499 and their application to the implementation of these functions.

[1] See http://www.holobloc.com for software tools, runtime platforms and up-to-date information on the IEC 61499 standard and its application to HMS.

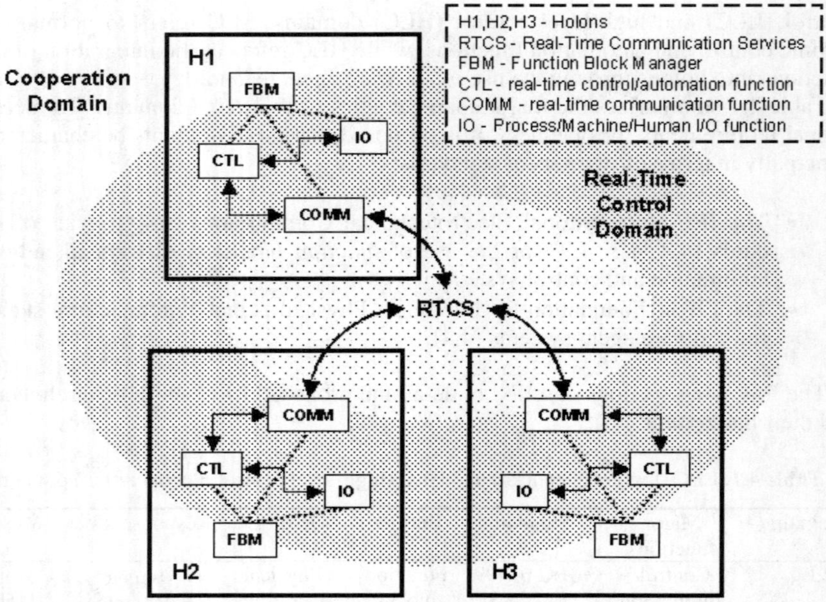

Fig. 4.3. Low-level control architecture

4.4.1 Basic Architectural Elements of IEC 61499

As shown in Fig. 4.4, the IEC 61499 model of a *system* consists of a number of *devices* (physically contained *functional units*). These devices can:

- communicate with each other over communication networks;
- interface to physical processes, equipment or machines; and
- serve as platforms for the execution of distributed applications.

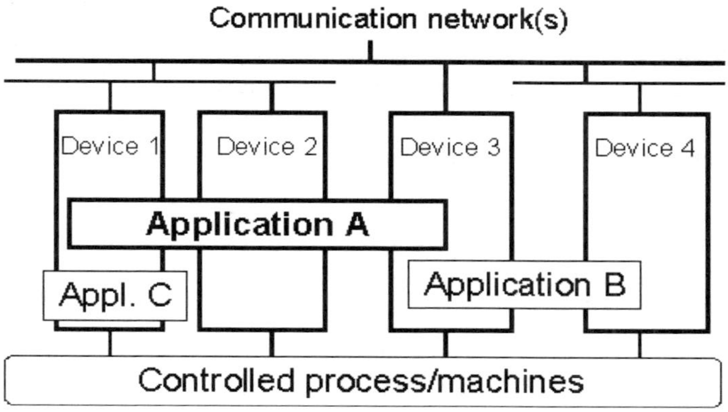

Fig. 4.4. IEC 61499 system model [4.2]

As shown in Fig. 4.5, the IEC 61499 model of an application comprises a network of function blocks interconnected by flows of events and data over event connections and data connections, respectively. As shown in Fig. 4.4, the elements of these applications are in principle distributable among multiple devices. The function blocks in turn are considered to be instances of function block types, which are specified in formal declarations using the means defined in IEC 61499-1 [4.2].

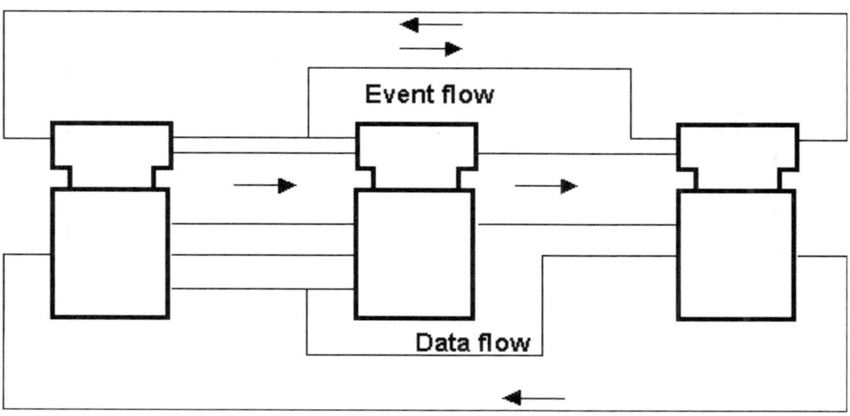

Fig. 4.5. IEC 61499 application model [4.2]

In IEC 61499 a device may consist of multiple resources, as shown in Fig. 4.6. These resources may share communication and process interfaces. Each resource may contain local applications, or the local parts of distributed applications, as shown in Fig. 4.7. In addition, the resource provides a platform for scheduling the execution of algorithms in function blocks, and for mapping underlying operating system services such as communications and machine/process I/O into service interface function blocks.

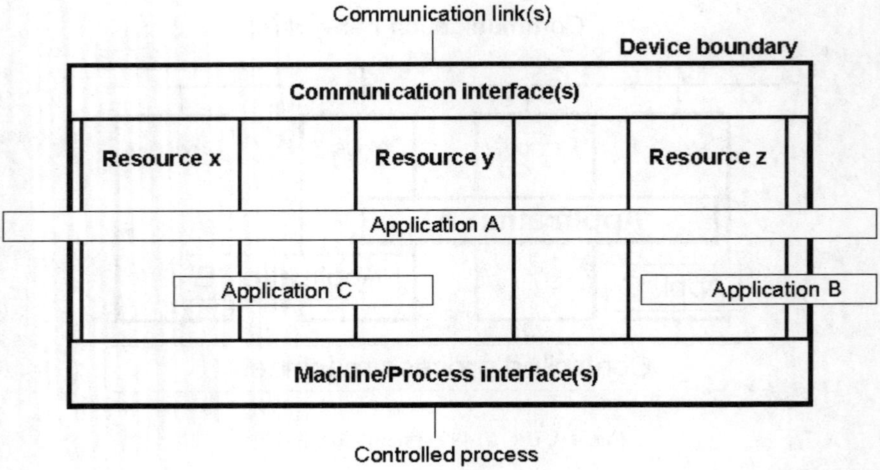

Fig. 4.6. IEC 61499 device model [4.2]

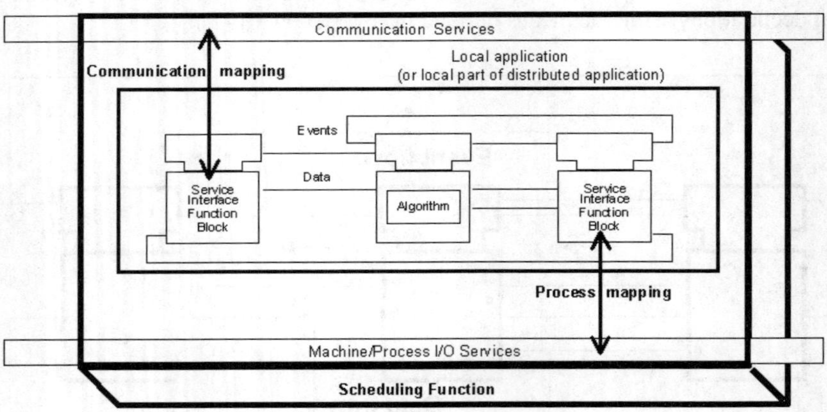

Fig. 4.7. IEC 61499 resource model [4.2]

4.4.2 Reference Example

The example shown in Fig. 4.8 will be used throughout this section to illustrate the application of the elements of IEC 61499 to meet the requirements of the LLC architecture. In this example, an actuator can move a workpiece along a slide in the "forward" direction at a velocity *VF* and in the "reverse" direction at a velocity *VR*. These velocities are characteristic of the particular physical actuator. In some mechanisms these velocities may even be time-varying according to a predetermined "velocity profile". The workpiece itself may be be another mechanism, for instance, an end effector such as a gripper.

Associated with the mechanism are two sensors: a HOME sensor, which is activated when the workpiece has moved to the end of the slide in the reverse direction, and an END sensor, which is activated when the workpiece has moved to the end of the slide in the forward direction. These sensors may exhibit hysteresis as a function of the sensor characteristics and their interaction with the mechanism.

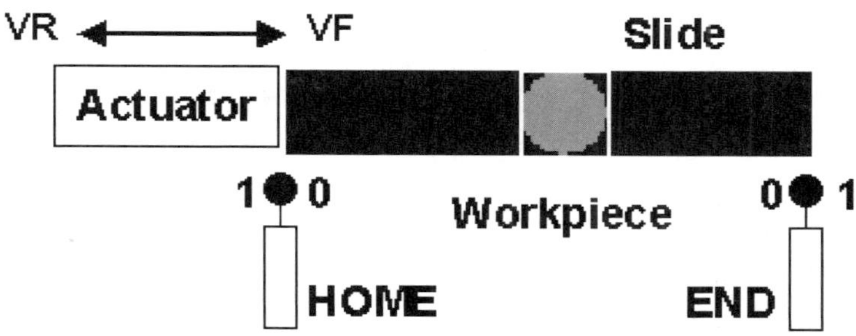

Fig. 4.8. A bidirectional mechanism

4.4.3 Interface Abstraction

IEC 61499-1 [4.2] defines the *adapter interface* construct to permit the abstraction of interfaces between function blocks. This has three major benefits, as will be seen later:

1. It permits the representation of multiple data and event connections between function blocks with a single line, thus helping to eliminate diagram clutter.
2. It enables the abstract representation of patterns of interaction between function blocks through the use of **service sequence** notation.
3. It facilitates the reuse of the interaction patterns for differing functional implementations , for instance in converting from a simulated to an actual physical system.

For the example of the bidirectional mechanism, this pattern is represented as shown in Fig. 4.9. Here a CMD event is used to convey Boolean values FWD and REV from a control algorithm, indicating that motion is commanded in the forward or reverse direction, respectively. Similarly, an IND event is used to convey changes in the boolean values of the FWD and REV sensors back to the control algorithm. Typical sequences of operation are shown in Figures 4.9b-d, and the corresponding values conveyed are shown in Table 4.2 in the informal notation of IEC 61499.

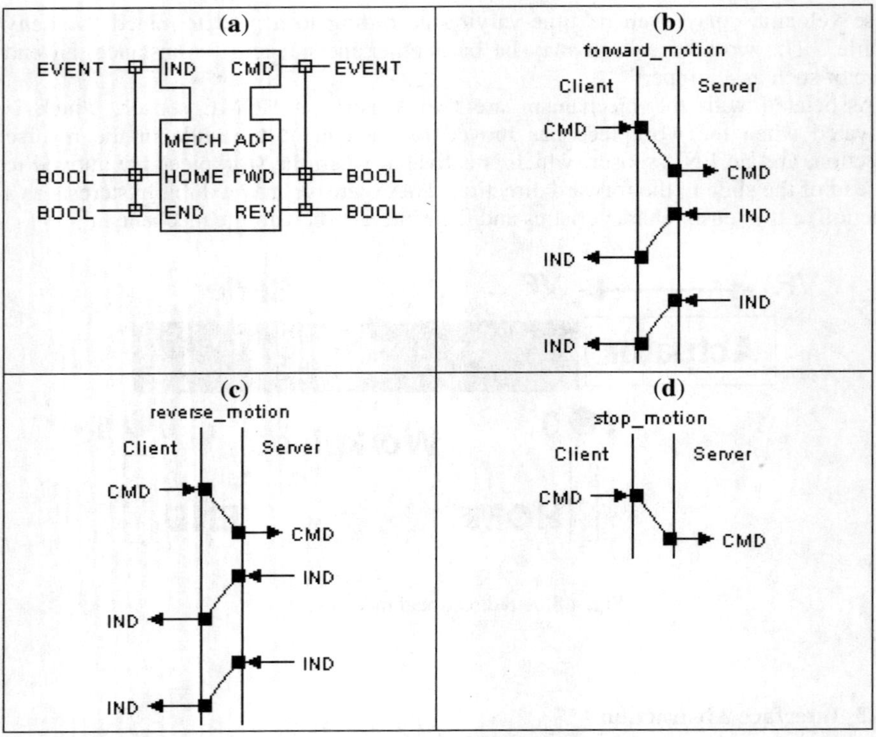

Fig. 4.9. Example of IEC 61499 adapter interface. (**a**) Interface. (**b**) Forward motion sequence. (**c**) Reverse motion sequence. (**d**) Stop motion sequence

Table 4.2. Operational sequences of mechanism interface

```
SERVICE Client/Server
SEQUENCE forward_motion
  Client.CMD(FWD=1,REV=0) -> Server.CMD(FWD,REV);
  Server.IND(HOME=0) -> Client.IND(HOME);
  Server.IND(END=1) -> Client.IND(END);
END_SEQUENCE

SEQUENCE reverse_motion
  Client.CMD(FWD=0,REV=1) -> Server.CMD(FWD,REV);
  Server.IND(END=0) -> Client.IND(END);
  Server.IND(HOME=1) -> Client.IND(HOME);
END_SEQUENCE

SEQUENCE stop_motion
  Client.CMD(FWD=0,REV=0) -> Server.CMD(FWD,REV);
END_SEQUENCE
END_SERVICE
```

4.4.4 Control and Automation Functions

In the IEC 61499 LLC architecture, combinational and sequential control functions are typically performed by *basic* function blocks, i.e., *instances* of basic function block *types*. In these types, the execution of control algorithms is under the control of event-driven state machines represented as *Execution Control Charts* (ECCs). As shown in Fig. 4.10, each *state* in the ECC is associated with one or more *actions*. Each action consists of zero or one *algorithm* to be executed and zero or one *output event* to be issued upon the completion of algorithm execution (or immediately if no algorithm is associated with the state). The algorithms are executed and output events issued as soon as possible after activation of the associated execution control state (EC state). Transition conditions between states are expressed as Boolean combinations of event inputs and Boolean expressions involving input, output and internal variables of the function block. A transition is *cleared*, that is, its predecessor state is deactivated and its successor state is activated when triggered by an associated event when its transition condition is true.[2]

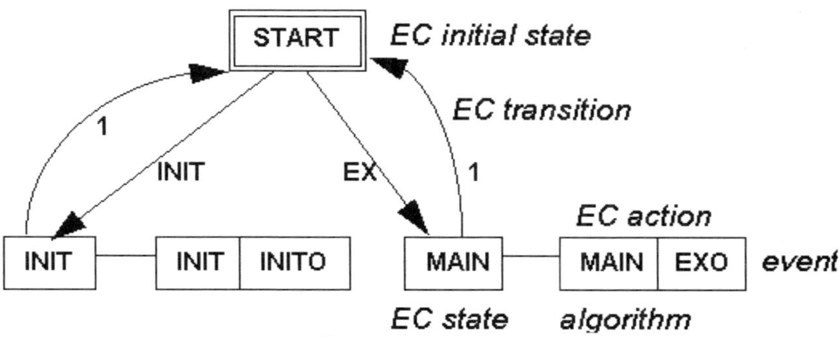

Fig. 4.10. Execution control chart (ECC) elements [4.2]

[2] Subclause 2.2.2.2 of IEC 61499 [4.2] provides a rather complex model for the operation of ECCs. A simpler model based on Harel state charts [4.6] is under consideration. In either case, transition conditions consisting of a single event, a pure Boolean condition not involving an event, or an AND combination of both will function as expected.

Figure 4.11 shows the usage of an ECC in a basic function block which provides cyclic operation of the bidirectional mechanism shown in Fig. 4.8. Operation for a single cycle only, or continuous cycling is provided.

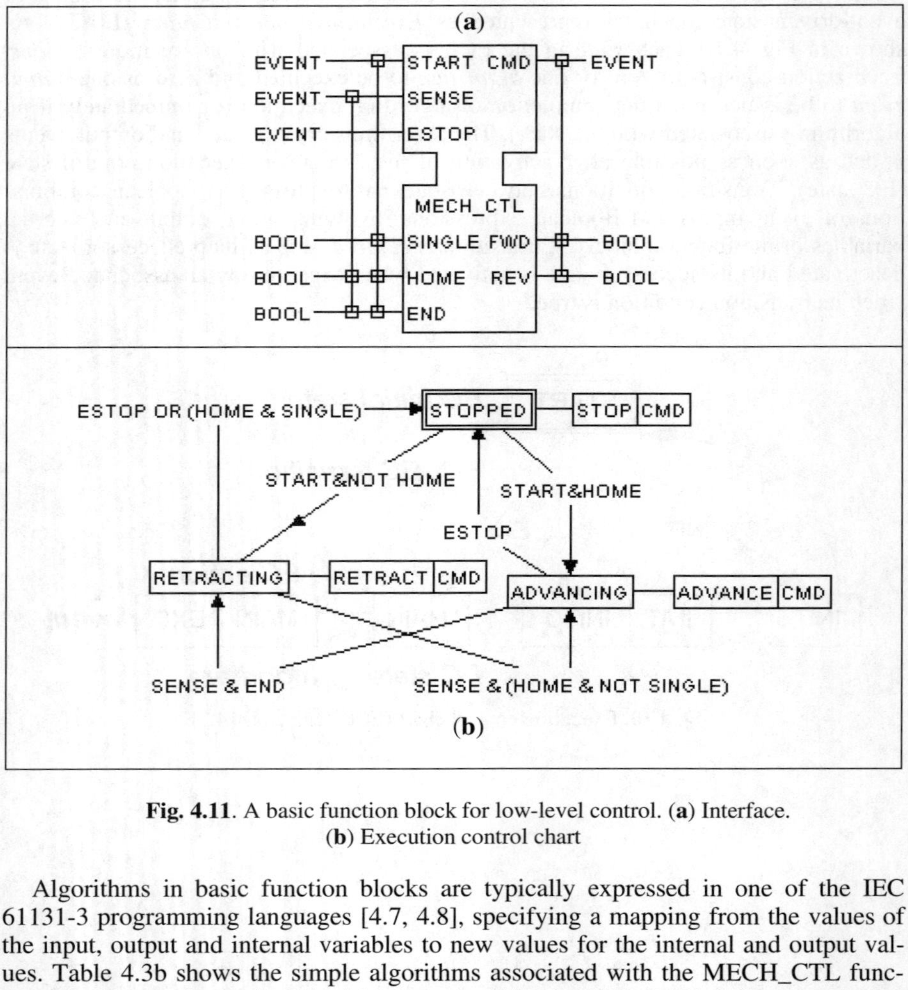

Fig. 4.11. A basic function block for low-level control. (**a**) Interface. (**b**) Execution control chart

Algorithms in basic function blocks are typically expressed in one of the IEC 61131-3 programming languages [4.7, 4.8], specifying a mapping from the values of the input, output and internal variables to new values for the internal and output values. Table 4.3b shows the simple algorithms associated with the MECH_CTL function block type shown in Fig. 4.11, expressed in the IEC 61131-3 Structured Text (ST) language [4.7]. Table 4.3a shows the informal semantics of the inputs and outputs of this function block type expressed in the textual syntax defined in IEC 61499-1 [4.2].

Table 4.3. Partial textual declarations for the MECH_CTL function block type. (**a**) Interface declarations. (**b**) Algorithms

(a)

```
EVENT_INPUT
  START WITH SINGLE, HOME, END; (* Start Cycle *)
  SENSE WITH HOME, END; (* Sensor Change Notification *)
  ESTOP; (* Emergency (Stop Immediately) *)
END_EVENT

EVENT_OUTPUT
  CMD WITH FWD, REV; (* FWD/REV Command *)
END_EVENT

VAR_INPUT
  SINGLE : BOOL; (* 1=Single Cycle, 0=Auto-Repeat *)
  HOME : BOOL; (* Tool at HOME position *)
  END : BOOL; (* Tool at END position *)
END_VAR

VAR_OUTPUT
  FWD : BOOL; (* Forward actuation *)
  REV : BOOL; (* Reverse actuation *)
END_VAR
```

(b)

```
ALGORITHM ADVANCE IN ST :
  FWD:=TRUE;
  REV:=FALSE;
END_ALGORITHM

ALGORITHM RETRACT IN ST :
  FWD:=FALSE;
  REV:=TRUE;
END_ALGORITHM

ALGORITHM STOP IN ST :
  FWD:=FALSE;
  REV:=FALSE;
END_ALGORITHM
```

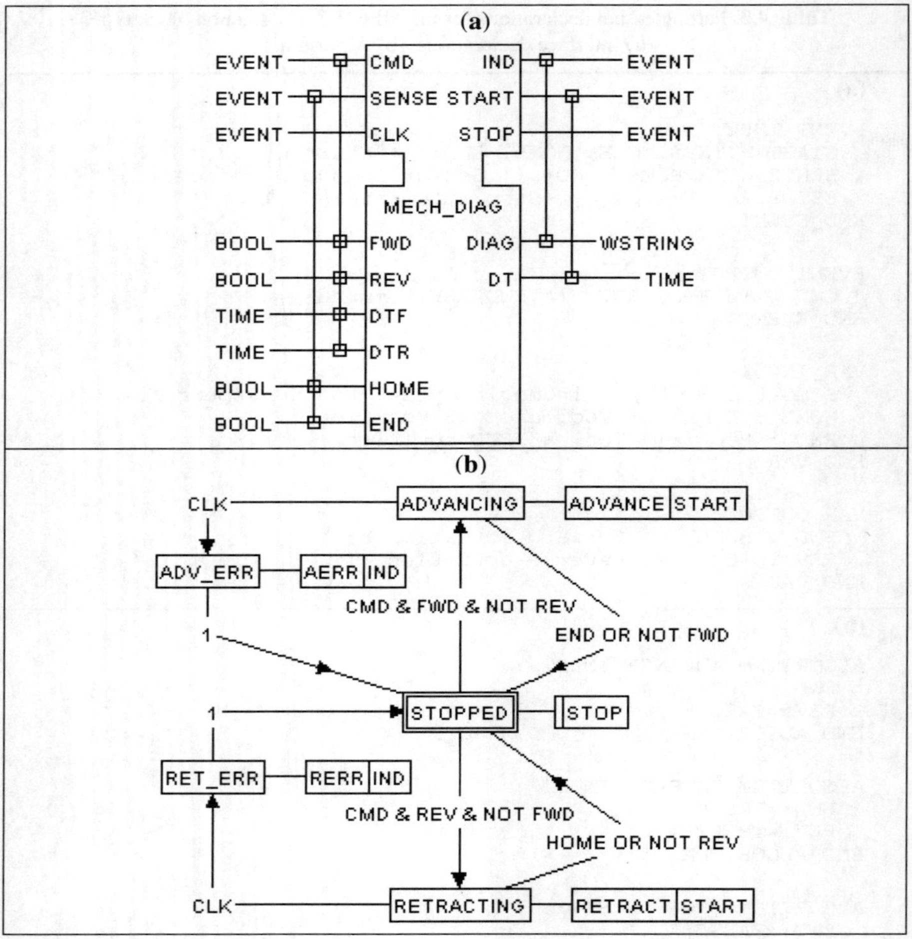

Fig. 4.12. A basic function block for diagnostics. (**a**) Interface. (**b**) Execution control chart

4.4.5 Diagnostic Functions

Figure 4.12 shows a function block type that may be used for model-based diagnostics of the operation of the bidirectional mechanism shown in Fig. 4.8. It is intended for use in conjunction with the operational control block MECH_CTL. shown in Fig. 4.11 and Table 4.3, and an external timer of the IEC 61499-1 E_DELAY type [4.2]. The function block's algorithms and the informal semantics of its inputs and outputs are shown in Table 4.4.

Table 4.4. Partial textual declarations for the MECH_DIAG function block type. (**a**) Interface declarations. (**b**) Algorithms

(a)

```
EVENT_INPUT
  CMD WITH FWD, REV, DTF, DTR; (* Motion Command *)
  SENSE WITH HOME, END; (* Sensor Change Notification *)
  CLK; (* External Timer Expired *)
END_EVENT

EVENT_OUTPUT
  IND WITH DIAG; (* Diagnostic Indication *)
  START WITH DT; (* Start External Timer *)
  STOP; (* Stop External Timer *)
END_EVENT

VAR_INPUT
  FWD : BOOL; (* Move in HOME -> END direction *)
  REV : BOOL; (* Move in END -> HOME direction *)
  DTF : TIME := t#5s; (* Maximum HOME -> END delay *)
  DTR : TIME := t#3s; (* Maximum END-> HOME delay *)
  HOME : BOOL; (* Tool at HOME position *)
  END : BOOL; (* Tool at END position *)
END_VAR

VAR_OUTPUT
  DIAG : WSTRING; (* Diagnostic String *)
  DT : TIME; (* Delay Period for External Timer *)
END_VAR
```

(b)

```
ALGORITHM ADVANCE IN ST :
DT:=DTF;
END_ALGORITHM

ALGORITHM RETRACT IN ST :
DT:=DTR;
END_ALGORITHM

ALGORITHM AERR IN ST :
DIAG:="TIMEOUT_ADVANCING";
END_ALGORITHM

ALGORITHM RERR IN ST :
DIAG:="TIMEOUT_RETRACTING";
END_ALGORITHM
```

4.4.6 Functional Composition

IEC 61499 provides for the construction of IP through the reuse and functional composition of lower-level encapsulated IP. This is done through the construction of *composite* function block types, whose bodies consist of networks of interconnected lower-level *component* function blocks. The execution control of the composite function blocks in an instance of this type is performed through the propagation of input events via the event connections of these networks. Therefore, a function block of this type does not directly utilize an ECC for execution control.

As an example of such encapsulation, consider the fact that the control and diagnostic functions for the bidirectional mechanism described earlier are closely related. Hence, it adds value for such functions to be "packaged" together in a composite function block as illustrated in Fig. 4.13. This provides the potential to make the mechanism and its associated control device more "intelligent", i.e. both automatically operating and self-diagnosing.

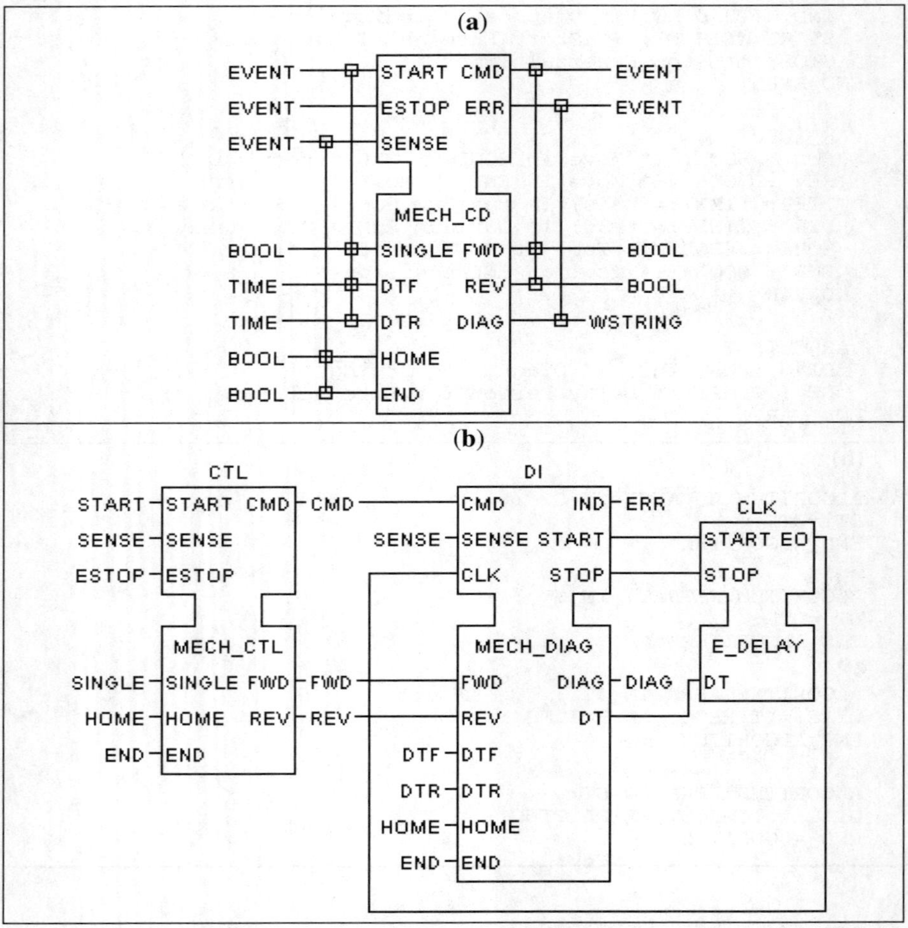

Fig. 4.13. A composite function block. (**a**) Interface. (**b**) Body

4.4.7 Human Interface Functions

Human interface services provided by an underlying operating environment may be encapsulated in IEC 61499 *service interface* function blocks. These blocks in turn may be combined into composite function blocks to deliver an appropriate set of human interface elements for a particular application. For instance, Fig. 4.14 shows the encapsulation of an appropriate set of human interface elements for the cyclic operation of the bidirectional mechanism described above.

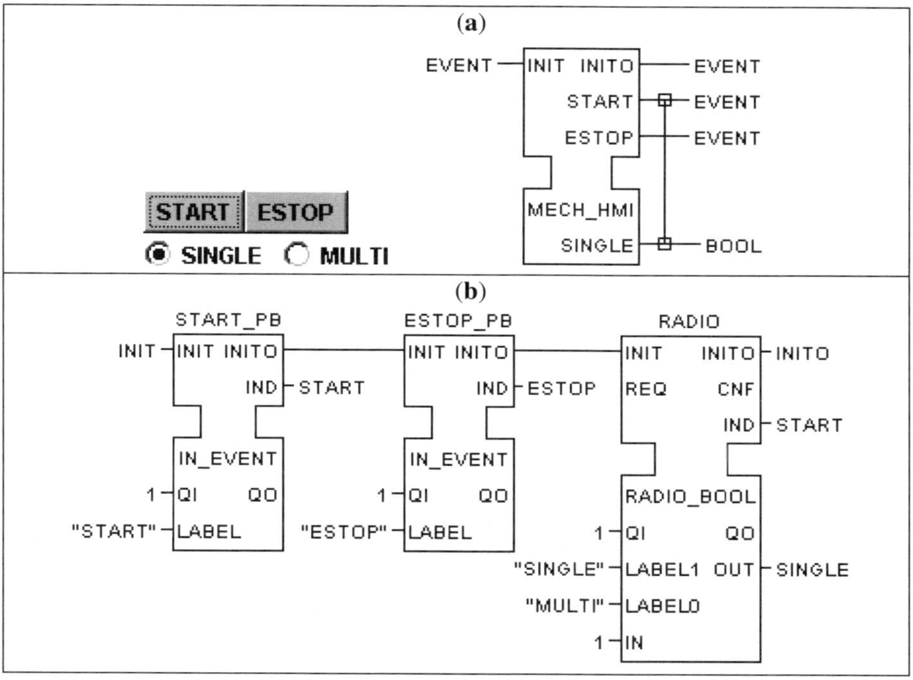

Fig. 4.14. A composite human-interface function block.
(**a**) Display and functional interface. (**b**) Body

4.4.8 Modeling, Simulation and Animation Functions

Modifications to the classical Model/View/Controller (MVC) design pattern [4.6] have been proposed [4.10] to permit the use of IEC 61499 for the purposes of modeling and simulation of intelligent devices and systems, and animated display of the results of such simulations. Inclusion of Diagnostic (D) and Adapter (A) elements leads to a proposed MVCDA design pattern.[3]

[3] See http://www.holobloc.com/doc/despats/ for more extensive descriptions of the use of design patterns for IEC 61499.

More recent experience[4] has indicated that a functional partitioning into Model/View/Adapter (MVA) and Control/Diagnostic/Adapter (CDA) groupings is more advantageous in the development and debugging of LLC systems. This modified design pattern is accordingly denoted MVA/CDA. The advantages of this partitioning are seen particularly in managing the transition from simulated to actual physical systems.

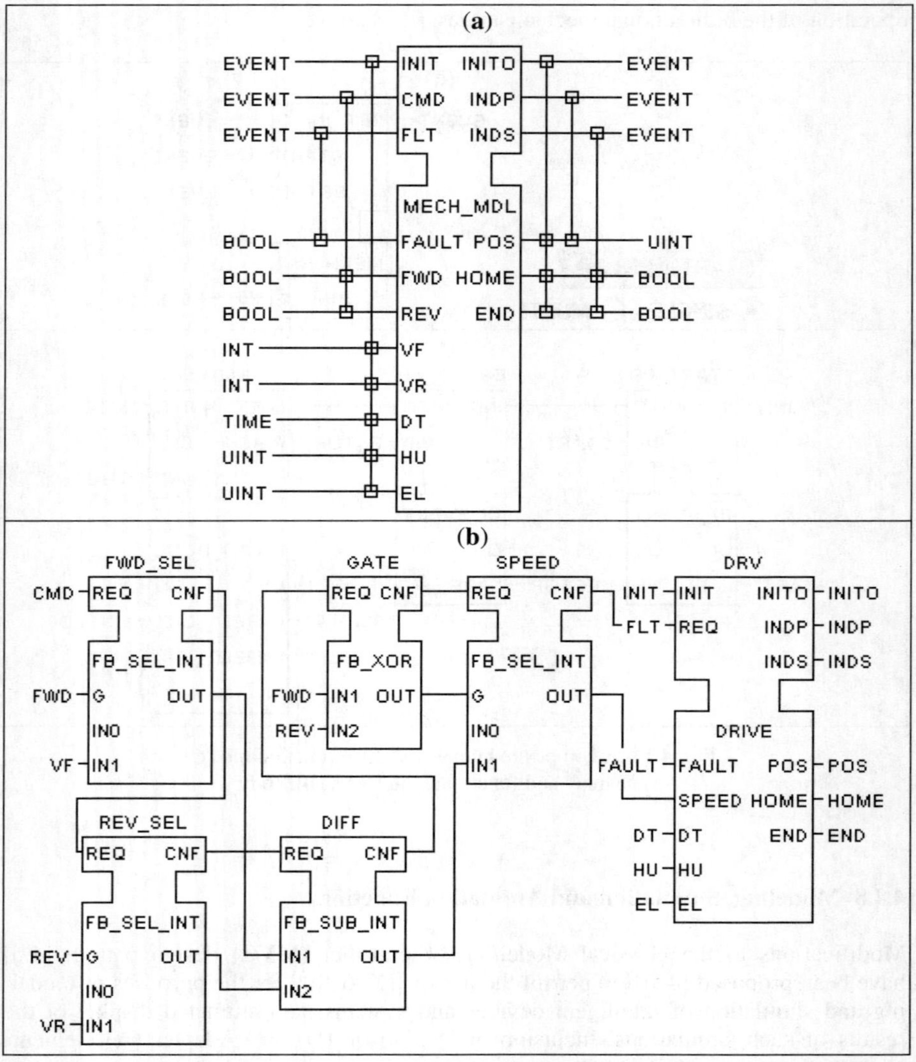

Fig. 4.15. A simulation model of the bidirectional mechanism. (**a**) Interface. (**b**) Body

[4] The author is indebted to Mr. Franz Auinger and Mr. Werner Rumpl of Profactor GmbH, Steyr, AT for their demonstration of the benefits of the MVA/CDA design pattern.

Figure 4.15 shows a composite function block type capable of simulating the operation of the bidirectional mechanism shown in Fig. 4.8. This utilizes standard component function blocks defined as specified in Annex D of IEC 61499-1 [4.2], in addition to a specialized service interface function block type (DRIVE) for simple motion simulation.[5] Table 4.5 provides documentation of the inputs and outputs of this composite function block type.

Table 4.5. Interface declarations for the MECH_MDL function block type

```
EVENT_INPUT
  INIT WITH VF, VR, DT, HU, EL;
  CMD WITH FWD, REV; (* Motion command *)
  FLT WITH FAULT; (* Change in FAULT status *)
END_EVENT

EVENT_OUTPUT
  INITO WITH POS, HOME, END;
  INDP WITH POS; (* Position change *)
  INDS WITH HOME, END; (* Sensor change *)
END_EVENT

VAR_INPUT
  FAULT : BOOL; (* 0=Enable,1=Simulate Fault(pause) *)
  FWD : BOOL; (* Move Forward at VF *)
  REV : BOOL; (* Move Back at VR *)
  VF : INT := 20; (* Forward speed in percent/sec *)
  VR : INT := 40; (* Reverse speed in percent/sec *)
  DT : TIME := t#250ms; (* Simulation clock interval *)
  HU : UINT := 10; (* Upper(advancing)limit of HOME LS *)
  EL : UINT := 90; (* Lower(retracting)limit of END LS *)
END_VAR

VAR_OUTPUT
  POS : UINT; (* Position, 0->100% *)
  HOME : BOOL := true; (* HOME position, 0->HU *)
  END : BOOL; (* END position, EL<-100 *)
END_VAR
```

Figure 4.16 illustrates the combination of the MECH_MDL function block described above with an XBAR_VIEW service interface function block displaying the progress of the simulation. Table 4.6 provides a description of the XBAR_VIEW inputs. A typical appearance of the XBAR_VIEW element is shown in Fig. 4.17. Note the small colored circle at the upper left of the display. The user can click on this subelement to toggle the FAULT output; this variable is used by the MECH_MDL function block to stop motion, thus simulating a simple fault. The colored circle changes from red when FAULT is true to green when FAULT is false, giving the user some feedback on the simulated FAULT state.

[5] See http://www.holobloc.com/doc/despats/mvc/DRIVE.htm for documentation of the DRIVE function block type.

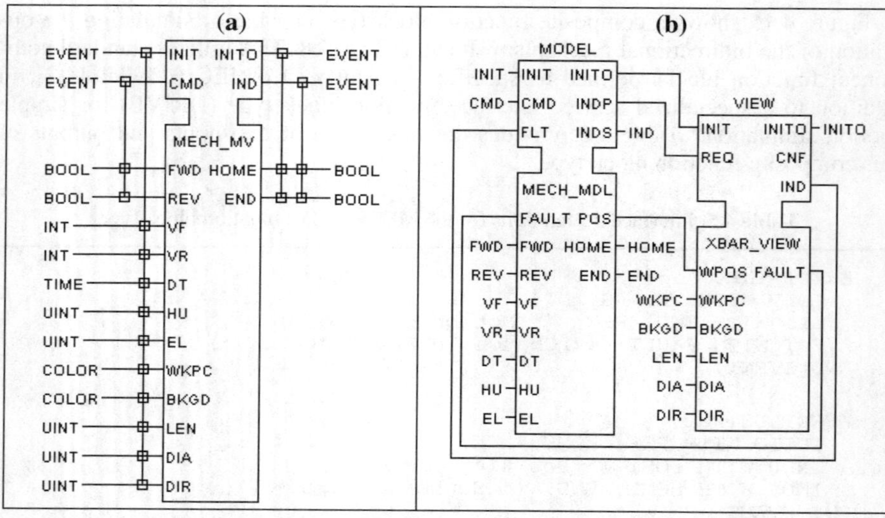

Fig. 4. 16. Composite function block for simulation and display of the bidirectional mechanism. (**a**) Interface. (**b**) Body

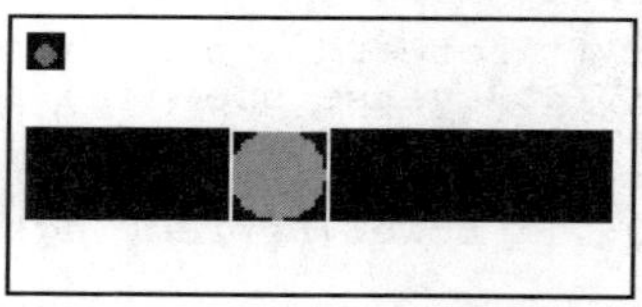

Fig. 4. 17. Appearance of the XBAR_VIEW element

Table 4.6. Partial interface declarations for the MECH_MV function block type

```
WKPC : COLOR := cyan; (* Workpiece Color *)
BKGD : COLOR := blue; (* Transfer Bar Color *)
LEN : UINT; (* Bar Length in Diameters *)
DIA : UINT; (* Workpiece diameter *)
DIR : UINT; (* Orientation:0=L/R,1=T/B,2=R/L,3=B/T *)
```

Adapter interfaces can be used to simplify the interconnection of control and diagnostic elements with simulation and display elements, and to facilitate the conversion from simulated to physical devices. The convention is that addition of adapters to a Model/View (MV) element converts it to a Model/View/Adapter (MVA) element, and addition of adapters to a Control/Diagnostic (CD) element converts it to a Control/Diagnostic/Adapter (CDA) element. Figure 4.18 illustrates the application of this principle to the previously developed MECH_MV and MECH_CD function block types.

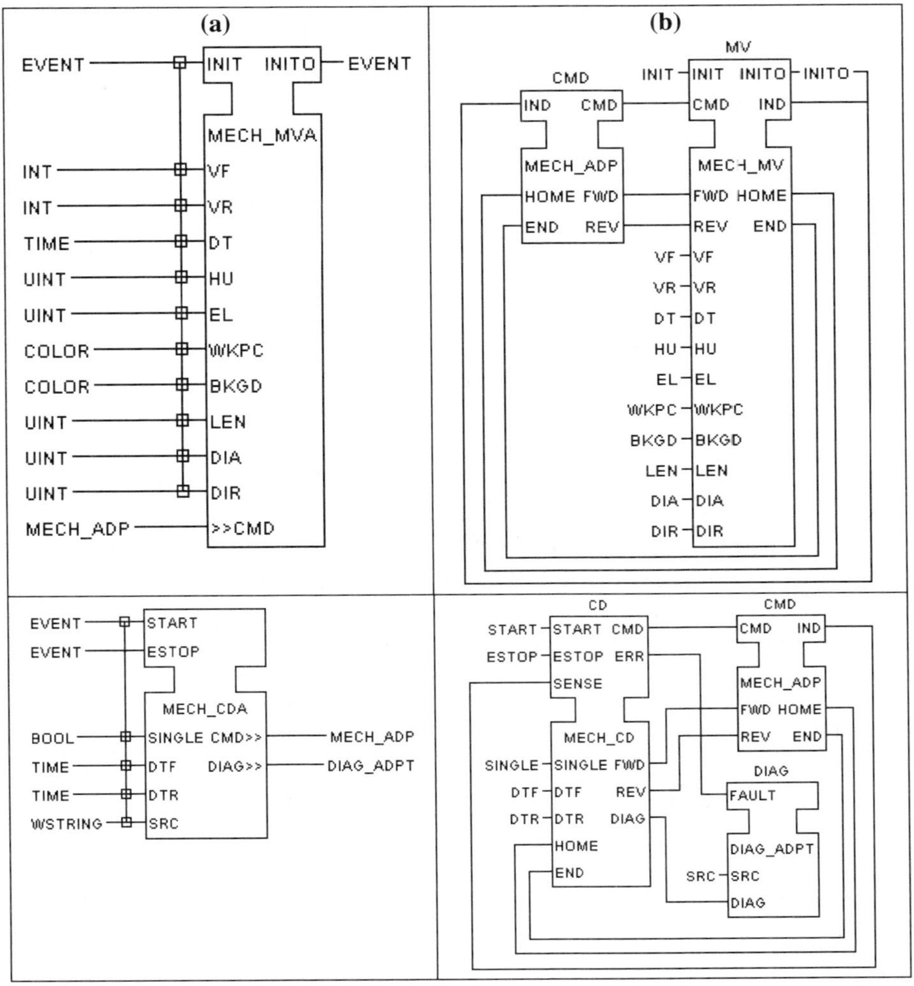

Fig. 4. 18. Addition of adapter interfaces. (**a**) External interface. (**b**) Body

4.4.9 System Configurations

A simple system configuration for testing the combined MECH_MVA and MECH_CDA functionality is shown in Fig. 4.19. This system contains a single device of type FRAME_DEVICE, which in turn contains a single resource of type PANEL_RESOURCE. These elements encapsulate the functionality of the classes Frame and Panel, respectively, in the Java AWT user interface [4.11]. The resource is populated with instances of MECH_CDA, MECH_MVA and associated user interface elements. An instance of the IEC 61499 standard type E_RESTART is used for initialization.

The appearance of the resulting user interface is shown in Fig. 4.20, and the composite user interface function block DIAG_HMI is shown in Fig. 4.21.

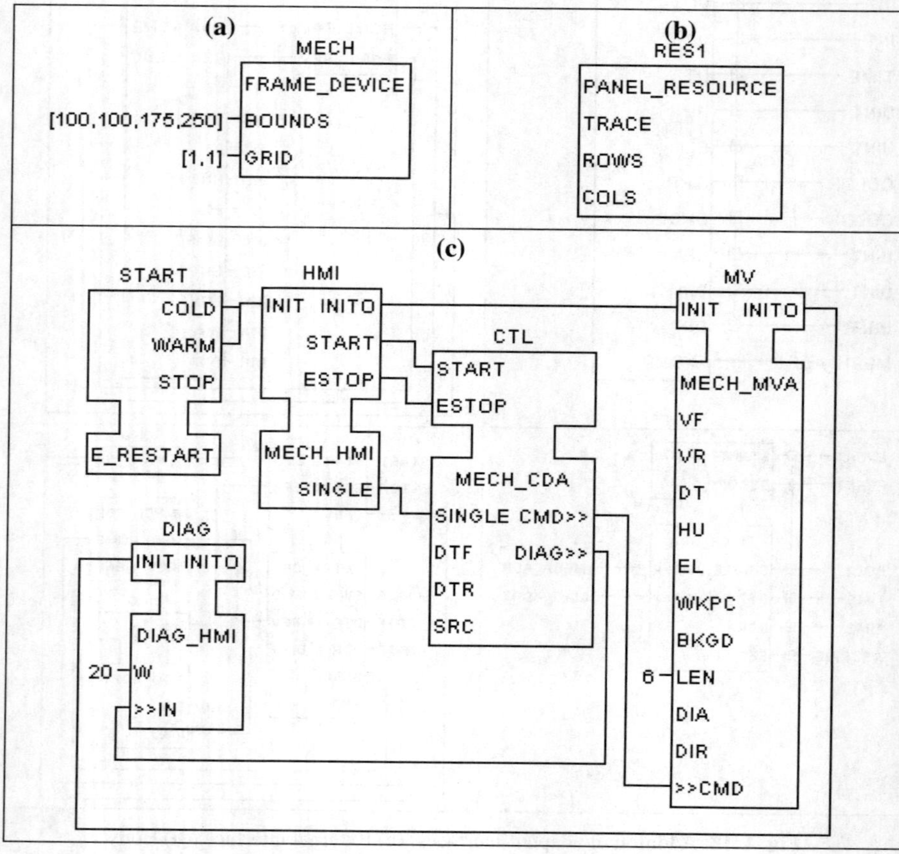

Fig. 4. 19. A simple system configuration. (**a**) System view. (**b**) Device view. (**c**) Resource view

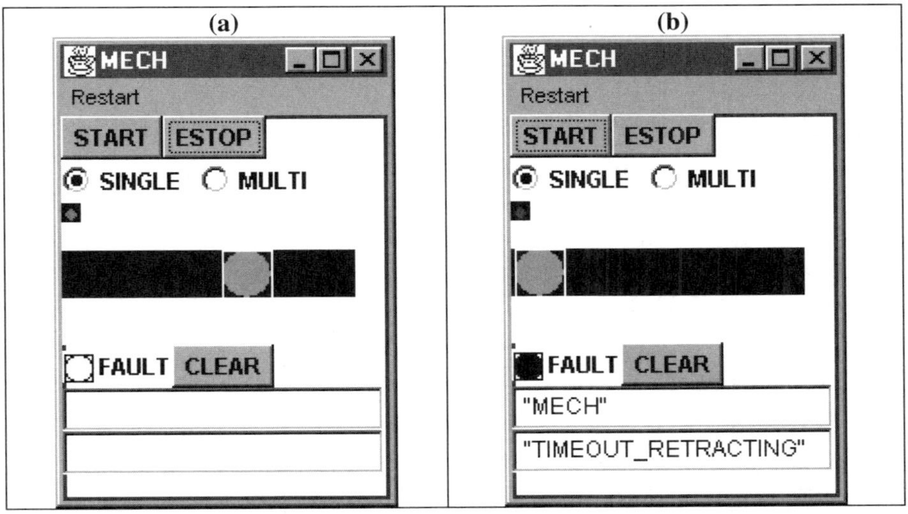

Fig. 4.20. The user interface to the simulation. (**a**) Normal operation. (**b**) Faulted operation

4.4.10 Communication Functions

Annex F of IEC 61499-1 [4.2] provides informative definitions for generic service interface function blocks to the most commonly used communication services in LLC, namely broadcast PUBLISH/SUBSCRIBE and one-to-one CLIENT/SERVER connections. In addition, IEC 61499-1 Annex F defines ASN.1 abstract syntax [4.12], and basic [4.13] and compact encoding rules for the IEC 61131-3 [4.7] elementary and derived data types used in IEC 61499.

Since this information is not normative, additional technical agreements, e.g. [4.15], must be employed to ensure interoperability of devices using these service interfaces.

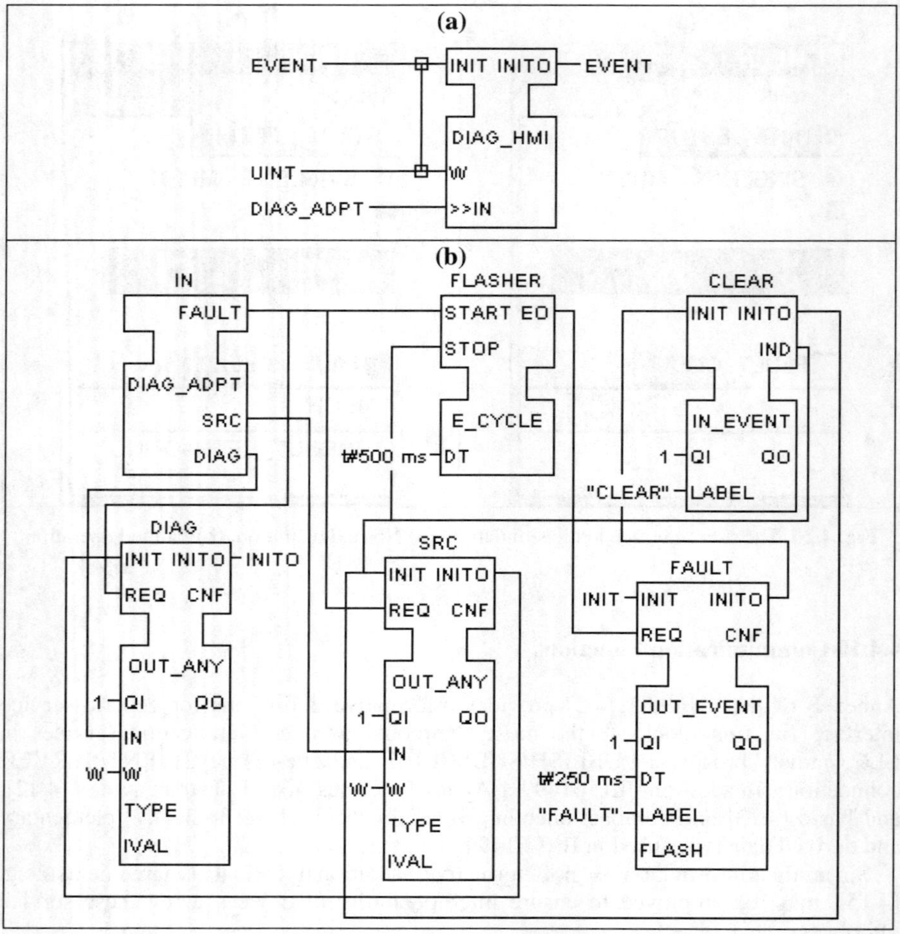

Fig. 4.21. A human interface for diagnostic messages. (**a**) Functional interface. (**b**) Body

4.4.11 Resource and Device Type Specifications

IEC 61499 provides for the definition of *device types* and *resource types*, for instance the previously illustrated FRAME_DEVICE and PANEL_RESOURCE types, respectively. Resources may be considered as providing "plug-in module" functionality within devices. To illustrate this point, consider the MECH_PNL resource type shown in Fig. 4.22. Note that the only interface defined for such types is a set of input variables that may be used for configuring instances of the type. Note also the provision of communication service interface function blocks to establish connectivity to the "outside world":

- An instance of the type SUBSCRIBE_1 receives the START event and value of the SINGLE variable for the MECH_CDA instance.
- An instance of the type SUBSCRIBE_0 receives the ESTOP event for the MECH_CDA instance.
- An instance of the composite DIAG_PUB type defined in Fig. 4.23 is used to publish the diagnostic information from the MECH_CDA instance. This publiched information is received by instances of the DIAG_SUB type also shown in Fig. 4.23.

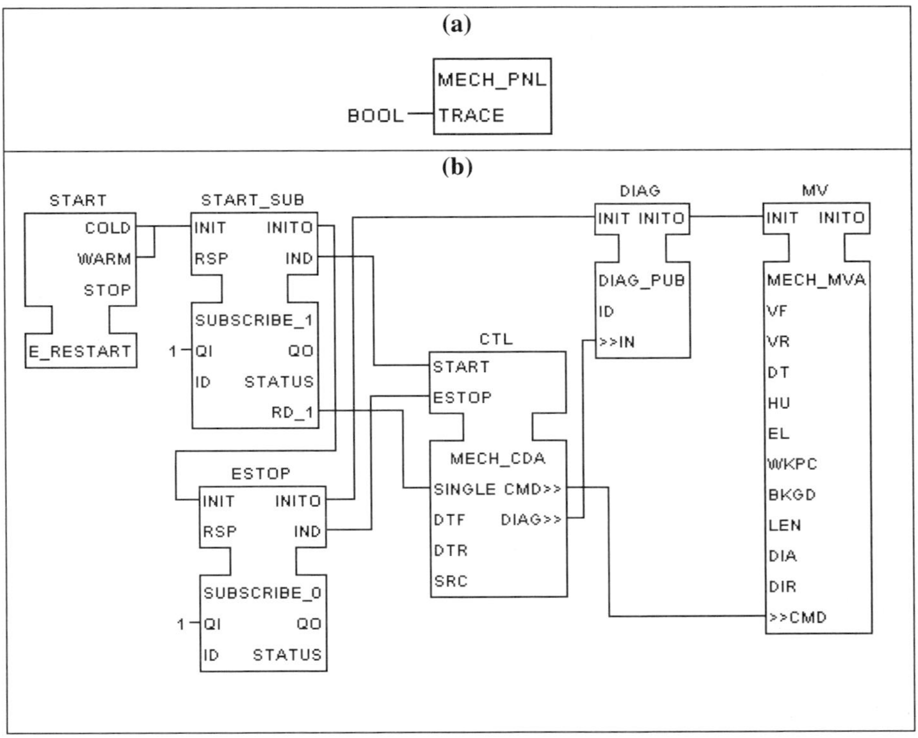

Fig. 4.22. A resource type definition. (**a**) Configuration interface. (**b**) Body

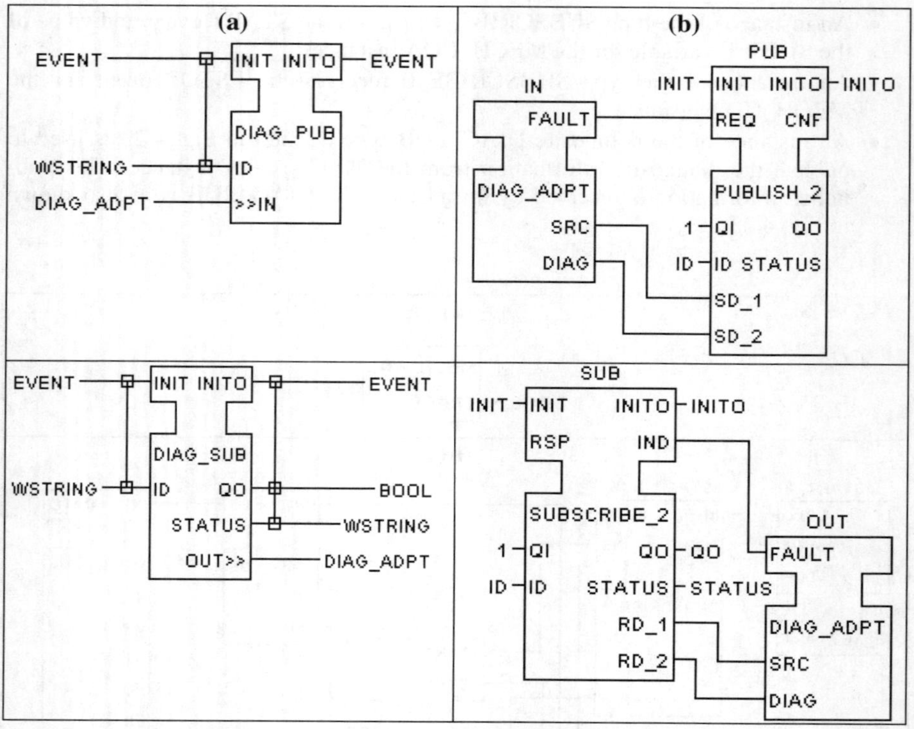

Fig. 4.23. Composite communication service interface types. (**a**) Functional interfaces. (**b**) Bodies

The system configuration shown in Fig. 4.24 presents a simple example of the integration of multiple distributed resources. The use of a grid layout illustrates the use of a device as a "rack" or "manifold" for multiple intelligent electomechanical devices, each of which is represented as an IEC 61499 resource within the manifold.

In this distributed system configuration, the operator interface is represented as a separate IEC 61499 device. Since the PUBLISH and SUBSCRIBE service interface function blocks in this implementation utilize multicast communications, a single operator interface can be used to control the operation of and receive diagnostic messages from multiple intelligent electromechanical devices of the same IEC 61499 resource type (in this example, the instances MECH1 and MECH2 of the MECH_PNL resource type in the MANIFOLD device). The OPERATOR.DISPLAY resource of this system configuration is shown in Fig. 4.24d.

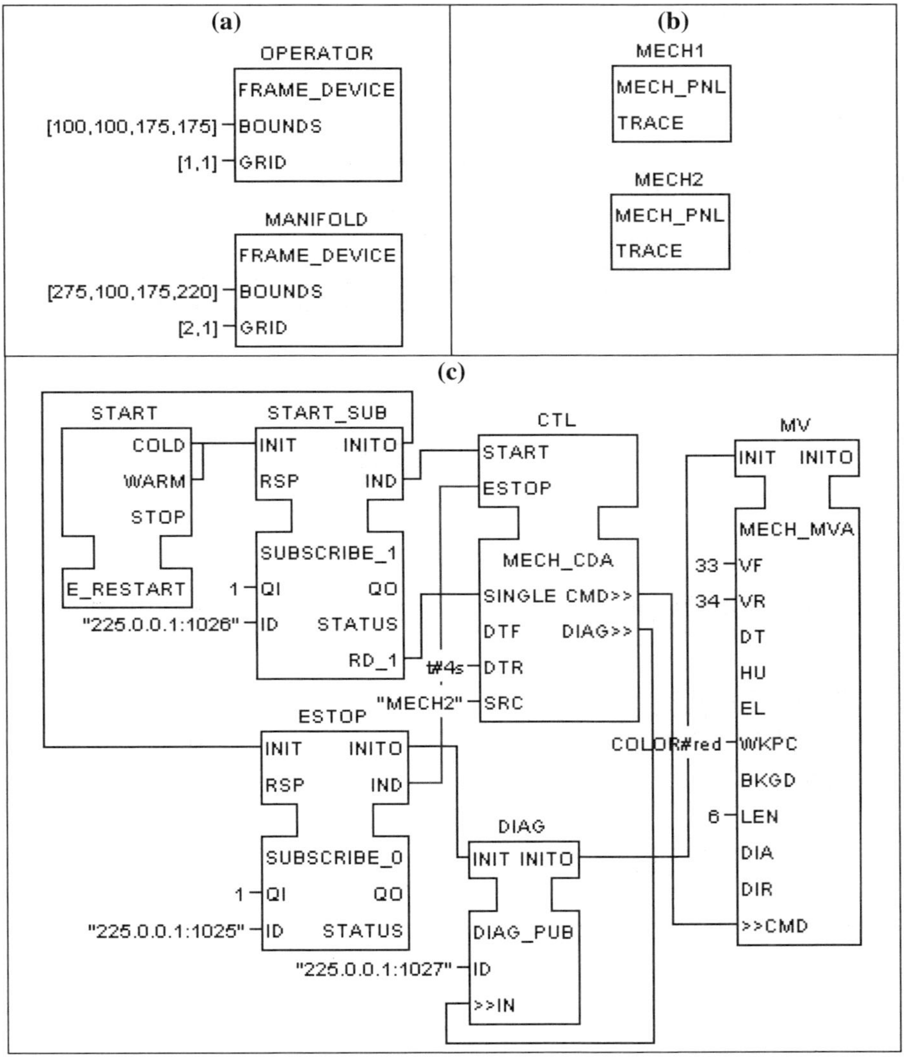

Fig. 4.24. A distributed system configuration. (**a**) System view. (**b**) MANIFOLD device view. (**c**) MANIFOLD.MECH2 resource view. (**d**) OPERATOR.DISPLAY resource view

Figure 4.25 shows the appearance of the operator interface, and of the manifold with two simulated electromechanical devices in a 2-row by 1-column grid, for the system configuration given in Fig. 4.24. Figures 4.25b,c show that the operator interface can receive error messages from either of the simulated electromechanical devices. Physical or simulated devices can easily be added to this system and be compatible with the same operator interface, as long as such devices receive their operative commands and send their diagnostic messages through the same UDP multicast socket addresses (for instance, they may receive the ESTOP command via a SUBSCRIBE_0 block with UDP socket address 225.0.0.1: 1025).

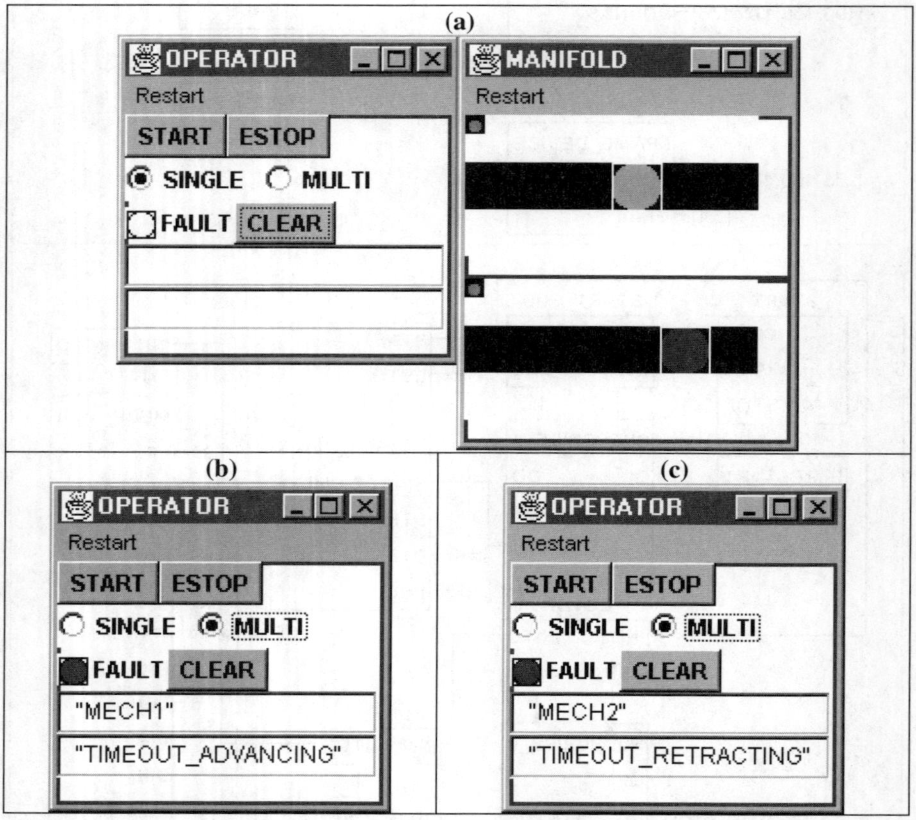

Fig. 4.25. The user interface to the simulation. (**a**) Normal operation. (**b**) MECH1 fault. (**c**) MECH2 fault

4.4.12 Software Portability

Portability of data types and other library elements (adapter types, function block types, resource types, device types and system configurations) is achieved in IEC 61499 via the use of XML [4.14] documents for exchange of library elements among software tools. The XML Document Type Definitions (DTDs) are given in IEC 61499-2 [4.3].

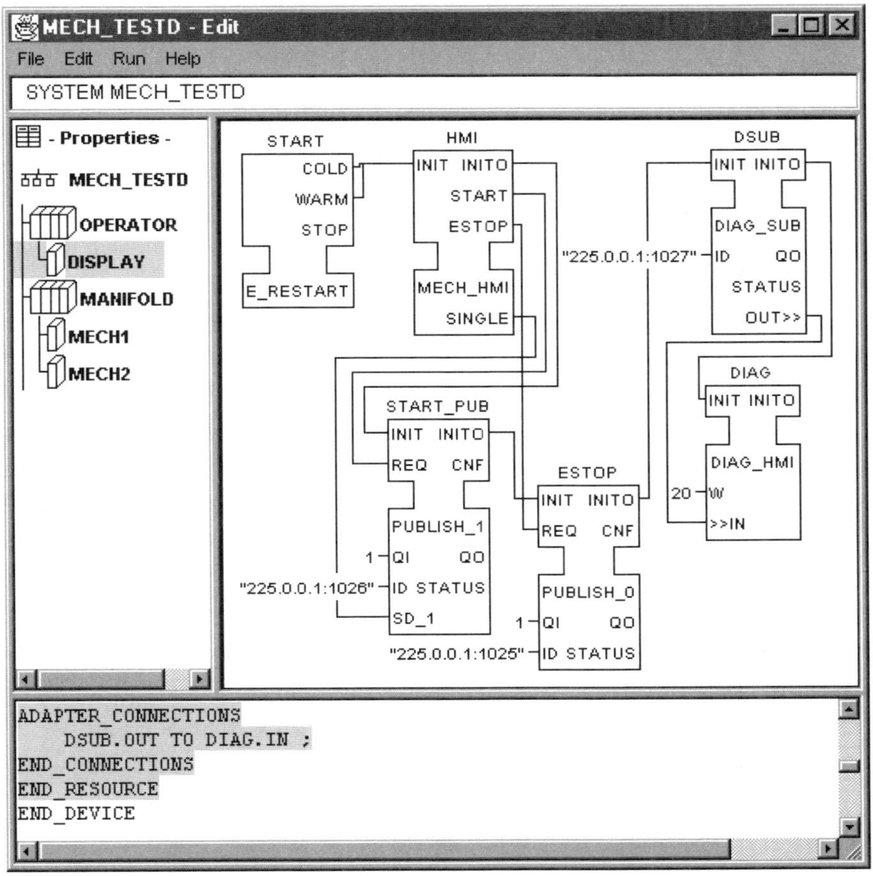

Fig. 4.26. An IEC 61499-2 compliant software tool.

The user interface of a prototype software tool[6] complying with the provisions of IEC 61499-2 is shown in Fig. 4.26. A human-readable textual representation of the library element being edited (in this case the system configuration previously shown in Fig. 4.24) is shown in the lower pane of the tool, in the syntax defined in IEC 61499-1. When a system configuration is being edited, the navigation tree in the upper left panel of this tool reflects the system/device/resource hierarchy of the IEC 61499 architecture, which is also reflected in the tree structure of the corresponding XML document. In this example, the OPERATOR.DISPLAY resource of the system configuration shown in Fig. 4.24 has been selected.

[6] Available at http://www.holobloc.com.

4.4.13 Device Configurability

To facilitate the configurability of devices and resources, subclause 4.3 of IEC 61499-1 [4.2] defines a management service interface function block type with standardized management commands for the dynamic creation and deletion of managed objects such as function block instances and event and data connections. However, the encoding of management commands and responses as well as the implementation of messaging services for remote configuration is given in non-normative Annexes of IEC 61499-1. Hence, as in the case of communication services, additional technical agreements must be employed to ensure configurability.

On the basis of practical experience, the previously mentioned technical agreement for feasibility demonstrations [4.15] proposes the use of an XML [4.14] DTD for encoding of management commands and responses. This agreement also specifies the combination of the normal IEC 61499 CLIENT/SERVER communication with the management service interface function block, to achieve configurability by any remote software tool or agent capable of encoding and decoding the appropriate XML commands and responses over a TCP/IP socket connection. This management service architecture is encapsulated in the device management kernel function block type, illustrated in Fig. 4.27.

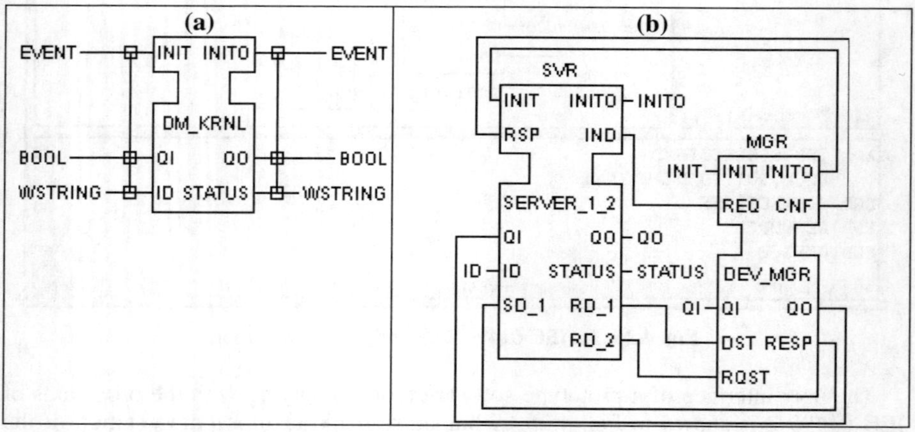

Fig. 4.27. Device management kernel. (**a**) Interface. (**b**) Body

4.4.14 LLC Architecture: Conclusion

Table 4.7 shows the partitioning of the major holonic system functions of *autonomy* and *cooperation* between the architectural levels of low-level control (LLC) and high-level control (HLC). From the preceding material in this section it can be seen that the IEC 61499 architecture addresses all of the LLC requirements as defined in Section 4.2. The extent to which this architecture can be used to address the remaining HLC requirements is addressed in the following section.

Table 4.7. Partitioning of holon functionality

Function		Architectural Level
Autonomous:		
	Task planning	HLC
	Task execution	LLC
	Fault detection	LLC
	Fault recovery	LLC/HLC
Cooperative:		
	Task scheduling	HLC
	Task coordination	LLC/HLC
	Fault detection	HLC
	Fault recovery	HLC

4.5 High-Level Control Architecture

As shown in Table 4.7, the HLC architecture addresses the functions associated with the domain of inter-holon *cooperation*, including negotiation and coordination of mutually agreed tasks and mutual action to recover from operational faults. Also included in this level of the HMS architecture are those intra-holon aspects of *autonomous* task planning and sequencing that contribute to the cooperative accomplishment of manufacturing tasks beyond the scope of LLC functionality.

4.5.1 HLC Structure

The overall structure of the HLC architecture is shown schematically in Fig. 4.28. HLC functions are performed within one or more *cooperation domains* which coexist with and are interfaced to the LLC functions of the real-time control domain. These HLC functions are built on an underlying set of *cooperation communication services*. The partitioning of functions within a cooperation domain is shown in Table 4.8.

Other chapters of this book provide further details on the requirements and implementation of HLC functions.

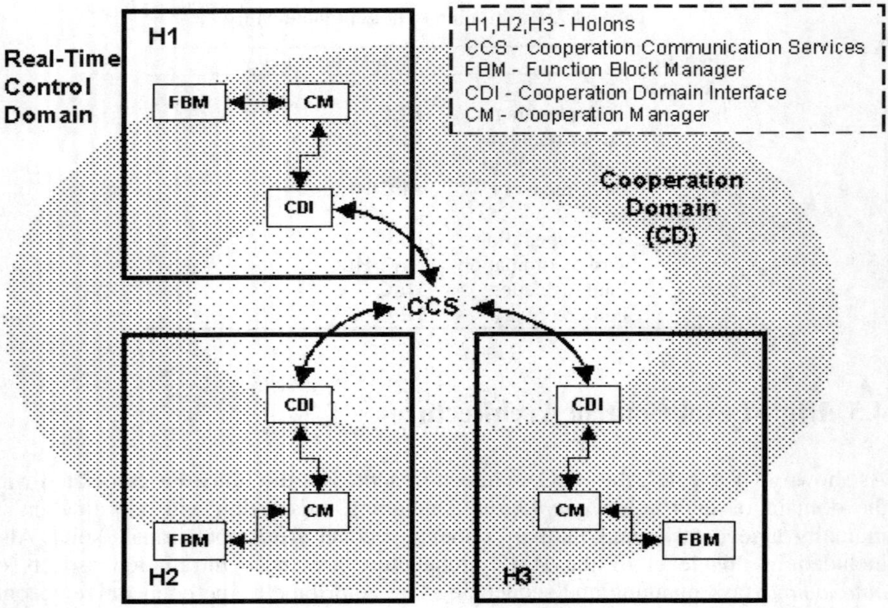

Fig. 4.28. High-level control architecture

Table 4.8. Partitioning of HLC functions

Module	Functions
CDI (Cooperation Domain Interface)	Locate, join, leave CDs Offer and find services ("Yellow Pages") Communicate with other agents
CM (Cooperation Manager)	Negotiate ontologies Maintain knowledge bases Negotiate tasks and schedules Generate/update LLC configurations Coordinate task performance
FBM (Function Block Manager)	Manage LLC configuration

4.5.2 LLC/HLC Integration

In accordance with the Open Systems Interconnection (OSI) model [4.16], the layered "onion skin" architectural models shown in Figs. 4.3 and 4.28 lead to the appearance of "service stacks" within individual holonic devices. Each layer of the stack provides value-added services to the layer above by utilizing the services of the layer below to interact with its distributed peers within the same layer. As shown in Fig. 4.29, the

LLC and HLC stacks coexist within the same device, with the FBM (function block management) services providing the interface between the LLC and HLC stacks.

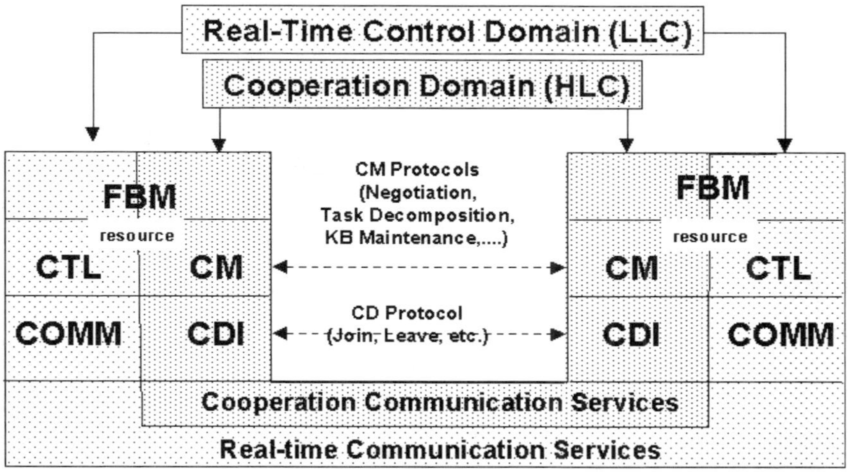

Fig. 4.29. LLC/HLC integration

4.5.3 Application of IEC 61499 to LLC/HLC integration

Since the proposed HLC architecture utilizes a layered service model, IEC 61499 *service interface* function block types can be used to model the encapsulation and integration of the services described in Table 4.8. Thus, an instance of the HMS_KERNEL type shown in Fig. 4.30 can be used in place of the FB_KERNEL type shown in Fig. 4.27 to transform an IEC 61499-compliant LLC device into a holonic device, provided that the device has sufficient resources to support the additional functionality. It is expected that the DEV_MGR functionality can be reused without modification, while the HMS_CM and HMS_CDI types will provided interfaces to the cooperation management (CM) and cooperation domain interface (CDI) services described above.

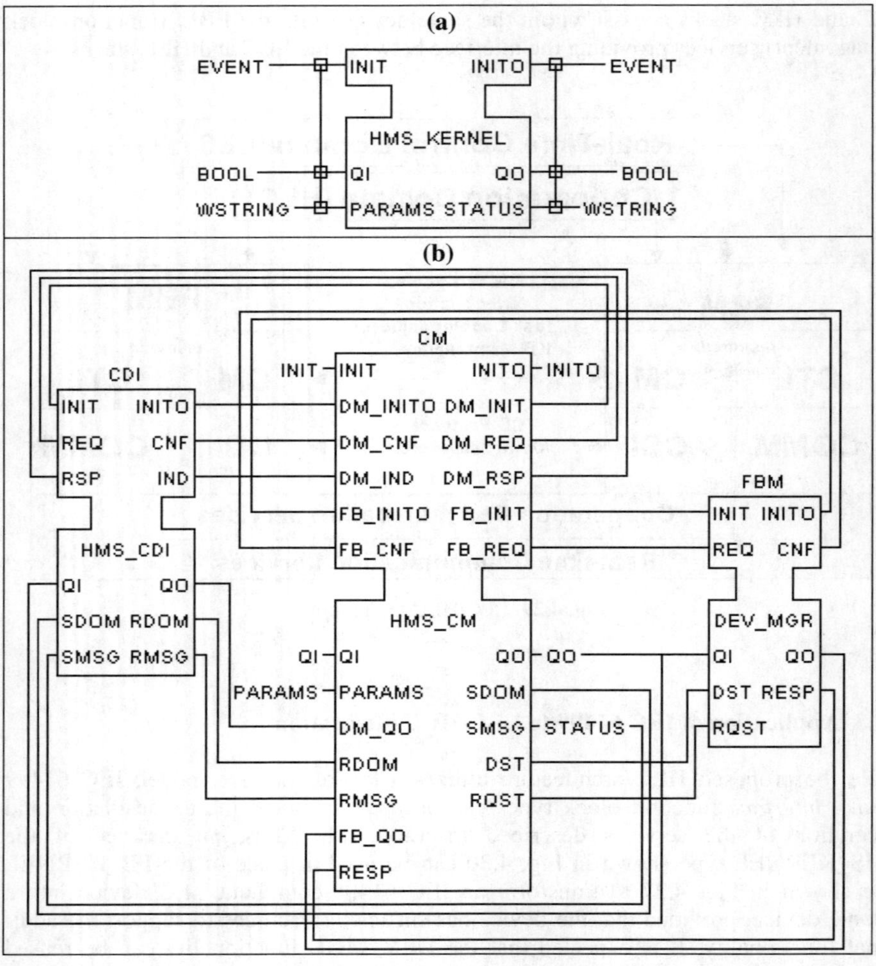

Fig. 4.30. The holonic kernel. (**a**) Interface. (**b**) Body

4.6 Conclusion

It has been shown that devices and software tools compliant with the IEC 61499 series of specifications [4.1, 4.2] and associated techical agreements are capable of meeting all the requirements for low-level control and its integration with high-level control functions in a holonic systems architecture. The LLC architecture has been extensively demonstrated in practical examples. The implementation of HLC functions within the IEC 61499 framework and their integration with LLC functions in practical physical equipment remain a topic of ongoing research.

References

[4.1] H. Shapiro and C. Varian:: *Information Rules: A Strategic Guide to the Network Economy,* Harvard Business School Press, Boston (1999).

[4.2] International Electrotechnical Commission: *Function Blocks, Part1 – Architecture*, IEC PAS 61499-1, Geneva (2000).

[4.3] International Electrotechnical Commission: *Function Blocks, Part 2 – Software tool requirements,* IEC PAS 61499-2, Geneva (2001).

[4.4] Foundation for Intelligent Physical Agents: *FIPA Abstract Architecture Specification*, http://www.fipa.org/specs/fipa00001/ (2001).

[4.5] M.Fletcher, E. Garcia-Herreros, J.H. Christensen, S.M. Deen and R. Mittmann: "An Open Architecture for Holonic Cooperation and Autonomy", in *11th International Workshop on Database and Expert Systems Applications (DEXA 2000)*, IEEE Computer Society, New York (2000).

[4.6] B.P.Douglass: *Real-Time UML: Developing Efficient Objects for Embedded Systems*, Addison Wesley Longman, New York (1997).

[4.7] International Electrotechnical Commission: *Programmable controllers, Part 3 – Programming languages,* IEC 61131-3, Geneva (1993).

[4.8] R.W. Lewis: *Modelling control systems using IEC 61499,* Institution of Electrical Engineers, Stevenage (2001).

[4.9] R.W. Lewis,: *Programming industrial control systems using IEC 61131-3.* Institution of Electrical Engineers, Stevenage (1995).

[4.10] J.H. Christensen: "Design patterns for systems engineering with IEC 61499", in *Fachtagung Verteilte Automatisierung – Modelle und Methoden fuer Entwurf, Verifikation, Engineering und Instrumentierung,* ed. Ch. Doeschner, Otto-von-Guericke-Universitaet, Magdeburg (2000).

[4.11] M. Campione and K. Walrath: *The Java Tutorial,* 2nd Edition, Addison-Wesley, New York (1998).

[4.12] International Electrotechnical Commission: *Information technology – Open Systems Interconnection – Specification of Abstract Syntax Notation One (ASN.1),* ISO/IEC 8824, Geneva (1990).

[4.13] International Electrotechnical Commission,: *Information technology – Open Systems Interconnection – Specification of Basic Encoding Rules for Abstract Syntax Notation One (ASN.1),* ISO/IEC 8824, Geneva (1990).

[4.14] W3 Consortium: *eXtended Markup Language (XML) Specification,* http://www.w3c.org/TR/1998/REC-xml-19980210 (1998).

[4.15] *Technical Agreement for IEC 61499 Feasibility Demonstrations,* http://www.holobloc.com/doc/ita/index.htm (2002).

[4.16] International Organization for Standardization: *Information technology – Open Systems Interconnection – Basic Reference Model: The Basic Model,* ISO/IEC 7498-1, Geneva (1994).

5. FIPA Standards and Holonic Manufacturing

Vladimír Mařík,[1,2] Michal Pěchouček,[1] Pavel Vrba[2] and Václav Hrdonka[2]

[1]Gerstner Laboratory, Department of Cybernetics,
Czech Technical University, Prague
Email:{marik|pechouc}@labe.felk.cvut.cz

[2]Rockwell Automation Research Center, Prague
Email:{vmarik|pvrba|vhrdonka}@ra.rockwell.com

Abstract. This paper documents the fact that the research in the two areas of holonic manufacturing systems and multi-agent systems contains substantial overlap. Many technologies and results achieved in the multi-agent area can be applied with advantage in the holonic field. This is especially true when talking about standardisation, which should enable interoperability of systems. The FIPA international consortium has introduced a systematic approach to development and maintenance of specifications and standards for the multi-agent domain. The principles of the FIPA Abstract Architecture are briefly described. Two examples of applications of FIPA standards in different areas within the scope of holonic systems (control, and production planning and supply chain management) are presented. FIPA standards have been recognised as a suitable candidate for ensuring complete interoperability in the holonic field.

5.1 Introduction

The increasing complexity of intelligent manufacturing systems, as well as the overall demands of flexible and fault-tolerant control on production processes, has stimulated development of two emerging technologies. These technologies will soon make an important breakthrough in the field of intelligent manufacturing and control. These two paradigms are the event-driven control strategy, typical of *holonic systems*, and distributed information processing resulting in *multi-agent systems*.

The research communities working in these two fields approach the problem of intelligent manufacturing from different viewpoints and nearly independently. They use their own specific terminologies and techniques.

5.1.1 Holons

The holonic system (HS) community is rooted in the concept of holons as presented by Koestler [5.20] and is strongly influenced by the requirements of industrial control. Holons are autonomous, cooperative units which can be considered as elementary building blocks of manufacturing systems with decentralised control [5.42]. They can be organised in hierarchical or heterarchical structures. Holons, especially those for real-time control, are usually directly linked to the physical hardware of the manufacturing facility and are able to physically influence the real world (e.g. they may be linked to a device, tool or other manufacturing unit, or to a transportation belt or a storage manipulator).

Holons for real-time control are expected to provide reactive behaviour rather than being capable of deliberative behaviour based on complex "mental states" and strongly proactive strategies. They are expected mainly to react to changes in the manufacturing environment (e.g. when a device failure or a change in the global plan occurs). Under "stable circumstances", during routine operation, they are not required to change the environment proactively. The reason for the prevalence of reactive behaviour of real-time control holons is that each of them is linked to a physical manufacturing facility/environment, changes to which are not very simple, cheap or desirable in a comparatively "stable" manufacturing facility. The physical linkage to physical equipment seems to be a strong limiting factor of the holons' freedom in decision-making.

The more generic holonic ideas and considerations have led to the vision of a "holonic factory" [5.6]. Here all the operations (starting from product ordering, planning, scheduling and manufacturing, to invoicing the customer) are based entirely on holonic principles. A holonic factory contains a group of principal system components (holons) that represent physical manufacturing entities such as machines, products and orders. The holons work autonomously and cooperate together in order to achieve the global goals of the factory. Thus, the factory can be managed towards global goals by the activities of individual autonomous holons operating locally.

The community of researchers trying to implement the vision of the holonic factory is well organised around the international HMS (Holonic Manufacturing Systems) consortium [5.17].

5.1.2 Agents

A comparatively much larger and more diverse community of researchers and developers is working in the multi-agent system (MAS) area. Their work is influenced by the ideas of highly distributed computing in computer networks, as well as by the ideas of distributed artificial intelligence. As the community is much more heterogeneous and agents have been applied to a wider spectrum of problems, there are different organisational frameworks in which the researchers are grouped. The European MAS researchers are organised into the AgentLink consortium [5.2]. Meanwhile, on the worldwide stage, they participate in IFMAS (International Foundation for MAS), the Agent Society, Agentcities [5.1] and, with an emphasis on setting industrial standards, FIPA (Foundation for Intelligent Physical Agents) [5.15]. Though the general paradigms of autonomous intelligent software agents and multi-agent systems have been documented over some considerable time, only recently have unifying principles been clearly specified [5.21].

Both the holonic system and multi-agent system metaphors share some ideas, but they differ on the other issues [5.21]. In general, the research topics of the HMS and MAS fields strongly overlap in the areas of communication protocols, reasoning, coordination and cooperation strategies, and planning and scheduling. Both research communities agree on the same, very fundamental attributes of holons and agents, e.g. they are *autonomous, cooperative* and *open*. Furthermore, the architectures of both holons and agents are multi-layered: a standardised head/kernel architecture in the case of holons and a body/wrapper architecture in the case of agents being the most popular at present. Both approaches use similar communication mechanisms, such as blackboards, matchmakers, brokers or mediators. Strong trends to standardisation are clearly visible in both areas (the IEC 61499 standard in the case of holonic systems, and the FIPA standards in the area of multi-agent systems).

The dual progress achieved by holonic systems and multi-agent systems, and the mutual convergence of these areas during the last few years, especially in the area of communication strategies and protocols, has been really significant.

5.1.3 Interoperability

The main communication issue in a distributed plug-and-play system is the problem of interoperability. Let us distinguish between the concept of agency (agent-hood) and the concept of interoperability, which people often mix up. The concept of agency stresses the autonomy of elements (agents or holons), which are expected to act collaboratively. The agents/holons can exchange information, contract with one another, form teams and coalitions, negotiate, and delegate responsibilities, while remaining fully autonomous. Interoperability is the means for achieving agency. Without the ability of agents/holons to communicate and interact on the physical level, we can hardly implement any aspect of agency. Agentification is the process of converting an old, "non-agent", "stand-alone" system into an agent/holon able to be integrated with a multi-agent community. This is very often implemented by means of a communication wrapper that enfolds the legacy application and provides the agent/holon with the sub-systems needed to achieve the required interoperability. Undoubtedly, interoperability is the key objective of standardisation efforts in all the application domains of multi-agent systems. It will necessarily play the dominant role, especially in holonic systems, in all the phases of implementation of the vision of the holonic factory, where integration of hardware and software components from a wide spectrum of vendors is expected.

Interoperability of agents/elements/modules is the main focus of the activities of FIPA. There should be several distinct levels of interoperability considered – interoperability of low-level near-to-hardware communication, interoperability in message syntax and semantics, interoperability in negotiation scenarios, and interoperability in frameworks for more general coordination and cooperation processes.

5.1.4 Structure of this Chapter

Section 5.2 briefly summarises the recent results in the MAS area, with the stress on FIPA-based solutions and tools applicable to the holonic manufacturing domain. In Section 5.3, the influence of MAS on standardisation efforts in the holonic field is briefly described. Complete information about the FIPA consortium, the FIPA Abstract Architecture as a general framework for achieving interoperability, the FIPA organisational structures and the FIPA specification life cycle is presented in Section 5.4. The Section 5.5 describes in detail the application of the FIPA approach in the material-handling domain. Another application, illustrating the viability of the FIPA architecture, is aimed at the tasks of production planning and supply chain management. This application (the ExPlanTech system) is described in detail in Section 5.6, which documents how interoperability among the planning and information components of the global solution can be easily achieved by applying the FIPA approach.

5.2 Recent Developments in the Multi-Agent Area

Unlike the holon, which is a methodological, design and organisational concept, an agent is an independent and autonomous computational (or physical to the extent needed) unit. Either hierarchical, heterarchical or partially structured communities of agents perform joint decision-making by means of communication, collaboration, negotiation and responsibility delegation that are based on agents' individual rationality and social intelligence. Agents group themselves into teams [5.40], coalitions [5.36], alliances [5.30], platforms, organisations, etc. In comparison with the cooperation domains in the holonic area, the coalitions and teams are much more flexible and allow agents to autonomously join and leave the coalition or the team.

In the area of MAS, the agent architectures have been studied from many different perspectives. An agent usually consists of the *agent's body* and the *agent's wrapper* [5.24]. The wrapper accounts for the inter-agent communication and real-time reactivity. The body is an agent's reasoning component. It very often happens that the agent's body is a pre-existing piece of software/hardware that is integrated by the agent's wrapper into the agent structure. From the outside, such a software system cannot be distinguished from an "ordinary" agent. This agentification activity provides an elegant mechanism for solving the migration problem, where new pieces of automation machinery must be integrated with the currently running information system. Recently, the agent integration architecture has been improved by formalising the concept of social intelligence [5.22, 5.23], which is represented by an *acquaintance model* that becomes a part of the agent wrapper.

The MAS research community provides various techniques and components for creating the architecture of an agent community. The crucial categories of agents, according to their intra- and inter-community functionalities (e.g. resource agents, order or customer agents and information agents), have been identified. Services have been defined for specific categories of agents (for example white pages, yellow pages, brokerage, etc). In some of the architectures, the agents do not communicate directly among themselves. They send messages via facilitators, which play the role of communication interfaces among collaborating agents [5.26]. Other architectures are based on utilisation of matchmakers, which proactively try to find the best possible collaborator, brokers which act on behalf of the agent [5.7], or mediators which coordinate the agents by suggesting and promoting new cooperation patterns among them [5.37]. Increasing attention has recently been paid within the multi-agent community to the concept of the meta-agent, which independently observes the inter-agent communication and suggests possible operational improvements [5.9, 5.22, 5.30]. Message exchange, interaction and negotiation scenarios are now widely accepted by organisations interested in agent technology. An important contribution of the agent community is their research work in the area of ontologies – definition of the domain-specific formal semantics of inter-agent communication [5.11].

There are *standardisation efforts* observable in the field. The FIPA consortium [5.15] plays the most important role here. It has created several *specifications* for agent communication, agent management, agent message transport, abstract architecture and application-specific standards. There are several open source platforms for implementing FIPA-compliant systems (for details see Section 5.4.6). Suitable knowledge ontologies for specific problem domains have been also defined within the FIPA effort.

In the past few years, the multi-agent research community has been much more focused on the investigation of specific models for representing knowledge about cooperating agents (acquaintance or social models) [5.23, 5.24], as well as on various negotiation scenarios, usually based on the contract-net protocol [5.38], auctions and market economies. Modal logic has also been widely applied, not only to represent the semantics of FIPA-compliant messages (often called "performatives" after the speech act theory of Searle [5.35]), but also to express and model agents' beliefs, desires and intentions, and other generic aspects of MAS [5.44]. The field is strongly influenced and motivated by the Rao and Georgeff's BDI (Beliefs-Desires-Intentions) model [5.33].

Meta-reasoning and (self-)reflectivity of a multi-agent system are other issues being intensively studied [5.31]. The coalition formation problem [5.40] and teamwork coordination [5.39] have become hot topics. In specific application domains, such as planning of humanitarian relief operations [5.30], separation of the private, semi-private and public knowledge owned by the agents has been required.

Besides the track record of multi-agent systems in production planning, scheduling and routing assistance, the theoretical backgrounds of multi-agency have penetrated to new application fields relevant to holonic manufacturing. This is the case for Internet business (especially the B2B area) and for various military applications (command support, coalition formation, detection of intrusion in information systems, etc.). The integration methodology for "gluing" sub-systems of diverse nature into a complex system (e.g. for diagnostics) has been explored more and more often.

The area of manufacturing and industrial control, especially when connected with the ambitious holonic visions, seems to be a very attractive and high-priority application field for MAS research.

5.3 Standardisation of Communication in Holonic Manufacturing Systems Influenced by MAS

The vision of the holonic factory covers several levels of information processing for manufacturing. We can distinguish at least three separate levels, namely

- *real-time control*, which is tightly linked with the physical manufacturing equipment;
- *production planning and scheduling*, both on the workshop and on the factory level;
- *supply chain management*, integrating a particular plant with external entities (supplier, customers, cooperators, sales network, etc.).

The specific research results in the holonic field are connected mainly with real-time control. In the other two subfields, the research remains on the philosophical or architectural level, and the specific implementations use MAS research results exclusively. But the real-time control architectures are also starting to be strongly influenced by the MAS results. Let us demonstrate how the holonic research into real-time control has naturally shifted its attention towards MAS architectures and solutions.

The holonic manufacturing community has fully realised that the function-block based control technology is mature for only very limited control tasks and that it is necessary to leverage the results achieved in the MAS field. Several general architectures for combining the function block and MAS technologies have been designed. For instance, the most popular new holon model encapsulates one or more function-block-oriented devices into a wrapper containing a higher-level software component (see Fig. 5.1).

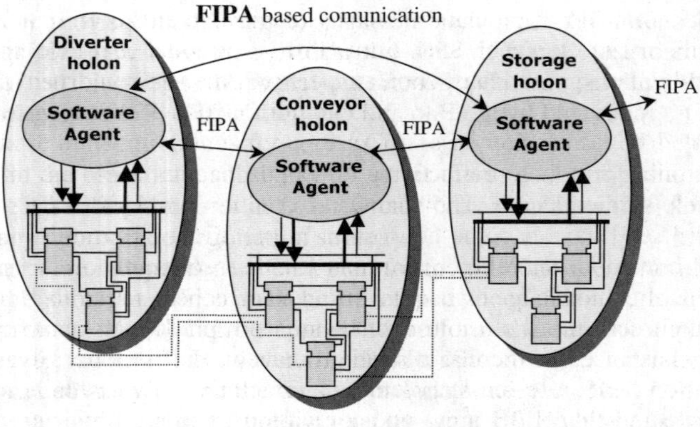

Fig. 5.1. The wrapper model

In such a holon equipped with a higher-level software component, three communication channels should be considered:

- *Intra-holon* communication between the function block part and the software agent component.
- *Inter-holon* communication that is aimed at communication among the agent-based parts of multiple holons – FIPA standards are used more and more often for this purpose.
- A *direct* communication channel between function block parts of neighbouring holons. If we are prepared to break the autonomy of an independent holon then this communication is standardised by IEC 61499 already; otherwise, a new type of real-time coordination technology is needed to ensure real-time coordination.

As a matter of fact, the holons defined in this way behave – on the level of inter-holon communication – like standard software agents. They can communicate widely among themselves, carry out complex negotiations, cooperate [5.8], develop manufacturing scenarios, etc. We can call them *holonic agents* (agentified holons), as they consist of both a holonic part connected with the physical layer of the manufacturing system (operating in hard real-time) and a software agent for higher-level, soft-real-time or non-real-time intelligent decision-making. It has already been mentioned that the inter-holon communication is usually standardised by the FIPA approach, and direct communication could be achieved by IEC standards (not necessarily IEC 61499). Let us stress that the FIPA standards are not applicable for the low-level real-time control purposes as they do not take account of the real-time control aspects.

The attention of system developers is currently directed mainly at the intra-holon communication, which is usually both application- and company-specific and is usually connected with the solution of the "migration problem" (the problem of exploring the classical real-time control hardware for holonic control). McFarlane et al. [5.25] and others (see articles reviewed in [5.6]) introduced a blackboard system for accomplishing the intra-holon communication, while others [5.14] proposed using a special management service interface function block.

The software components of holons help to solve the migration problem as well; for example it is possible to locate several software components on one PLC (Programmable Logic Controller) and thus to create several holons. The software component can even be located on two or more PCs and transferred to the PLC when necessary. The "boundary" between holons running on the same platform or several platforms is mainly determined by the organisation of the intra-holon and inter-holon communication channels [5.5].

The encapsulation approach can be viewed as just a bottom-up attempt to extend the capabilities of the low-level, hard real-time control elements by adding some higher-level software components. This might help in solving certain categories of tasks where the groups of cooperating holons (interacting via the *cooperation domains* in HMS terminology) are not too complex and are clearly separated. For more complicated tasks, a more general, top-down analysis of the control and decision-making specifications is required.

It is expected that the communication among holonic agents will be standardised, for many reasons. One reason is that these holonic agents should be involved in global communities of company agents, where they can directly participate in supply chain management negotiations or contribute to virtual enterprise simulation games, etc. The FIPA communication standards are considered preferable for implementing the inter-holon communication. To develop these standards, the HMS community must declare messages, define their semantics and develop the appropriate knowledge ontologies. This seems to be quite a demanding task, as manufacturing, material-handling, production planning and supply chain management requirements differ significantly between different industries and between different types of production.

There are many *descriptive frameworks* which describe and organise agent behaviour. Some of them represent just a rough, mainly philosophical description of different aspects of the life of a multi-agent community. They describe how the agents should behave, how they are expected to cooperate, how the organisation of the community should look and which types of services and communication are recommended. These frameworks are far from becoming widely accepted standards that would ensure agents' interoperability. KQML [5.10] and the FIPA standards represent the best developed and the most precisely defined descriptive frameworks currently available. Moreover, the FIPA specification contains a *consistent formal model of semantics*, which is defined using a multi-modal logic – Semantic Language (SL). Semantic Language can be considered as a tool for formalising the BDI-oriented communication among the agents.

From a wider perspective, for the FIPA standards to be applicable to holonic manufacturing, they should take account of the pre-existence and coexistence of other standards. In manufacturing industry, there is STEP (Standard for the Exchange of Product Model Data), which is a comprehensive ISO standard (ISO 10303) that is used for representation and exchange of engineering product data and specifies the EXPRESS language for product data representation in any kind of industry. Integration of these widely accepted concepts with the HMS and FIPA effort seems to be of high importance. On the level of physical interoperability, there are various standards, such as the TCP/IP and UDP protocols, CORBA (Common Request Broker Architecture), DCOM (Distributed Common Object Model) and others. Similarly, use of higher-level cooperation standards in the area of multi-agent systems such as KQML (Knowledge Query Manipulation Language), FIPA and JINI is inevitable for dynamic, flexible and reconfigurable manufacturing enterprises.

Owing to its scalability, the concept of agency brings advantages to the implementation of holonic visions in both big companies and small and medium enterprises. This article aims to provide the reader with an insight into one particular, but well-acknowledged and extremely important set of multi-agent and interoperability standards – the FIPA standards.

5.4 FIPA – Foundation for Intelligent Physical Agents

The Foundation for Intelligent Physical Agents is a non-profit association registered in Geneva, Switzerland, founded in December 1995. The main goal of FIPA is to maximise interoperability across agent-based applications, services and equipment. This is done through FIPA specifications.

FIPA provides specifications of basic agent technologies that can be integrated by agent systems developers to make complex systems with a high degree of interoperability. FIPA specifies the set of interfaces which the agent uses for interaction with various components in the agent's environment, i.e. humans, other agents, non-agent software and the physical world. It focuses on specifying *external communication among agents* rather than the internal processing of the communication at the receiver.

FIPA produces mainly two kinds of specifications: (i) *normative specifications* that mandate the external behaviour of an agent and ensure interoperability with other FIPA-specified subsystems, and (ii) *informative specifications* of applications for guidance to industry on the use of FIPA technologies.

FIPA standards attempt to be high-level, *neutral abstractions*. The core FIPA specifications are neutral with respect to both (a) the application area and (b) the hardware and software platforms to be used. Of course, the existing software infrastructures should be considered when designing the standards and leveraging the advantages of the existing, non-agent software technology.

The *FIPA Abstract Architecture* (Fig. 5.2) defines a high-level organisational model for agent communication and core support for it [5.32]. It is neutral with respect to any particular network protocol for message transport or any service implementation. This abstract architecture cannot be directly implemented; it should be viewed as *a basis or specification framework for the development of particular architectural specifications*. Concrete implementations can cover part or all of a particular specification.

The FIPA Abstract Architecture contains agent system specifications in the form of both descriptive and formal models. It covers three important areas, namely

a) agent communication
b) agent management
c) agent message transport.

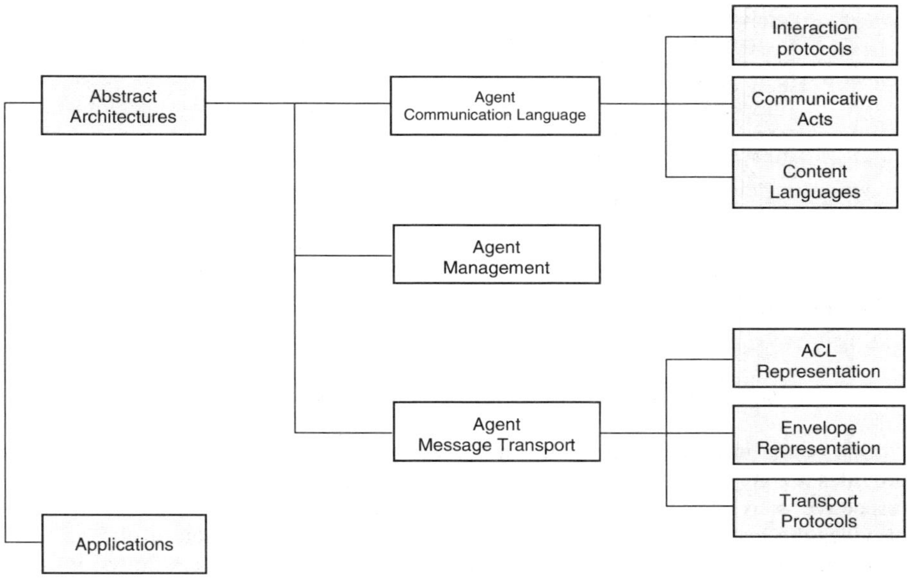

Fig. 5.2. The FIPA Abstract Architecture

5.4.1 Agent Communication and Agent Communication Language

FIPA provides standards for agent communication languages. The messages exchanged among the agents must comply with a FIPA-ACL (Agent Communication Language) specification. FIPA-ACL is based on speech act theory [5.35] and resembles KQML (Knowledge Query Manipulation Language). Each message is labelled by a *performative*, denoting a corresponding *communicative act* (such as “inform” or “request”). Along with the performative, each message bears information about its sender and receiver, a conversation ID, the content of the message, a content language specification, an ontology identifier, information about the conversation protocol which applies to the current message, and the time-out period during which the sender will wait for a reply. An example of a message is shown in Fig. 5.3.

```
(request
  :sender (agent-identifier :name the-sender)
  :receiver (set (agent-identifier :name the-receiver))
  :content
   ((action (agent-identifier :name the-receiver)
    (deliver box017 (loc 12 19))))
  :protocol fipa-request
  :language FIPA-SL
  :reply-with order567
)
```

Fig. 5.3. Example of a request type of message: the agent "the-sender" requests the agent "the-receiver" to deliver a specific box to a specific location, using the FIPA-SL content language and the FIPA-Request conversation protocol

A complete list of all the FIPA communicative acts [5.13] is presented in Table 5.1.

One of the interesting properties of a FIPA-ACL message is its ontology, which provides a high-level description of a concept that organises meanings (syntax) of the respective conversation. Sometimes a dedicated agent, the Ontology Agent (OA), provides the rest of the community with services such as translation of different ontologies, maintenance of public ontologies and selecting a shared ontology for communication. Sending a message is usually part of a conversation that is ruled by a conversation scenario called a *conversation* or *interaction protocol* – the name of the corresponding protocol is contained in a separate slot of the FIPA-ACL message. FIPA provides standards for several protocols, such as simple request/query interaction and more sophisticated negotiation protocols, e.g. contract-net protocol (CNP), iterated CNP, English or Dutch auctions, and brokering or recruiting conversation protocols.

The core of the FIPA ACL message is its content, which is encoded in one of the *FIPA-compliant content languages* (denoted in the "language" slot of the message). All possible FIPA content languages are listed in the FIPA Content Language Library (FIPA-CLL). FIPA-CLL provides a specification for the FIPA Semantic Language (FIPA-SL), which is a general-purpose knowledge representation formalism for different agent domains. This formalism maps each agent message type (performative) to an SL formula that defines constraints (called feasibility conditions) that the sender has to satisfy, and to another formula which defines the rational effect of the corresponding action. Thus, it is possible to verify whether each message sent conforms to the FIPA formal-logic specifications, i.e. whether it satisfies both the feasibility and rational-effect conditions. As a consequence, standard axiomatic proofs can be applied to determine the behaviour of the agents. There are subsets of the FIPA-SL with reduced expressivity:

- SL0 (SL minimal representational form) allows representation of actions and the result, as well as the completion of an action, a term representing a computation, and simple binary propositions.
- SL1 extends SL0 by adding Boolean connectives to represent propositional expressions.
- SL2 allows first-order predicate and modal logic expression representation with restriction of required decidability (Fig. 5.4).

Table 5.1. A complete list of FIPA communicative acts (conversation protocols) [5.13]

Accept Proposal:	The action of accepting a previously submitted proposal to perform an action.
Agree:	The action of agreeing to perform some action, possibly in the future.
Cancel:	The action of one agent informing another agent that the first agent no longer has the intention that the second agent perform some action.
Call for Proposal:	The action of calling for proposals to perform a given action.
Confirm:	The sender informs the receiver that a given proposition is true, where the receiver is known to be uncertain about the proposition.
Disconfirm:	The sender informs the receiver that a given proposition is false, where the receiver is known to believe, or believe it likely, that the proposition is true.
Failure:	The action of telling another agent that an action was attempted but the attempt failed.
Inform:	The sender informs the receiver that a given proposition is true.
Inform If:	A macro action for the agent of the action to inform the recipient whether or not a proposition is true.
Inform Ref:	A macro action for sender to inform the receiver the object which corresponds to a descriptor, for example, a name.
Not Understood:	The sender of the act informs the receiver that it perceived that it performed an action, but did not understand what the sender just did.
Propagate:	The sender intends that the receiver treat the message as sent directly to the receiver, and wants the receiver to identify the agents denoted by the given descriptor and send the received propagate message to them.
Propose:	The action of submitting a proposal to perform a certain action, given certain preconditions.
Proxy:	The sender wants the receiver to select target agents denoted by a given description and to send an embedded message to them.
Query If:	The action of asking another agent whether a given proposition is true.
Query Ref:	The action of asking another agent for the object referred to by an referential expression
Refuse:	The action of refusing to perform a given action, and explaining the reason for the refusal.
Reject Proposal:	The action of rejecting a proposal to perform some action.
Request:	The sender requests the receiver to perform some action.
Request When:	The sender wants the receiver to perform some action when some given proposition becomes true.
Request Whenever:	The sender wants the receiver to perform some action as soon as some proposition becomes true and thereafter each time the proposition becomes true again.
Subscribe:	The act of requesting a persistent intention to notify the sender of the value of a reference, and to notify again whenever the object identified by the reference changes.

```
(forall ?situation
 (or
  (B agent-1 (planned ?situation))
  (B agent-2 (manageable ?situation))))
```

Fig 5.4. Example of the SL2 encoding of the content of a message: for any situation that may happen, it is considered as planned by agent-1 or manageable by agent-2. The model operator B expresses an agent's belief in a given formula

Another option is to use a FIPA Constraint Choice Language (FIPA-CCL), which has been designed primarily for enabling communication between agents about multiple interrelated choices. The choice problems are defined as constraint satisfaction problems. The content of the FIPA-ACL message may be also encoded in the well-known Knowledge Interchange Format (KIF) that was defined by the Knowledge Sharing Effort (KSE) initiative and is often used within KQML. FIPA also supports the Resource Description Framework (RDF) for representing the content of the communication.

5.4.2 Agent Management

The FIPA Agent Management Specification (AMS) provides a specification of how the members of the multi-agent community shall be registered, organised and managed. According to the FIPA philosophy, the agents are grouped into high-level organisational structures called Agent Platforms (APs). Members of each platform are usually geographically "close" – for example, they may be run on one computer or be located in a local area network. Each platform must provide its agents with several mandatory services that are made available by three mandatory components: the Agent Management System (AMS), the Directory Facilitator (DF) and the Agent Communication Channel (ACC).

The Agent Management System (AMS) administers a list of agents registered with the platform. This component implements the creation and deletion of the running agents and provides the agents with a "white-page-list" type of service, e.g. a list of all agents accommodated on the given agent platform and their addresses. Unlike the AMS, the Directory Facilitator supplies the community with a "yellow-page-list" type of service, where each agent can register its services. The Agent Communication Channel is a component that is in charge of message transport among agents.

The mandatory naming convention assigns a GUID (Global Unique Identifier) that looks like the following:

```
<name>@<hostname>:<port> / <target>
```

where "name" is a unique expression for an agent within a single AP, "hostname" is the IP address of the AP, "port" is the port number of that host on which the ACC is listening, and "target" is an object key which is used to identify the receiver of the message. This convention enables communication between agents resident on different agent platforms.

5.4.3 Message Transport Service

The Agent Communication Channel provides *message transport services* (MTS) that implement the infrastructure for message transport within one agent platform, and a *message transport protocol* (MTP) that is responsible for interoperability between agents from different agent platforms. For a platform to be FIPA-compliant, its ACC must have at least one interface which supports a FIPA MTP, such as IIOP (Internet Inter-Orb Protocol [5.29]), WAP [5.43] or HTTP [5.18].

5.4.4 Existing FIPA Specifications

Since its initiation, FIPA has produced three sets of specifications: FIPA 97, FIPA 98, and FIPA 2000. These standards were gradually improved and adjusted to the practical requirements of the industrial applications. Currently, the FIPA 2000 set of specifications (Table 5.2) contains 39 items.

Table 5.2. FIPA 2000 list of specifications

Abstract Architecture:	
XC00001	FIPA Abstract Architecture Specification
PC00089	FIPA Domains and Policies Specification
Agent Communication Languages:	
XC00061	FIPA ACL Message Structure Specification
Interaction Protocols:	
XC00025	FIPA Interaction Protocol Library Specification
XC00026	FIPA Request Interaction Protocol Specification
XC00027	FIPA Query Interaction Protocol Specification
XC00028	FIPA Request When Interaction Protocol Specification
XC00029	FIPA Contract Net Interaction Protocol Specification
XC00030	FIPA Iterated Contract Net Interaction Protocol Specification
XC00031	FIPA English Auction Interaction Protocol Specification
XC00032	FIPA Dutch Auction Interaction Protocol Specification
XC00033	FIPA Brokering Interaction Protocol Specification
XC00034	FIPA Recruiting Interaction Protocol Specification
XC00036	FIPA Propose Interaction Protocol Specification
Communicative Act:	
XC00037	FIPA Communicative Act Library Specification
Content Languages:	
XC00007	FIPA Content Languages Specification
XC00008	FIPA SL Content Language Specification
XC00009	FIPA CCL Content Language Specification
XC00010	FIPA KIF Content Language Specification
XC00011	FIPA RDF Content Language Specification
Agent Management:	
XC00023	FIPA Agent Management Specification
PC00087	FIPA Agent Management Support for Mobility Specification
Agent Message Transport:	
XC00067	FIPA Agent Message Transport Service Specification
PC00093	FIPA Messaging Interoperability Service Specification
ACL Representations:	
XC00069	FIPA ACL Message Representation in Bit-Efficient Specification
XC00070	FIPA ACL Message Representation in String Specification
XC00071	FIPA ACL Message Representation in XML Specification
Envelope Representations:	
XC00085	FIPA Agent Message Transport Envelope Representation in XML
XC00088	FIPA Agent Message Transport Envelope Representation in Bit Efficient Encoding
Transfer Protocols:	
XC00075	FIPA Agent Message Transport Protocol for IIOP Specification
XC00076	FIPA Agent Message Transport Protocol for WAP Specification
XC00084	FIPA Agent Message Transport Protocol for HTTP Specification
Applications:	
XC00014	FIPA Nomadic Application Support Specification
XC00079	FIPA Agent Software Integration Specification
XC00080	FIPA Personal Travel Assistance Specification
XC00081	FIPA Audio-Visual Entertainment and Broadcasting Specification
XC00082	FIPA Network Management and Provisioning Specification
XC00083	FIPA Personal Assistant Specification
PC00092	FIPA Message Buffering Service Specification

5.4.5 FIPA-Approved Applications

There are currently seven approved informative specifications, usually called FIPA applications:

- The *FIPA Nomadic Application Support Specification* provides agent middleware support for a mobile computing environment. This specification standardises the functionality of the Monitor Agent (MA) and Control Agent (CA) and an appropriate ontology for nomadic application support.
- The *FIPA Agent Software Integration Specification* provides a specification of agentification technologies enabling the integration of non-agent software into a multi-agent community.
- The *FIPA Personal Travel Assistance Specification* is an informative specification for the application of agents in the travel industry.
- The *FIPA Audio-Visual Entertainment and Broadcasting Specification* deals with information filtering and retrieval, information customisation, user-friendliness, education and entertainment, and information compatibility.
- The *FIPA Network Management and Provisioning Specification*.
- The *FIPA Personal Assistant Specification* addresses construction of secretary-like software agents that act on behalf of users.
- The *FIPA Message Buffering Service Specification*.

Currently, no application aimed at manufacturing systems has been approved or even discussed.

5.4.6 FIPA Agent Development Platforms

While there is an extensive range of both proprietary and open source agent development platforms [5.2], there are just a few that are fully FIPA-compliant: April Agent Platform (Fujitsu Labs of America) [5.3], FIPA-OS (Emorphia), Grasshopper (IKV++), JADE (CSELT) and Zeus (British Telecom) [5.27]. Probably the best-known platform is FIPA-OS, provided by Emorphia (http://www.emorphia.com; formerly provided by Nortel Networks). FIPA-OS is a component-based toolkit enabling rapid development of FIPA-compliant agents. FIPA-OS complies with the majority of the FIPA specifications. It is fully implemented in Java and has two distributions: *Standard FIPA-OS*, one version running under JDK1.2 and another running under JDK1.1, and *MicroFIPA-OS*, which is an extension to the JDK 1.1 version of FIPA-OS, designed primarily for execution agents on PDAs.

The main competitor to FIPA-OS is JADE (Java Agent DEvelopment Framework), which has been implemented by Telecom Italia Labs (http://sharon.cselt.it/ projects/jade/). JADE is also implemented in Java and requires JDK 1.2. JADE is distributed as open source software under the terms of the LGPL (Lesser General Public License), version 2. As in the case of FIPA-OS, the agent platform can be distributed across machines with different operating systems (if required). The configuration, which is controlled via a remote GUI, can be even changed at run-time by moving agents from one machine to another, as and when required. JADE provides the backdrop for development of LEAP (Lightweight Extensible Agent Platform), which allows FIPA-compliant agents to run on PDAs and cell phones (http://leap.crm-paris.com/).

We should mention that talking about FIPA-compliant solutions means that the FIPA Abstract Architecture generic specifications were respected when the particular architecture implementation was designed and developed. FIPA-compliance does not mean full interoperability. Unfortunately, there is currently no formal procedure available to determine the interoperability of different FIPA-compliant solutions. There is no other way than just practical testing of interoperability. Some interoperability tests of the most important FIPA-compliant platforms (JADE, FIPA-OS and ZEUS) were carried out during the FIPA meeting in London in April 2001. Such tests are of crucial importance for making the specifications more precise.

5.4.7 FIPA Organisational Structure

The FIPA activities are coordinated by the FIPA Architectural Board and the Board of Directors. The technical work is carried out by the FIPA Technical Committees (TCs) – currently the Architecture TC, which maintains the abstract architecture, the Gateways TC working in the area of wireless and wireline domains, and the Ontologies TC, which standardises methods for knowledge sharing. There are also several FIPA Working Groups (WGs) organised to deal with aspects of the FIPA-related work which are not necessarily based on technology. There are currently the Security WG and the Product Design and Manufacturing (PDM) WG. The latter WG [5.11], which is the most relevant to the manufacturing domain, is working on the three following tasks: (i) supply chain management, (ii) planning and scheduling, and (iii) holonic control devices. Some other tasks that fall within FIPA's scope of interest are carried out by Special Interest Groups (SIGs), such as the Agentcities SIG, the Liaison SIG, the Ontology SIG and the Semantics SIG.

5.4.8 FIPA Specification Life Cycle

The procedure for getting specifications approved is well organised. Each FIPA specification is given a status which clearly defines the position of the specification in the FIPA specification life cycle (Fig. 5.5). The first draft of a specification starts the life cycle when it is accepted by a Technical Committee. When the Technical Committee confirms that the specification is ready to be implemented, Experimental status is assigned to it, and the specification is frozen for a period of 24 months. If experimental implementations and practical trials document the suitability and viability of the specification, the specification can be promoted to Standard status. Approval by the FIPA Architectural Board (FAB), as well as ratification by the FIPA members, is required for this promotion. A specification with Standard status is expected to be mature and stable. If the specification becomes superseded by another application, it is given Deprecated status for a fixed period of six months. The Deprecated status warns the FIPA users that a more valuable substitution is expected and provides them with time to change their implementations accordingly. After six months, a specification with Deprecated status retires and is given the Obsolete status.

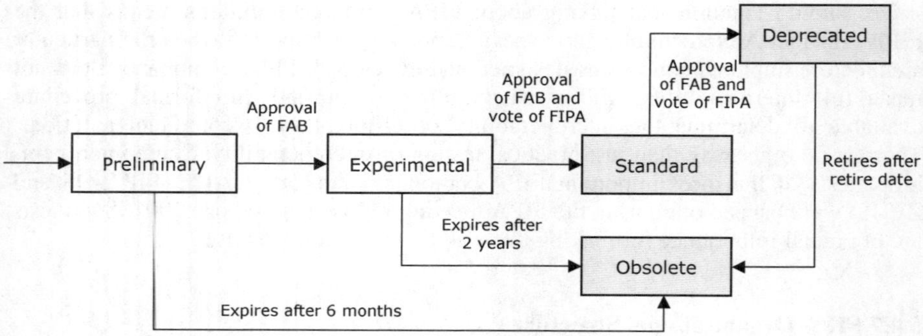

Fig. 5.5. The life cycle of a FIPA specification

5.5 A FIPA-Compliant System for Control of Material Handling

5.5.1 Problem specification

Let us consider a material-handling problem, with a focus on its transportation subtask. We consider a simple material transportation and sorting system consisting of several conveyors, diverters and storage units. As shown in Fig. 5.6, the problem can be represented by a graph, where each node or arc represents a particular component of a material-handling system (MHS), such as a storage unit, diverter or conveyor belt.

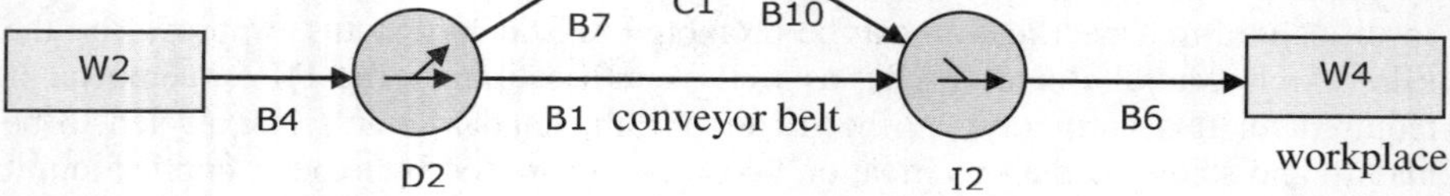

Fig. 5.6. Transport problem of a material-handling system treated as a graph

From the nature of the graph, we can distinguish two main types of components: *node components* representing the nodes of the graph and *transport components* representing its arcs. Basically, the node components constitute (static) places in an MHS between which *workpieces* are transported, where by *workpieces* the subjects of a transport operation, such as pallets, semi-products, parcels and pieces of raw material, are meant. As depicted in Fig. 5.6, the node components can be separated further into two different classes. First class can be called a *workplace* (the rectangles W1, ..., W4), which could be, for example, a storage/retrieval system, a CNC machine or any other component in an MHS able to receive, store/process and resend workpieces. The second class is the node component that we call a *crossing* (circles in Fig. 5.6), the function of which is to receive workpieces coming from several input transport components (see below) and reroute them (according to their destination) to several different output transport components.

As Fig. 5.6 shows, we can distinguish two specialised subtypes of crossing, (i) a *diverter*, which receives workpieces from one input and reroutes them to two outputs (D1 and D2 in Fig. 5.6), and (ii) an *intersection*, which merges workpieces coming from two inputs into one output (I1 and I2).

The transport components, i.e. the arcs of the graph, are those capable of the physical transport of workpieces between node components. For simplicity, let us consider only *conveyor belts* (B1, ..., B10) capable of unidirectional or bidirectional transportation between one input and one output node component. However, the proposed structure can easily be extended by other transportation facilities, such as an AGV (automated guided vehicle) or a robot arm. In any case, we presume that each of them transports for a prespecified *cost* (where less means better), which is then used as a criterion in the finding of optimal routes between workplaces.

In such a discrete system we assume that the basic transportation task is the delivery of a workpiece from the *start* to the *destination* workplace via transport and crossing components. If there are several possible routes for making the delivery, the optimal one must be found; this should be done during the transportation by on-line minimisation of a criterion computed from the sum of the transport component costs.

5.5.2 Agent Implementation

In the multi-agent implementation of the material-handling problem specified above, an agent part must be introduced into each material-handling component. Each of these components is then an *autonomous* unit able to solve local tasks (e.g. rerouting transported items and maintaining the list of reachable destinations in the case of a crossing), and also *cooperative* unit able to send messages to other units in order to solve global-level tasks, e.g. to deliver items between workplaces.

As already mentioned, we decided to follow the FIPA specifications, and used two different open source platforms, FIPA-OS and JADE. Not only for the content language [5.12] of the FIPA-compliant messages sent among agents but also as a general format for the ontology knowledge representation, it was decided to use Extensible Markup Language (XML) to be consistent with current industrial trends.

Workplace Agent. This agent represents a node component, e.g. a storage unit or CNC machine, i.e. a place in an MHS capable of receiving, storing and/or processing and sending out workpieces. From the transport problem point of view, the workplace plays a role of a starting place and a destination place in the system, between which the workpieces are transported.

For these purposes we suppose that the workplace has one transport component connected to it as an input (i.e. one input conveyor belt in this stage), with a sensor which detects the workpieces arriving at it, and one output conveyor by which the workpieces can be sent out to the next workplaces. Moreover, we suppose that the input sensor is equipped with a bar code reader that can read workpiece IDs (see below), so that the workplace knows which item is currently arriving at it.

In the following text, the names of the FIPA performatives (messages) are given in capital letters. One of the messages the workplace agent can receive is a REQUEST message to send a specified workpiece, which is assumed to be present in that workplace (particularly in the case of a storage unit). Let us assume that the workplace W3 in the example in Fig. 5.6 requires the workplace W2 to send it a workpiece with the ID "wp1245". The message sent to W2 looks as follows:

```
(REQUEST
   :sender W3
   :receiver W2
      :content (<sendWorkPiece>workPieceInXML</sendWorkPiece>)
      :language XML
      :ontology material-handling
 )
```

where `workPieceInXML` specifies the item to be sent using the format:

```
<workPiece ID = "wp1245" destinationWorkPlace = "W3" />.
```

Take note of the attribute `destinationWorkPlace`, which indicates the final destination to which the workpiece should be delivered (workplace W3 in this case). This information is used during the transportation of the workpiece through the whole system, especially when the workpiece passes crossings, which have to switch it to the appropriate direction in order to reach the intended destination.

When the workplace W2 receives the message mentioned above, it looks to determine whether the item with ID "wp1245" is really stored in it. If there is no such item available, it replies to W3 with a REFUSE message. If the required workpiece is present in W2, it takes the following sequence of actions: first, it replies AGREE to W3 and second, it sends the workpiece to its output conveyor (B4 in this example), i.e. it sends an INFORM message to the conveyor B4 with content `<workPiece-Sent>workPieceInXML</workPieceSent>`.

Crossing Agent. The crossing agent is one of the "smartest" agents in our system. Each crossing maintains knowledge about which destination workplaces are reachable by each of its output conveyors, together with the delivery costs. Whenever a workpiece comes to a crossing from one of its input conveyors (indicated by an input sensor as in the case of a workplace), the knowledge base is searched by this crossing in order to find the appropriate output conveyor, to which the workpiece should be switched so that it finally reaches its required destination at minimal cost.

We have found this solution to be similar to the ideas in [5.41] of applying *stigmergy* (techniques inspired by the behaviour of social insects) to manufacturing control. The important capability of this approach is that global information (i.e. how to reach the destination workplace, in the case of a material-handling problem) is made available locally (i.e. in crossings, which know what direction the workpiece should be switched to). Let us note that this distributed knowledge is kept up to date by the material-handling environment itself, i.e. it is dynamically changed as various workplaces become unreachable when failures of system components (e.g. conveyors) are detected (see Sect 5.5.4).

5.5.3 Proposal for a Material-Handling Ontology

The ontology plays a significant role in inter-agent communication because it represents a structure providing agents with the semantics of the terms (or symbols) used in the messages [5.28]. Thus, if an agent sends a message to another agent using a specified ontology, it can be sure that the other agent (if, of course, it shares the same ontology) will understand the message.

The key ingredients that make up an ontology are a vocabulary of basic terms and a precise specification of what those terms mean and what their relationship to other terms is. Let us note that our choice of XML as the main data representation format simplifies the definition and further utilisation of the knowledge ontology. XML enables one to define custom tags with a desired syntactic structure via its DTD (Document Type Definition), which can be viewed as an ontology specification according to [5.28].

Some of the ontology elements have already been mentioned in this article, e.g. the specification of the transported item, which we call a *workpiece*. Its XML description for transportation purposes may look like the following:

```
<workPiece ID = "wp1245" destinationWorkPlace = "W3" />
```

The "ID" attribute is used for the unique identification of the workpiece in the material-handling system, whereas the "destinationWorkPlace" attribute denotes the destination workplace to which this workpiece should be delivered.

Another important ontology element has been introduced for the description of material-handling system components, such as the crossings, conveyor belts and workplaces already mentioned. The XML description we propose looks like

```
<component type = "WorkPlace" name = "W1" />
```

The "type" attribute denotes the type of a component (one of `WorkPlace`, `Crossing`, `Diverter`, `Intersection` or `ConveyorBelt`), and the "name" attribute is used to uniquely identify the appropriate agent in the multi-agent community.

If we look at the Fig. 5.6, we can see that the workplace component W1 (described earlier) is connected to the diverter component D1 by the conveyor belt B1. Let us suppose that the cost it which this conveyor delivers is 12. The XML specification of this conveyor component will look as follows:

```
<component type = "ConveyorBelt" name = "B1">
   <connection from = "W1" to = "D1" defaultCost = "12" />
</component>
```

It is obvious that this approach makes our solution open to the future introduction of new material-handling components, such as AGVs, robots and CNC machines.

The last ontology term that we would like to mention here we call `reachableWorkPlaces`. This one is used by agents while looking for the optimal routes through the system during initial configuration and also during reconfiguration as a consequence of a failure detection. The use of this ontology term will be described in the following section.

5.5.4 Dynamic Reconfiguration Capabilities

It is obvious that in any particular system there can be more than one possible route from the start to the destination and that the crucial components which have to contribute to the optimal-delivery solution are the crossings. As mentioned in Section 5.5.2, crossing component has more than one output conveyor and therefore must decide which of them to send the arriving workpiece to so that it finally reaches its destination. Moreover, this decision must be optimal, i.e. the "best one" has to be chosen so that the delivery cost of the remaining route (i.e. from this crossing to the destination workplace) is minimal.

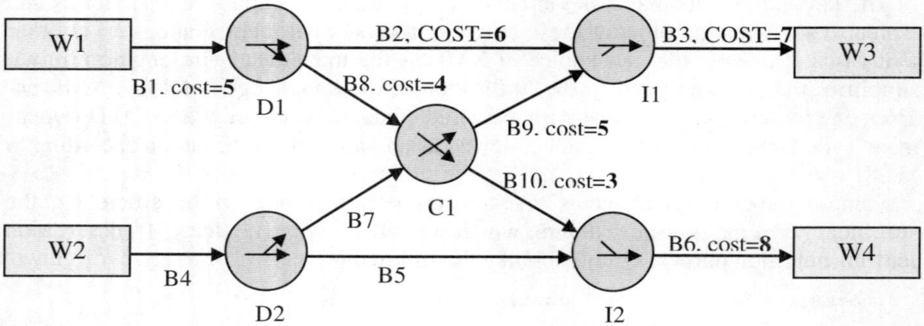

Fig. 5.7. Example of a material-handling system. Each conveyor belt delivers at a certain cost, where "less means better". Crossing components (D1, D2, C1) maintain knowledge about optimal routes to the destination workplaces (W3, W4) using `reachableWorkPlaces` ontology term

In the case of the system mentioned earlier (depicted once more in Fig. 5.7), the diverter D1, for example, can use both the B2 and B8 output conveyors in order to reach the destination W3. Thus if the cost of the route to W3 using conveyor B2 is less then the cost using B8, the diverter should prefer B2 to B8.

To perform such a task, each crossing agent holds, for each output conveyor, a list of names of all reachable destination workplaces together with the costs of delivery. Whenever a workpiece enters the crossing, the agent searches through these lists and finds which one contains the workpiece's destination at the lowest cost. Finally, the workpiece is sent out to the appropriate output conveyor.

In this example, the diverter D1 maintains the necessary knowledge (using the `reachableWorkPlaces` ontology term), which looks like the following:

For output conveyor B2:

```
<reachableWorkPlaces>
  <workPlace name="W3" cost="13" />
</reachableWorkPlaces>
```

For output conveyor B8:

```
<reachableWorkPlaces>
  <workPlace name="W3" cost="16" />
  <workPlace name="W4" cost="15" />
</reachableWorkPlaces>
```

Whenever a workpiece with the destination W3 enters D1, this diverter knows that if it uses output conveyor B2, the cost of the remaining route to W3 is 13, whereas if it uses B8 the cost is 16. Its obvious that it sends this workpiece out to B2.

The information about the destinations accessible through a particular output conveyor is obtained by the crossing agent from this conveyor, which propagates such a list received from its output node backward to its input node while adding its cost to the costs of all destinations in the list.

When the crossing receives such a list, it stores it for the purposes mentioned above and then merges it with the other lists received from other output conveyors.

The result is then sent by the crossing back to all of its input conveyors (which propagate it back again in the same way) so that other components in the system will be informed about the accessibility of destinations. Its obvious that such a cascade of messages is originally triggered by the destination workplaces themselves when each of them sends a list (including itself as a reachable workplace) to its input conveyor. In other words, the agents use a *backward propagation* algorithm, i.e. from destinations to sources, for optimal route-finding.

In the example in Fig 5.7, the destination workplaces W3 and W4 send messages to their input conveyors (B3 and B6). These conveyors then hold the following knowledge:

B3:
```
<reachableWorkPlaces>
    <workPlace name="W3" cost="0" />
</reachableWorkPlaces>
```

B6:
```
<reachableWorkPlaces>
    <workPlace name="W4" cost="0" />
</reachableWorkPlaces>
```

These conveyors add their costs (7 and 8, respectively) and send their lists to their input nodes, i.e. to intersections I1 and I2. Thus, these intersections hold another piece of knowledge:

I1:
```
<reachableWorkPlaces>
    <workPlace name="W3" cost="7" />
</reachableWorkPlaces>
```

I2:
```
<reachableWorkPlaces>
    <workPlace name="W4" cost="8" />
</reachableWorkPlaces>
```

Both intersections propagate their knowledge back to all of their input conveyors, i.e. I1 to B2 and B9 and I2 to B5 and B10; they add their costs and propagate the knowledge again. The crossing agent C1 then receives and stores this knowledge:

C1 for output B9:
```
<reachableWorkPlaces>
    <workPlace name="W3" cost="12" />
</reachableWorkPlaces>
```

C1 for output B10:
```
<reachableWorkPlaces>
  <workPlace name="W4" cost="11" />
</reachableWorkPlaces>
```

As already said, the crossing C1 merges these lists and sends them to all of its input conveyors. When B8 receives the resulting list, adds its cost (4) and sends it to the diverter D1, its clear that the knowledge of reachable destinations held by this diverter will be as shown earlier in this section, i.e.

D1 for output conveyor B2:
```
<reachableWorkPlaces>
    <workPlace name="W3" cost="13" />
</reachableWorkPlaces>
```

D1 for output conveyor B8:
```
<reachableWorkPlaces>
  <workPlace name="W3" cost="16" />
  <workPlace name="W4" cost="15" />
</reachableWorkPlaces>
```

The diverter D1 combines this knowledge and sends the result to its input conveyor. Eventually, B1 propagates the knowledge back to the workplace W1, which is then equipped with the following knowledge base:

```
W1:
<reachableWorkPlaces>
    <workPlace name="W3" cost="18" />
    <workPlace name="W4" cost="20" />
</reachableWorkPlaces>
```

The key benefit of this solution is the capability of the system to react safely to the failure of any of its components, i.e. what we call *lightweight* dynamic reconfiguration. In a situation when a low-level control code of some component detects a physical failure (e.g. a conveyor belt stops for some reason), the "failure" method of the agent part of a holon is called up. This causes the agent to send an *empty* list of reachable destinations, because none of the destinations is reachable through this component now. This information is propagated backwards to the other agents in the same way that described above so that other agents, especially crossing agents, know that the routes to some destinations do not exist or have changed.

Another benefit of similar importance is the capability for what we call *heavyweight* dynamic reconfiguration. This allows the physical reconstruction of the system "on the fly", i.e. removing components from the system and/or incorporating new components into it at run-time without the need to stop the system or to make changes to the software. Considering the proposed architecture, such a capability was easy to implement. Whenever a component is removed from the system, it sends a `<disconnected/>` message to its neighbours. These react to it as they would if they were to receive an empty list of reachable destinations. If a new component is added to the system, it sends a `<connected>` message to its neighbours and, as a consequence, receives the `reachableWorkPlaces` list, which propagates further in the same manner.

5.6 ExPlanTech: A FIPA-Compliant System for Production Planning and Supply Chain Management

ExPlanTech is a FIPA-compliant multi-agent system that has been implemented for providing decision support in project-oriented production planning, as well as for supply chain management purposes. The individual agents model the factory production units (e.g. workshops, departments and machines). They are used for preparing various courses of production, identifying relationships among different parts of the manufacturing process, and detecting and solving conflicts in sharing resources (which are very often unique). ExPlanTech is based on the ProPlanT technology [5.22] and has been implemented within the EC-funded project IST-1999-20171. The system is fully integrated into the company information system. Instead of forcing the factory management to adopt the multi-agent system as an integral and unifying framework and agentify (and integrate) all existing pieces of software, ExPlanTech has been encapsulated within the ERP (Enterprise Resource Planning) system (see Fig. 5.8). All the production data are specified in the ERP system, and are then sent to ExPlanTech, where a production plan is elaborated. ExPlanTech provides the user with graphical and numerical representations of the resulting plan. There are various means for the user to revise the plan and to solve possible conflicts (priority handling, increasing workshop capacity, changing project definition, etc.). The result of planning is propagated back to the ERP system.

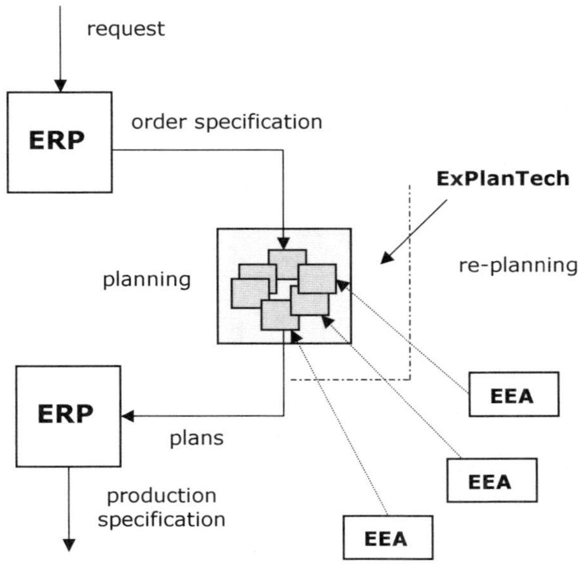

Fig. 5.8. Role of ExPlanTech in ERP

5.6.1 ExPlanTech Architecture

The system architecture is composed of two different classes of agents – *intra-enterprise* and *extra-enterprise* agents. The intra-enterprise community consists of a configurator agent (CA), several workshop agents (WAs) and, finally, a database agent (DBA). The main role of the DBA is to provide other agents with production data (orders, pre-orders, calendar and capabilities of the workshops). The DBA is a communication bridge between ExPlanTech and ERP system (see Fig. 5.9). The CA coordinates the group of WAs and provides them with production data about orders. The WAs simulate workshops and departments within the factory, and their main role is to create plans for workshops. There are two extra-enterprise agents – the monitor and resource agents. The monitor agent serves customers of the factory, who can access the status of a production order using an Internet browser. The monitor agent can be also used by the factory management to inspect activities of particular workshops of the factory while they are off site. The resource agent works on the side of suppliers, cooperators or contractors (in the case of outsourcing) and allows the building and managing of a fast and efficient supply chain.

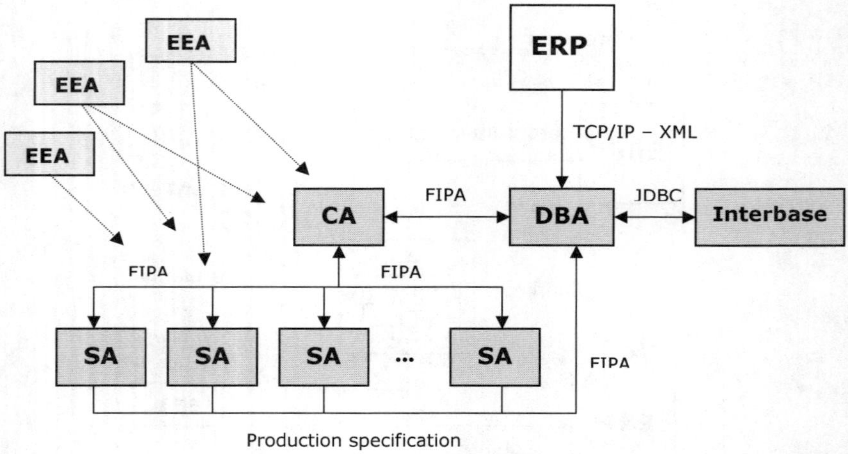

Fig. 5.9. Architecture of the ExPlanTech multi-agent system

As we have already mentioned, the ExPlanTech system is based on the ProPlanT technology [5.22]. Originally, ProPlanT was implemented in C++, KQML was used as an inter-agent communication language, the content language was KIF and a proprietary ProPlanT message transport protocol was applied. Within the ExPlanTech project, there were strict calls for standardisation of the system's operation. As a natural choice, the FIPA specification was adopted, as it provides both the communication standards and the agent management standards. FIPA provided a reference implementation, enabling the developers to build their own implementations.

5.6.2 FIPA-Compliant Communication in ExPlanTech

As already noted, the agents in ExPlanTech exchange FIPA-ACL messages with a rather limited vocabulary of performatives (speech acts). The context-sensitive action specification is always a part of the content of the message, which is encoded in XML. Examples of such actions include DeletePlan, MoveTask, MakePlan and CapacityRequest. Apart from agent-management-related protocols (registration, deregistration, etc.), ExPlanTech uses primarily the FIPA-Request protocol. A conversation following this protocol is initiated by a message with the "request" performative:

```
(request
  :sender operator@mas0:1099/JADE
  :receiver workshop@mas0:1099/JADE
  :language XML
  :ontology plan-task-ontology
  :protocol fipa-request
  :content
    (<?xml version="1.0" encoding="UTF-8"?>
       <plan-task>
              <task-id>Z_211-0242</task-id>
              <time>10</time>
              <day-load>8</day-load>
              <priority>1</priority>
              <start-day>20011026T000000000</start-day>
```

```
                <dead-line>20011120T000000000</dead-line>
                <strategy>E</strategy>
        </plan-task >)
  )
```

The agent `operator@mas0:1099/JADE` sent this message to the agent `workshop@mas0:1099/JADE` requesting it to perform the action `<plan-task>` with parameters specifying the properties of the task to be planned (such as ID, deadline and strategy). This message includes information about the content language used, the conversation protocol and the ontology as well.

The receiver replies to the message with an "agree" speech act:

```
(agree
  :sender workshop@mas0:1099/JADE
  :receiver operator@mas0:1099/JADE
  :language XML
  :ontology plan-task-ontology
  :protocol fipa-request
  :content
    (<?xml version="1.0" encoding="UTF-8"?>
        <plan-task>
                  <task-id>Z_211-0242</task-id>
                  <time>10</time>
                  <day-load>8</day-load>
                  <priority>1</priority>
                  <start-day>20011026T000000000</start-day>
                  <dead-line>20011120T000000000</dead-line>
                  <strategy>E</strategy>
        </plan-task >)

    )
  )
```

Here the agent `workshop@mas0:1099/JADE` informs the agent `operator@mas0:1099/JADE` about its agreement to the suggested request. This message is sent as soon as possible and it follows only the agent's decision to accept the request.

After this, the agent `workshop@mas0:1099/JADE` starts to work on planning the requested task. Upon completing this operation, it sends a message with an "inform" speech act:

```
(inform
  :sender workshop@mas0:1099/JADE
  :receiver operator@mas0:1099/JADE
  :language XML
  :ontology plan-task-ontology
  :protocol fipa-request
  :content
    (<?xml version="1.0" encoding="UTF-8"?>
        <plan-task >
                  <task-id>Z_211-0242</task-id>
                  <start-day>20011026T000000000</start-day>
                  <end-day>20011029T000000000</end-day>
        </plan-task>
    )
  )
```

In this message, the agent `workshop@mas0:1099/JADE` informs the agent `operator@mas0:1099/JADE` about having managed to plan the requested task. In the content of the message, there are properties of the resulting plan.

Apart from this classical protocol, ExPlanTech relies strongly on the subscribe-inform communication mechanism, for which there is no specific conversation protocol defined by FIPA. In such a case, a sender subscribes to a receiver for some type of information. The receiver replies with an appropriate "inform" message. Upon each update of the truthfulness of its previous reply, the receiver sends an updated "inform" message to the sender. The following is en example of such a conversation mechanism:

```
(subscribe
  :sender workshop@mas0:1099/JADE
  :receiver DBA@mas0:1099/JADE
  :language XML
  :ontology capacity-ontology
  :content
    (<?xml version="1.0" encoding="UTF-8"?>
       <change-capacity/>
    )
)
```

The agent `workshop@mas0:1099/JADE` subscribes to the agent `DBA@mas0:1099/JADE` for information about new capacities and capacity changes that is propagated from the ERP system (such as extra shifts or holidays) automatically. This happens at the beginning of the community life cycle. Upon such an event (a change of the capacity of the workshop) the agent `DBA@mas0:1099/JADE` automatically sends an "inform" message to the agent `workshop@mas0:1099/JADE`:

```
(inform
  :sender DBA@mas0:1099/JADE
  :receiver workshop@mas0:1099/JADE
  :language XML
  :ontology capacity-ontology
  :content
    (<?xml version="1.0" encoding="UTF-8"?>
<request-change-capacity-message>
    <day-capacity-list>
    <day-capacity>
      <day>20011120T000000000</day>
      <capacity>10</capacity>
    </day-capacity>
  </day-capacity-list>
</request-change-capacity-message>)
 )
```

There are several ontologies in the system which specify the semantics of inter-agent communication. Some of them are simple and some of them are complex, but their structure and the way in which they are defined complies with FIPA specifications. Below is a complete list of ontologies implemented:

```
ExPlanTechOutputOntology
BackupWorkshopAgentOntology
CapacityOntology
CapacitySynchronizationOntology
PlanOntology
ChangePlanTaskOntology
DeletePlanTaskOntology
PlanTaskOntology
TaskMoveOntology
TaskOntology
TaskPlanOntology
TaskRemoveOntology
ExPlanTechServiceOntology
TaskUpDateOntology
OperOntology
ExPlanTechConfigurationOntology
```

The method that grants access to the unique instance of the PlanTask ontology is presented below. It returns a <code>Ontology</code> object, containing the concepts of the PlanTask ontology.

```
public class PlanTaskOntology {
 public static final String NAME = "plan-task-ontology";
 private static Ontology theInstance = new DefaultOntology();
 public static Ontology instance() {
         return theInstance;
 }
static { initInstance(); }

private static void initInstance() {
  try {
    theInstance.addRole("request-plan-task-message", new SlotDescriptor[5.] {
      new SlotDescriptor("task-id", Ontology.PRIMITIVE_SLOT, Ontology.STRING_TYPE, Ontology.M),
      new SlotDescriptor("time", Ontology.PRIMITIVE_SLOT, Ontology.INTEGER_TYPE, Ontology.M),
      new SlotDescriptor("day-load",Ontology.PRIMITIVE_SLOT, Ontology.INTEGER_TYPE, Ontology.M),
      new SlotDescriptor("priority",Ontology.PRIMITIVE_SLOT, Ontology.INTEGER_TYPE, Ontology.M),
      new SlotDescriptor("start-day", Ontology.PRIMITIVE_SLOT, Ontology.DATE_TYPE, Ontology.M),
      new SlotDescriptor("dead-line", Ontology.PRIMITIVE_SLOT, Ontology.DATE_TYPE, Ontology.M),
      new SlotDescriptor("strategy", Ontology.PRIMITIVE_SLOT, Ontology.STRING_TYPE, Ontology.M)
    }, RequestPlanTaskMessage.class);

    theInstance.addRole("inform-plan-task-message",new SlotDescriptor[5.] {
      new SlotDescriptor("task-id", Ontology.PRIMITIVE_SLOT, Ontology.STRING_TYPE, Ontology.M),
      new SlotDescriptor("start-day", Ontology.PRIMITIVE_SLOT, Ontology.DATE_TYPE, Ontology.M),
      new SlotDescriptor("end-day",  Ontology.PRIMITIVE_SLOT, Ontology.DATE_TYPE, Ontology.M)
    }, InformPlanTaskMessage.class);

  } // End of try
  catch(OntologyException oe) {
     oe.printStackTrace();
  }
}
```

5.6.3 Agent Management in ExPlanTech

The ExPlanTech system is built on top of the JADE framework [5.4]. The agent platform can be distributed across machines (which do not even need to share the same operating system) and the configuration can be controlled via a remote GUI. The configuration can even be changed at run-time by moving agents from one machine to another, as and when required.

Apart form the community of extra-enterprise agents and intra-enterprise agents, the following components are used (see Fig. 5.10):

- The *Agent Management System* (AMS) component that maintains the white-page list of agents and allows registration, deregistration and migration of agents.
- The *Directory Facilitator* (DF) that administers the yellow-page list and allows searching for a service.
- The *Agent Communication Channel* (ACC) facilitating the physical transport of messages among the agents.

The extra-enterprise agents are connected to the community via servlets and a special monitor agent that is a member of the community. JADE has several interesting features that make the process of implementation easier. One of these features is the Sniffer agent, which enables the user to observe the message flow among agents.

The first full implementation [5.34] of the ExPlanTech system was carried out for the purposes of both production planning and supply chain management purposes for the Czech company Modelarna LIAZ s.r.o., an important producer of component patterns for the automotive industry. The system is currently in routine use.

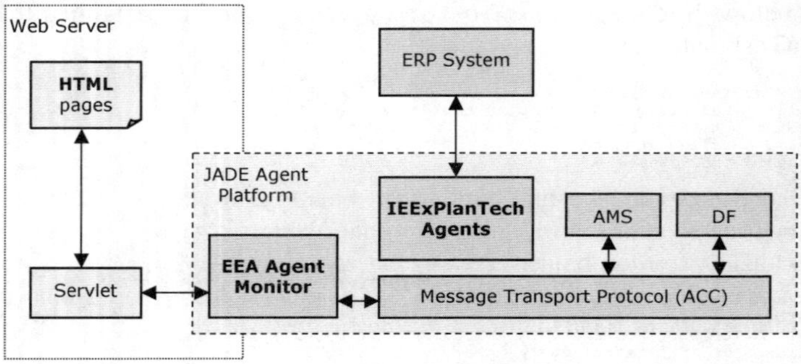

Fig. 5.10. Architecture of the ExPlanTech implementation

5.7 Conclusions

The extension of the original holonic visions (aimed at real-time low-level control) towards (production) planning and scheduling, and supply chain management issues covers exactly the same area as the current MAS research. The same ideas but different terms are used. But at this higher level of a manufacturing business, the holonic community is currently interested in the general ideas (fully compliant or even identical with the MAS ideas), whereas the MAS research is already providing a theoretical background and particular technical solutions. It seems to be quite natural that the holonic visions could and should be implemented through exploring the MAS technology.

The global holonic ideas represent a more flexible alternative to computer-integrated manufacturing (CIM), and provide principles to ensure a higher echelon of responsiveness and handling of system complexity. But holonics is not a new technology, it is a system-wide philosophy for developing, configuring, running and managing a manufacturing business. The area of multi-agent systems is a promising technology provider for accomplishing the ambitious holonic visions. This is especially true in the tasks of (ontology) knowledge representation, communication architecture, negotiation, coordination and cooperation principles, and algorithms, but most of all in the field of *standardisation*.

In contrast to traditional CIM approaches, a holonic manufacturing system is constructed in a bottom-up fashion by integrating holons in such a way that they collaborate to provide an array of system-wide characteristics. These behavioural attributes include flexibility, robustness, self-similarity and openness. The appearance and the whole existence of holons is tightly connected with the requirement of system reconfigurability to support manufacturing agility, where holons are considered as the lowest level of granularity in the reconfiguration tasks. A high degree of interoperability based on widely accepted standards is the main requirement for successful reconfiguration activities.

Holonic terminology has been used mainly in the engineering of control systems, for a varied set of diverse manufacturing domains. Several industrial holonic-based solutions have already been proposed [5.16, 5.19]. The only standard developed in the holonic manufacturing area is the IEC 61499 standard, aimed at the low-level real-time function-block-oriented control accomplished by PLCs. But the importance and applicability of this standard are very often overestimated by the holonic manufacturing community. Special products to support holonic manufacturing are still missing from the commercial automation and control marketplace.

One of the main reasons for this lack is the fact that, within the holonic community, there do not exist adequate standardisation activities aimed at implementation levels above hard real-time control. On the other hand, the FIPA consortium, which was established recently to ensure full interoperability of multi-agent systems, provides standards which might be immediately explored by the holonic community. The FIPA Abstract Architecture can be considered as an ideal basis for holonic software solutions. The FIPA performatives and other features of FIPA standardisation can be easily applied to control problems, production planning and supply chain management tasks. It is just necessary to develop a particular implementation by following the FIPA guidelines. Several very efficient open source FIPA-compliant platforms are available. Thus, the application of the FIPA standards is quite easy; the messages and ontologies can be developed in a quite natural way after a detailed analysis of the problem at hand. We have documented this fact by presenting two examples from different areas in detail (material-handling control and production planning and supply chain management).

Real interoperability of different holonic components will be achieved when informative FIPA specifications of ontologies for different control and decision-making areas have been developed and widely accepted.

To join the FIPA efforts and to explore widely the FIPA specifications is the only useful way to ensure full interoperability of holonic and information management subsystems in the area of industrial manufacturing.

Acknowledgements

The research activities described here were funded in part by the European Commission through the Trial IST project No. IST-1999-20171, by the Rockwell Automation Comp. and by Grant No. MSMT 212300013 from the Ministry of Education, Youth and Sports of the Czech Republic.

References

[5.1] Agentcities, a new initiative aiming to build a worldwide, publicly accessible, testbed for the deployment of dynamic, composed, agent-based services, http://www.agentcities.org/ (2002).

[5.2] AgentLink, Europe's network of excellence for agent-based computing, http://www.agentlink.org/ (2000).

[5.3] April Agent Platform: web site http://sourceforge.net/projects/networkagent/ (2002).

[5.4] F. Bellifemine, G. Rimassa and A. Poggi: "JADE: A FIPA-Compliant agent framework", *Proceedings of the 4th International Conference on the Practical Applications of Intelligent Agents and Multi-Agent Technology*, London (1999), pp. 97–108.

[5.5] O. Bochmann, P. Valckenaers and H. Van Brussel: "Negotiation-based manufacturing control in holonic manufacturing systems", *Proceedings of the ASI'2000 Conference*, Bordeaux (2000).

[5.6] J.L. Chirn and D.C. McFarlane: "Building holonic systems in today's factories: a migration strategy", Journal of Applied Systems Studies, **2**(1) (2001).

[5.7] K. Decker, K. Sycara and M. Williamson: "Middle agents for internet", *Proceedings of the International Joint Conference on Artificial Intelligence* 97, Nagoya (1997) Vol. 1, pp. 578–583.

[5.8] S.M. Deen: "A co-operation framework for holonic interactions in manufacturing", *Proceedings of the 2nd International Conference on Cooperating Knowledge Based Systems*, Keele (1994), pp. 103–124.

[5.9] J. Dix, V.S. Subrahmanian and G. Pick: "Meta agent programs", Journal of Logic Programming, **46**(1–2), 1–60 (2000).

[5.10] T. Finin, R. Fritzson, D. McKay and R. McEntire: "KQML as an agent communication language", *Proceedings of the 3rd International Conference on Information and Knowledge Management* CIKM'94, Gaithersburg, Maryland (1994), pp. 456–463.

[5.11] FIPA's Working Groups: web site http://www.fipa.org/activities/groups.html (2002).

[5.12] FIPA organization: *FIPA Content Languages Specification*, document identifier XC00007, http://www.fipa.org/specs/fipa00007 (2002).

[5.13] FIPA organization: *FIPA Communicative Act Library Specification*, document identifier XC00037, http://www.fipa.org/specs/fipa00037 (2002).

[5.14] M. Fletcher and R.W. Brennan: "Designing a holonic control system with IEC 61499 function blocks", *Proceedings of the International Conference on Intelligent Modeling and Control*, Las Vegas (2001).

[5.15] Foundation for Intelligent Physical Agents: web site http://www.fipa.org/ (2002).

[5.16] T. Hasegawa, L. Gou, S. Tamura, P.B. Luh and J.M. Oblak: "Holonic Planning and Scheduling Architecture for Manufacturing", *Proceedings of the 2nd International Conference on Cooperating Knowledge Based Systems*, Keele (1994).

[5.17] Holonic Manufacturing Systems (HMS) consortium: website http://hms.ifw.uni-hannover.de/ (2002).

[5.18] The Interent Society: *Hypertext Transfer Protocol – HTTP/1.1. Request for Comments*, http://www.ietf.org/rfc/rfc2616.txt (1999).

[5.19] Institute for Manufacturing, University of Cambridge: web site http://www-mmd.eng.cam.ac.uk/menulist.htm (2002).

[5.20] A. Koestler: *The Ghost in the Machine*, Arkana, London (1967).

[5.21] V. Mařík, M. Fletcher and M. Pěchouček: "Holons & agents: recent development and mutual impacts", in *Multi-Agent Systems and Applications II*, Lecture Notes in Artificial Intelligence No. 2322, eds. M. Luck, V. Mařík, O. Štěpánková and R. Trappl, Springer, Berlin, Heidelberg (2001), pp. 223–259.

[5.22] V. Mařík, M. Pěchouček, O. Štěpánková and J. Lažanský: "ProPlanT: multi-agent system for production planning", Applied Artificial Intelligence Journal, **14**(7), 727–762 (2000).

[5.23] V. Mařík, M. Pěchouček and O. Štěpánková: "Acquaintance models in replanning and reconfiguration", in *Advances in Networked Enterprises*, Kluwer Academic Publishers, Boston (2000), pp.175–186.

[5.24] V. Mařík, O. Štěpánková and M. Pěchouček: "Social knowledge in multi-agent systems", in *Multi-Agent Systems and Applications*, Lecture Notes in Artificial Intelligence No. 2086, eds. M. Luck, V. Mařík, O. Štěpánková and R. Trappl, Springer, Berlin, Heidelberg (2001), pp. 211–245.

[5.25] D.C. McFarlane, M. Kollingbaum, J. Matson and P. Valckenaers: "Development of algorithms for agent-based control of manufacturing flow shops", *Proceedings of the IEEE International Conference on Systems, Man and Cybernetics*, Tucson (2001).

[5.26] J. McGuire, D. Kuokka, J. Weber, J. Tenebaum, T. Gruber and G. Olsen: "SHADE: technology for knowledge-based collaborative engineering", Concurrent Engineering: Research and Applications, **1**(3), 137-146 (1993).

[5.27] H. Nwana, D. Ndumu, L. Lee and J. Collis: "ZEUS: a tool-kit for building distributed multi-agent systems", Applied Artificial Intelligence Journal, **13**(1), 129–186 (1999).

[5.28] M. Obitko: *Ontologies: Description and Applications*, Research Report No. GL 26/01, Gerstner Laboratory for Intelligent Decision Making and Control, Czech Technical University, Prague (2001), http://cyber.felk.cvut.cz/gerstner/reports/GL126.pdf.

[5.29] Object Management Group: *Common Object Request Broker Architecture specification*, http://www.omg.org/technology/documents/formal/corba_iiop.htm (2002).

[5.30] M. Pěchouček, V. Mařík and J. Bárta: "Acquaintance models in coalition planning", in *Intelligent Agent Technology – Research and Development*, World Scientific, Singapore (2001), pp. 434–443.

[5.31] M. Pěchouček and D. Norrie: *Knowledge Structures for Reflective Multi-Agent Systems: On Reasoning about Other Agents*, Research Report 538, Department of Mechanical and Manufacturing Engineering, University of Calgary, Canada (2001).

[5.32] S. Poslad, P. Buckle and R. Hadingham: "The FIPA-OS agent platform: open source for open standards", *Proceedings of the 5th International Conference on the Practical Applications of Intelligent Agents and Multi-Agent Technology*, Manchester (2000), pp. 355–368.

[5.33] A.S. Rao and M.P. Georgeff: "An abstract architecture for rational agents", *Proceedings of Knowledge Representation and Reasoning KR&R-92* (1992), pp. 439–449.

[5.34] A. Riha, M. Pěchouček, J. Vokřínek and V. Mařík: "ExPlanTech: exploitation of agent-based technology in production planning", in *Multi-Agent Systems and Applications II*, Lecture Notes in Artificial Intelligence No. 2322, eds. V. Mařík, O. Štěpánková, H. Krautwurmová and M. Luck, Springer, Berlin, Heidelberg (2002), pp. 300–314.

[5.35] J.R. Searle: *Speech Acts*, Cambridge University Press (1969).

[5.36] O. Shehory and S. Kraus: "Methods for task allocation via agent coalition formation", Artificial Intelligence, **101**(1–2), 165–200 (1998).

[5.37] W. Shen, D.H. Norrie and J.A. Barthes: *Multi-Agent Systems for Concurrent Intelligent Design and Manufacturing*, Taylor and Francis, London (2001).

[5.38] R.G. Smith and R. Davis: "Negotiation as a metaphor for distributed problem solving", Artificial Intelligence, **20**, 63–109 (1983).

[5.39] P. Stone and M. Veloso: "Task decomposition, dynamic role assignment and low bandwidth communication for real time strategic teamwork", Artificial Intelligence, **110**(2), 241–273 (1999).

[5.40] M. Tambe: "Towards flexible teamwork", Journal of Artificial Intelligence Research, **7**(1), 83–124 (1997).

[5.41] P. Valckenaers, H. Van Brussel, M. Kollingbaum and O. Bochmann: "Multi-agent coordination and control using stigmergy applied to manufacturing control", in *Multi-Agent Systems and Applications*, Lecture Notes in Artificial Intelligence No. 2086, eds. M. Luck, V. Mařík, O. Štěpánková and R. Trappl, Springer, Berlin, Heidelberg (2001), pp. 317–334.

[5.42] E.H. Van Leeuwen and D. Norrie: "Intelligent manufacturing: holons and holarchies", Manufacturing Engineer, **76**(2), 86–88 (1997).

[5.43] WAP Forum: *Wireless Application Protocol Specification Version 1.2.* http://www.wapforum.org/what/technical.htm (2002).

[5.44] M. Wooldridge: *Reasoning about Rational Agents*, MIT Press, Cambridge, MA (2000).

6. Towards a formalised HMS model

C. A. Johnson

Department of Computer Science
University of Keele
Staffordshire, ST5 5BG
United Kingdom[1]

Abstract. An abstract model of holonic systems is presented using concepts of cooperating knowledge-based systems. The model is shown to be fault tolerant and to handle failures as locally as is possible. The model is sufficiently abstract to enable theoretical properties to be derived. In particular it is shown that if a task is reported as being completed correctly, then this is indeed the case. Tasks that are not explicitly aborted by the system are guaranteed to terminate successfully.

6.1 Introduction

Holonic manufacturing systems (HMS) is a new area of engineering research in which "intelligent" autonomous units, called *holons*, cooperate to provide a flexibile (agile), but largely unmanned environment for low-volume, high-variety production and assembly. A holonic system comprises a manufacturing environment (i.e. a shop floor) and numerous holons which are responsible for product ordering, artefact production (e.g. welding machines) and inspection, transportation (e.g. automated guided vehicles (AGVs)), information processing, etc. Holonic manufacturing systems operate under real-time conditions and are inherently distributed and dynamic, requiring a high degree of flexibility and fault tolerance which is not provided by traditional approaches based upon static and hierarchical control.

The central concepts employed within an HMS – autonomous units, cooperation, negotiation and distributed functionality – are standard terms in agent-based research, and in [6.11] an architecture for HMS is described using an agent-based approach. This architecture is not, however rigorous, and this together with the combination of autonomy and distribution raises some important concerns about the overall system properties of an HMS. These properties might include the following (without being limited to them):

1. *Correctness*: If a task is reported as being successfully completed, can we be certain that this is indeed the case?
2. *Termination and completeness*: Is every sub-task either successfully completed, or aborted (say in response to some failure-driven recovery)?
3. *Recovery*: Is every failed sub-task properly recovered?

---

[1] Present address: School of Computing, University of Plymouth, Plymouth, PL4 8AA, United Kingdom. Email: *chrisj@soc.plymouth.ac.uk*

4. *Stability*: Are local failures and perturbations likely to have global consequences? Do processes (e.g. for negotiation) converge to a stable point, or is oscillation possible?

An attempt to show that holonic systems are well behaved will certainly require that the concepts of holonic systems are made precise. However, given the diversity of views which currently exist with respect to these concepts, there seems little chance of achieving this without raising the level of abstraction. This then points to the need for a high-level model which extracts the essential properties of (and can be mapped onto) holonic systems. The problem of demonstrating that holonic systems possess the above-mentioned properties then reduces to showing that the same is true of the high-level model.

Of course, at the concrete, detailed level, we must expect that manufacturing systems and processes will be arbitrarily complex and ill prescribed. This too is another justification for the need to raise the level of abstraction, in that a high-level model then imposes minimal restrictions on the target systems.

In this chapter we present such a high-level model using an agent-based approach known as cooperating knowledge-based systems (CKBS) [6.5]. The specific properties that we concentrate on are 1–3 above. Stability is the topic of ongoing research. The CKBS approach differs from the more traditional approaches to agent-based research found in distributed artificial intelligence, in that the emphasis is on solving real-world distributed problems, as opposed to simulating human or social behaviour. In this respect, CKBS has been said to be an *engineering* approach to agent-based systems, and is part of an ongoing research effort known as DREAM [6.8].

6.2 CKBS Model Overview

In the CKBS approach, an *agent* is a large-grain entity having a compulsory software component and an optional hardware component. We assume that agents are cooperative (i.e. they participate in joint task processing) but that they are autonomous, in the sense that agents decide independently which joint tasks to participate in, and negotiate their own involvement in such task processing. An agent may also include a human role, although this issue is not considered here in any depth. We make no assumptions about the internal components and behaviour of agents.

We assume that a finite number of *task classes* are known to the system. Agents possess skills, and for each task class there are potentially two *skills*, one for processing and another for coordinating tasks belonging to that class. An agent may have both coordinating and processing skills for the same and/or differing task classes. For each skill, we assume the existence of a *minder* [6.7] – itself an agent – which monitors agents possessing the minder's skill. In particular, if an agent fails, then we assume that the relevant skill minders are employed to replace it with other agents covering the skills that the agent was currently applying.

In order to process a request for task execution, the task is first allocated to a coordinator agent, which forms a *cooperation block*, this being a temporary and dynamic alliance between agents allied to the processing of the given task request. Negotiation involves firstly the *decomposition* of the task into (possibly interdependent) subtasks and the allocation of these subtasks to agents (in this context referred to as *cohorts*) with the appropriate skill to carry out (or coordinate) subtask execution. Following this allocation, a further decomposition phase decomposes these subtasks still further in an attempt to generate a *cooperation plan* which determines how the cohorts will coordinate their individual executions in order to ensure that the relevant interdependencies are satisfied.

Any cohort may, in turn, form a lower-level cooperation block with other agents for the execution of a subtask. In this case it acts as a cohort in the upper-level block and as the coordinator of the lower-level block. Cooperation blocks thus form hierarchies, which parallel the hierarchy of task decompositions.

These concepts are defined formally in Sections 6.3–6.6. In Section 6.7 we then go on to discuss failure and recovery, and in Section 6.8 we present a computational model for CKBS systems. In Section 6.9 we present some formal properties of CKBS systems such as correctness, termination, completeness and fault tolerance. Section 6.10 reviews the ways in which our model allows CKBS systems to amend their behaviour dynamically, and presents some suggestions as to the manner in which such behaviour could be extended. In Section 6.11 we discuss the mapping of our model onto the HMS model, and we finally conclude with a review of related work.

6.3 Tasks

Tasks and task decompositions form the basis for our architecture. We need to distinguish between

(a) the general description of a task, which we refer to as a *task schema*;
(b) the format of a task request, which will include the task schema, but not necessarily full details of the actual task (e.g. input values), as these may not be available until the start of task execution; and
(c) the task itself.

We assume that a finite number of *task schemata* are known to the system, where a task schema describes the task procedure, the task parameters, and the pre/post-conditions that are inherent (i.e. which apply to every instance of the schema). We assume that agents with the appropriate skill know how to process (or coordinate) the task procedure, and moreover are able to report whether they were successful in applying that skill via a *success-parameter*.

A (task) *parameter* is any value associated with the execution of the task procedure which is externally visible. An *input parameter* is a task parameter whose value must be set before execution can start (possibly using defaults). The values of other parameters may be set either prior to or during execution. Any parameter value can, in theory, be output from one task, and input into another. Three obvious parameters are *start-time*, *end-time*, and the success-parameter.

Pre-conditions may include a *system pre-condition* (which tests that resources are not down), a *skill pre-condition* (which tests that the allocated agent does indeed have the required skill to carry out the actual task with the given input values), a *state pre-condition* (i.e. a Boolean condition over the input values and other (assigned) parameters), a *temporal pre-condition* (e.g. *start-time* ≤ *time*), etc. Similarly, post-conditions may include a *quality check*, a test on the success-parameter, a *state post-condition* (as above), a *temporal post-condition* (e.g. *end-time* ≤ *deadline*), etc. For both pre- and post-conditions, the above are intended to be suggestive rather than exhaustive.

The skill pre-condition and the post-conditions guard against the fact that in real applications – particularly those requiring physical processing – task specification and planning might not be so accurate as to guarantee success, even if the plan is executed without an observed failure.

A *task request* (for a particular schema) identifies the task schema, values for some (possibly all or none) of the task parameters, and additional pre- and post-conditions. A task is assumed to be fully described (at least from the external viewpoint) by its task request and a fully instantiated set of task parameter values (which must clearly be consistent with the task request's parameter values and pre-conditions). We refer to the task as being an *instance* of the task schema and of the task request. Clearly the task parameters must satisfy the post-conditions if the task instance is to be regarded as successful.

A *task decomposition,* intuitively breaks a task into subtasks. Note, however, that an agent might wish to decompose a task well before execution (either whilst considering a proposal for a task, or whilst planning the task), and thus it is actually the task *request* that is to be decomposed (into subtask requests). In general the subtask requests within a decomposition will be interdependent, these interdependencies being expressed by *dependency rules* – Boolean conditions defined on their task parameters. A decomposition should *satisfy* its parent task request in that successful instances of the subtask requests satisfying the dependency rules should result in the successful execution of an instance of the parent request.

Note that a decomposition is a logical specification for accomplishing a task request. It does not prescribe how the individual subtasks are to be carried out so as to satisfy the dependency rules: such details are contained in the *cooperation plan* (see Section 6.4).

A task request is *sparse* iff there are no input values and no additional pre- or post-conditions. Decompositions might be derived in a number of ways. The simplest approach would be to assume a finite set of predetermined decompositions (including decompositions for all sparse task requests), from which all other decompositions are to be derived. Note that there are infinitely many task requests (since the task parameter values and additional pre- and post-conditions are unconstrained), and hence predetermining decompositions for all task requests is impossible.

6.4 Cooperation

Cooperation, in the present context, is the process by which the agents in a cooperation block accomplish a task through all its processing stages, using negotiation and coordination as appropriate. These stages include decomposition, planning, scheduling and execution.

A *cooperation block CB* is a temporary and dynamic alliance between agents allied to the processing of a specific task request. *CB* will have a coordinator agent, *co-*

ord(*CB*), which is (at least initially) the agent allocated the task request. We can identify the following phases and events in the life of a cooperation block (Fig. 6.1):

a) cooperation block creation;
b) pre-invocation (static) negotiation;
c) invocation (of the underlying task);
d) (dynamic) negotiation;
e) task execution;
f) log results and commit;
g) cooperation block destruction.

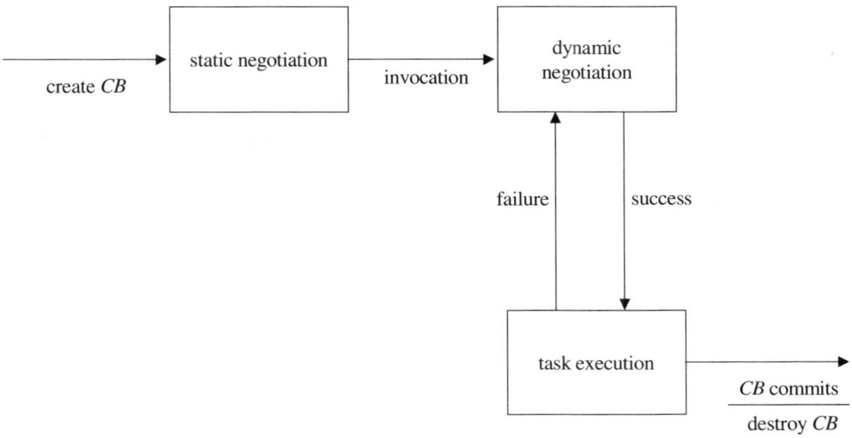

Fig. 6.1. Cooperation block life cycle

As we shall see later, cooperation blocks will form hierarchies, and a cooperation block will persist through the above phases unless it is aborted by the coordinator of its parent block (for example in response to a failure). Communication with the parent block is via the child block's coordinator. Phases (d) and (e) may be iterated in response to failure.

Static/dynamic negotiation may be absent. These forms of negotiation are *internal* to the block, as opposed to *external* negotiation between the block and its parent block (via the child block's coordinator). External negotiation might take place at any point within phase (b), (d) or (e). During the life of the cooperation block, the underlying task may be amended as a result of external negotiation, in which case a cooperation block *CB* whose currently assigned task request is t will be written as CB_t. All of these issues are expanded upon in Sections 6.5–6.8.

Agents may join a cooperation block subject to the consent of the coordinator, and may leave if they have no unfulfilled contract with the block.

In the present context, *negotiation* is the process whereby the agents in a cooperation block CB_t arrive at a decomposition, a (cooperation) plan and a schedule for the task request t. If the negotiation is unsuccessful, then *coord*(CB_t) reports the failure to the coordinator of its parent block (see later). Negotiation aims to ensure subsequent *coordination*, which we define as the process whereby the execution of subtasks by different agents is synchronised so as to satisfy the dependency rules and to prevent undue delays and deadlocks. In this sense coordination will be a system property, rather than an activity of the coordinator as such. Such synchronisation is defined by the cooperation plan (see below): the full world state does not exist in any particular agent, but is a distributed system property, its correctness being ensured by the coordination of the agents in the cooperation block.

The first phase in negotiation is for $coord(CB_t)$ to invite a subset of the other agents to participate in the block, and subsequently admit them. A decomposition D_t is chosen for t, and each subtask request in D_t is allocated to an agent in CB_t. The allocation may involve any appropriate negotiation strategy.

If the subtasks are independent, then a plan for the full task is effectively the union of plans for the subtasks, which may be developed independently. In this case the contract net protocol [6.34] can be applied to perform the subtask allocation, with agents bidding for subtasks. However, if the subtasks are interdependent, then the contract net protocol may not be the most appropriate method. The reason for this is that an agent may not be able to predict exactly what work will be involved in the processing of a subtask until the development of the plan – which occurs after subtask allocation. Thus, in this case, negotiation requires a multi-stage protocol.

A *cooperation plan* for CB_t defines how the agents in CB_t intend to cooperate and coordinate their activities in an attempt to achieve both their respective subtasks and the given dependencies. The development of such a plan involves each agent A_i in the block decomposing its allocated subtask request(s) t_i still further yielding u_i, v_i, w_i, etc. These further decompositions are then used to create, via negotiation, a network of lower-level task requests, with the cooperation plan defining how these lower-level task requests fit together, i.e. what inputs and outputs each expects to receive from, and deliver to, the others. We shall refer to each of these lower-level task requests, together with the information about their expected inputs and outputs, as an *action element*. Their essential characteristic is that each action element in CB_t's plan should be a black box *as far as CB_t is concerned.*

Clearly the action elements in the plan might well include the lower-level subtask requests u_i, v_i, w_i, ... generated in the second-level decomposition, but these may be duplicated, or additional action elements may be included as well. The intention here is to allow the plan to meet the dependency rules and to also handle some of the possible eventualities (e.g. a failure to complete some subtask as specified) without resorting to replanning (see later). Figure 6.2 illustrates this two-stage decomposition and planning process. Note that the cooperation plan is not cyclic.

A *schedule* is formed by adding timing constraints to some (possibly all or none) of the action elements within a cooperation plan. We tentatively regard decomposition, planning and scheduling as three distinct phases, in which case it is clear that we need to allow iterative feedback, e.g. a failure to produce a schedule would feed back into the reformulation of an alternative cooperation plan. This is illustrated in Fig. 6.3. Details of the initial and final transitions are given in Section 6.8.

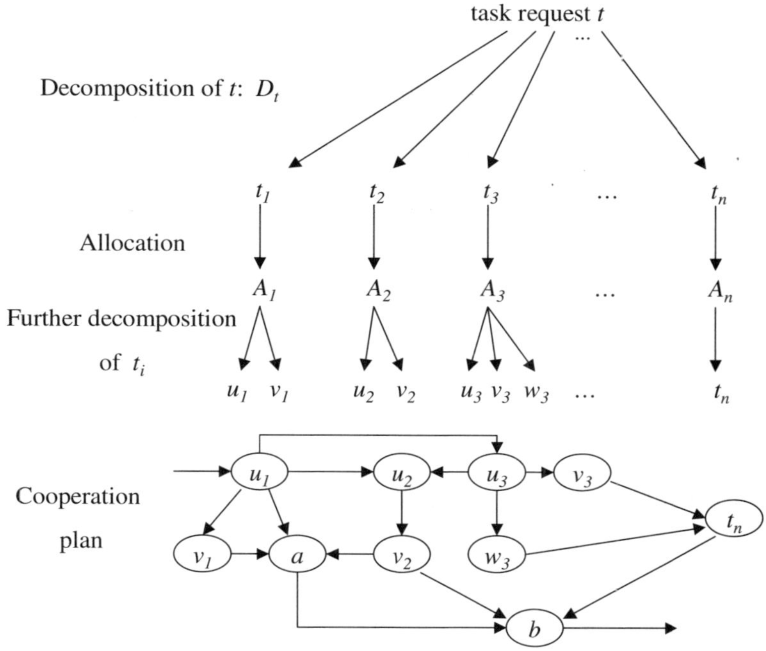

Fig. 6.2. Two-stage task decomposition and cooperation plans

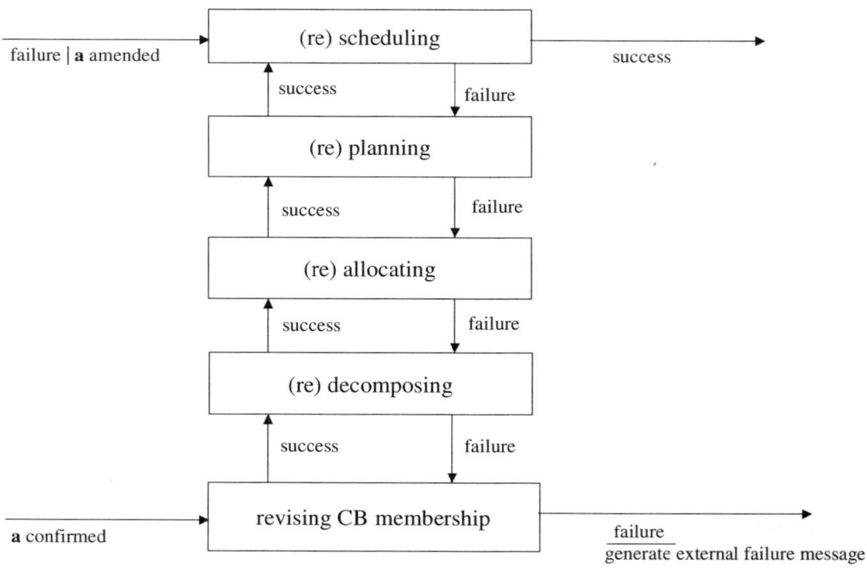

Fig. 6.3. Internal negotiation

Notice that the formulation of a plan is, for the most part, a logical activity. Implementation issues are considered to the extent that the formulation is constrained by agent (i.e. skill) availability, and this availability is input into the plan formulation by the allocation of subtasks to agents at the decomposition stage. This then results in a plan with implementation pre-conditions such as the system and skill pre-conditions. The timing constraints added during scheduling are of course further implementation conditions.

We also note that we could define planning to be an entirely logical activity, involving the development of a *logical plan* (in which pre- and post-conditions are purely logical), and then regard the development of an implementation plan, i.e. the allocation of agents and the addition of implementation pre- and post-conditions (including timing constraints), as belonging to a separate phase. Of course, the downside of such a strategy would then be that agent availability would not be taken into consideration until later in the negotiation, whence allocation failure would be far more likely.

6.5 Action Elements and Cooperation Plans

An *action element* consists of a task request a set of *input rules*, and a set of *output rules*. If a is an action element with the underlying task request t, then we write a_t. An input rule specifies either

(a) that the arrival of an input is a pre-condition for the task request; or
(b) an assignment of a value to a task parameter in terms of the input values and defaults.

Notice that the specification of a parameter as an input parameter (see Section 6.3) can be viewed as an input rule. The input rules and task request should together ensure that every input parameter is defined.

Output rules have the form

if ⟨condition⟩ *then send* ⟨task-parameter⟩ *to* ⟨action element⟩

where ⟨condition⟩ is a Boolean condition over the task parameters. A post-condition *PC* can be thought of as an output rule of the form

if ⟨not *PC* ⟩ *then send* ⟨“PC failure”⟩ *to* ⟨coordinator⟩

6.5.1 Example

Suppose that a task involves polishing a piece of metal to a certain level of smoothness. The obvious post-condition (a quality check) is that the required level of smoothness has been achieved (and clearly this requirement would be input into the planning of the process).

We might allow – within our plan – for the possibility that the first attempt might not succeed, whence a second "reworking" would be undertaken. Of course if the first attempt did succeed then the reworking action element would not be invoked at all. The plan thus has the form indicated in Fig. 6.4.

Notice that in the first action element, the test "smoothness OK" is a condition within an output rule (and whose negation invokes repolishing), whereas in the second action element it is a post-condition (whose failure invokes a failure message to the block's coordinator).

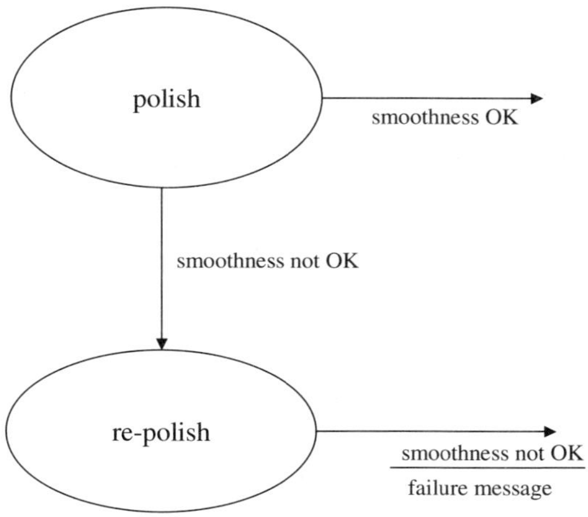

Fig. 6.4. Post-conditions and output rules

The life-cycle of an action element a_s within a cooperation block CB_t thus contains the following phases and events:

(a) invocation;
(b) pre-condition phase (in which the input rules and then the pre-conditions are checked);
(c) task processing (which terminates when CB_s commits);
(d) post-condition phase (the output rules are fired, output messages are logged and sent);
(e) acknowledgement receipt; and
(f) termination.

These phases are illustrated in more detail in Section 6.8. The invocation time specifies when the pre-condition phase begins. The two obvious possibilities are as follows.

(a) Invocation occurs when *time* > *start-time*, and the coordinator agent's internal schedule allows. In this case we need a specification as to what happens if the essential inputs are not present.

(b) Invocation occurs when all the essential inputs have arrived, and the coordinator agent's internal schedule allows.

A *cooperation plan* P for D_t is a finite set of action elements. As in Section 6.3, P should *satisfy* D_t in that successful execution of P (in which each of its action elements terminates successfully) should imply that each of the subtask requests in CB_t has been successfully executed according to the dependency rules in D_t (which in turn implies successful execution of t).

If a and b are action elements in P, then we write a $\Rightarrow_P$ b if the arrival of output from a is an essential input in b. Clearly the $\Rightarrow_P$ relation should not be cyclic. In addition, P needs to be internally consistent, in that if $a \Rightarrow_P b$, then the output rules in a should guarantee that the required inputs to b will indeed be sent if a terminates successfully.

A *schedule* S for the cooperation plan P consists of P together with the assignment of timing constraints to some (possibly all or none) of the action elements in P. Clearly the timing constraints should respect the $\Rightarrow_P$ relation. Timing constraints are of the form (*start-time*, *duration*), where either *start-time* and/or *duration* might be absent. We shall view these timing constraints as additional pre- and post-conditions.

6.6 Cooperation Block Hierarchies

In accepting the allocation of a subtask t_i in D_t, an agent A in CB_t is also implicitly accepting the role of coordinating (within a lower-level cooperation block) each action element a_s within the further decomposition of t_i. However, if A has the required skill, then A might process a_s itself, but we still regard the processing of s as taking place via a cooperation block. We refer to this as a *trivial* cooperation block (Fig. 6.5), with a trivial decomposition $D_s = \{s\}$, and a single action element within CB_s's plan being allocated back to the coordinator.

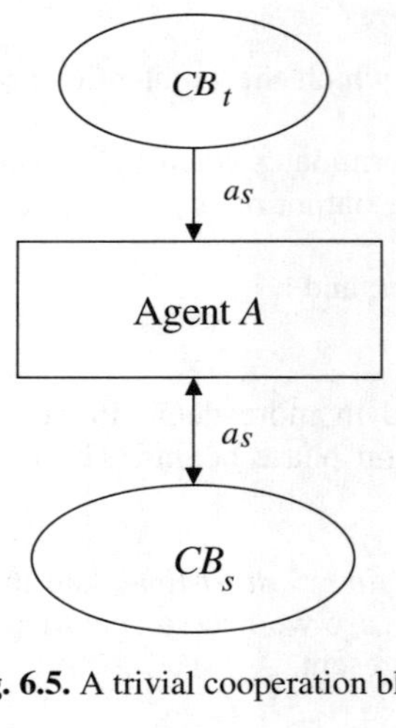

Fig. 6.5. A trivial cooperation block

The insistence upon using a cooperation block is motivated by the requirements of *autonomy*: if some failure occurs within A, then the block may need to reconfigure itself in order to process s in some other way (in which case the block may change from being trivial to non-trivial).

Suppose that A is a cohort in a non-trivial cooperation block CB_t. During planning, if an action element a_s is *proposed* to A, then A generates a cooperation block CB_s (of which it becomes the coordinator) in order to decide whether it can accept a_s and if so, to subsequently process it. Static negotiation within CB_s during the consideration of the proposal might well be total, in that a complete plan/schedule could be developed for a_s before acceptance or rejection was given to the proposal. More generally however, in order to consider the proposal of a_s, CB_s might employ only partial negotiation, possibly with the consent of critical lower-level agents and components. The response from the consideration of the proposal is sent back from *coord*(CB_s) to *coord*(CB_t) which then (using information about other acceptances/rejections) decide whether to *confirm* a_s to *coord*(CB_s), or to *abort* it (thus resulting in the destruction of CB_s and any child blocks). If a_s is confirmed, then it becomes part of the upper block's plan. Following confirmation, further negotiation might then take place within CB_s, either before (static) or after (dynamic) invocation. These issues are discussed further in Section 6.7.

A *cooperation block hierarchy* (Fig. 6.6) is a finite non-empty tree, in which each node is labelled with an action element and the underlying cooperation block (a_t, CB_t). If CB_t has committed, then in reality CB_t has been destroyed, although for the sake of formal simplicity we continue to write (a_t, CB_t). If (a_s, CB_s) is a child node of (a_t, CB_t), then we may informally say that a_s or CB_s is a child node of a_t or CB_t. Figure 6.6 thus illustrates the features of a cooperation block hierarchy.

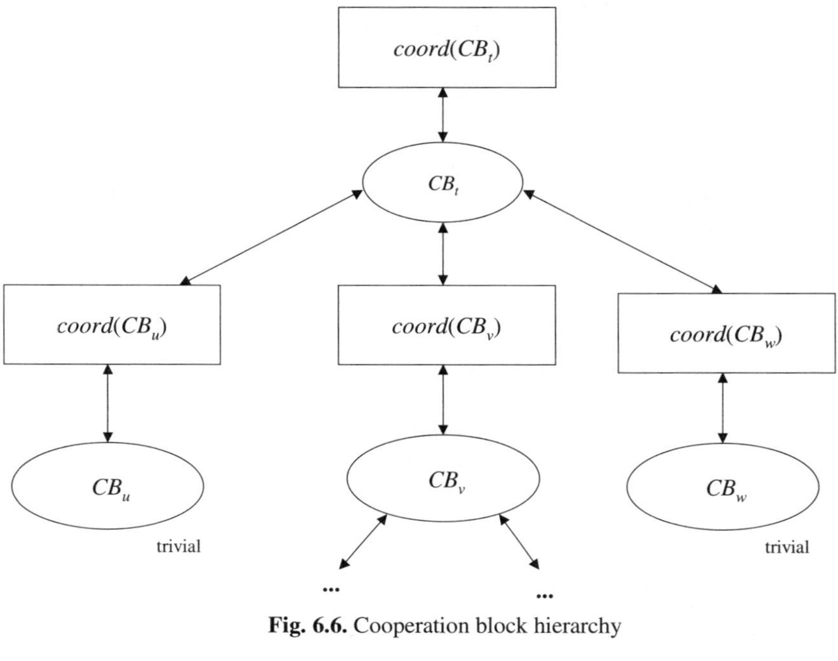

Fig. 6.6. Cooperation block hierarchy

If the action element a_s is proposed within a non-trivial block CB_t, then (a_s, CB_s) becomes part of the hierarchy (as a child of (a_t, CB_t)). Once added to the hierarchy (a_s, CB_s) is removed if (and only if) the action element is aborted.

If (a_s, CB_s) is a child of (a_t, CB_t) and a_s has not terminated, then coord(CB_s) is a cohort in CB_t, and is responsible for communication with CB_t, in particular, for

(a) receiving inputs for a_s from other action elements in CB_t;
(b) sending outputs from a_s to other action elements in CB_t; and
(c) negotiation with CB_t on behalf of a_s/CB_s, in response to failures within a_s/CB_s and proposals from CB_t.

As such, we also regard *coord*(CB_s) as the coordinator of a_s, and write *coord*(a_s). We shall assume that the plan *P* of a cooperation block CB_s always contains an action element i_P whose role is to pass the inputs (derived from (a) above) from *coord*(a_s) to the relevant cohorts in CB_s, and an action element o_P whose role is to collect outputs from the cohorts in CB_s on behalf of *coord*(a_s) in order to allow *coord*(a_s) to perform the post-condition phase, and hence send the outputs (as in (b) above). Clearly both i_P and o_P are allocated to *coord*(a_s), and their underlying cooperation blocks are trivial.

6.6.1 User Tasks and Agents

If (a_t, CB_t) is the root node of some cooperation block hierarchy, then *t* is said to be a *user task*, and *coord*(CB_t) is a user agent. We suggest [6.7] that the user should be allowed involvement in the top-level cooperation block, specifying for example (partial) details of the decomposition or plan (or parameters to be used in the development of the plan), the form of negotiation to be employed (e.g. a variant of the contract net protocol), and strategies to be employed in the event of certain types of failure. The user may limit the properties of the action elements: for example by insisting that certain action elements are (or are not) dynamically scheduled. The user would also certainly need to be involved in negotiation if the user task fails, so as to agree a new deadline for example. In contrast, since lower-level cooperation blocks are supposedly black boxes, the negotiation and strategy within such blocks will not be the concern of the user. We thus view the user as being a component of the user agent.

6.7 Failure and Recovery

If a_s is an action element in the cooperation block CB_t, then *coord*(CB_s) can report a_s as being *failed*, this failure message being sent to *coord*(CB_t). Such a failure message is said to be *internal* to CB_t but *external* to CB_s. We can identify the following types of failure:

(a) failure to receive an essential input (on time);
(b) pre-condition failure;
(c) post-condition failure;
(d) failure to receive an acknowledgement (to an output message);

(e) resource failure at *coord*(CB_s);
(f) resource failure at a cohort in CB_s (other than coord(CB_s));
(g) failure within CB_s (other than (e) or (f)), e.g. negotiation failure.

Note that failures of type (a)–(d) are internal to CB_t, but external to CB_s. Failures of type (f) and (g) are internal to CB_s. A failure of type (e) is internal to both parent and child blocks. Network failure manifests itself as message failure (see below). Note also that one type of failure may cause another: for example, a failure to receive an essential input would probably result in an internal timing failure.

If the coordinator agent *coord*(CB_s) itself fails, then, as stated in Section 6.2, we assume that the agent is replaced with another coordinator agent from the same skill class [6.7]. The coordinator may also act as a cohort (i.e. as a processing agent), and thus the failure of an agent may require the use of several replacements, depending upon the number of skills the failed agent was employing.

Failure to receive an essential input obviously causes the recipient to delay the start of its execution. We assume that the sender adopts some protocol for resending messages until either an acknowledgement arrives or a timeout occurs, the latter resulting in a failure message as above (see Fig. 6.10).

In Section 6.4 we mentioned that negotiation would take place either statically (before invocation) or dynamically (after invocation), and that static negotiation may well be partial or total. We shall, somewhat arbitrarily, assume that dynamic negotiation is total in that it is required to generate a complete plan. (An alternative to this approach is discussed briefly in Section 6.10.) Dynamic negotiation is also required to resolve failed child nodes using the techniques described below. We thus make the following definition.

Suppose that (a_s, CB_s) is a child node of (a_t, CB_t). (a_s, CB_s) is *consistent* iff either

(a) a_s has not yet been invoked, or
(b) CB_s's current plan satisfies s and no child node of (a_s, CB_s) (within the hierarchy) is currently failed.

Dynamic negotiation has the goal of restoring consistency by recovering failed child nodes and by negotiating an amended plan (see below). If this internal negotiation is unsuccessful, then *coord*(CB_s) reports a_s as being failed to *coord*(CB_t). If *coord*(CB_s) is the top-level user agent then the user component in *coord*(CB_s) allows the user involvement in replanning and/or negotiating an amendment (such as a new deadline), or tidy-up (using rollback or abort as described below) for the user task. Thus we see that failures are handled as locally as possible within the cooperation block hierarchy, this being an important aspect of fault tolerance and autonomy.

6.7.1 Recovery of Failed Action Elements

Suppose that a_s is a failed action element in CB_t's plan; then *coord*(CB_t) can instruct a_s to either abort, amend or rollback.

If *coord*(a_s) receives an *abort*(a_s) instruction from *coord*(CB_t), then *coord*(a_s) issues abort instructions to each of the action elements in CB_s, and (upon receiving acknowledgements) then ceases processing of a_s at the current position. The agents in CB_s take no recovery action, other than that which is required to enable them to reset themselves for other work. Any inputs and (partial) results of the action element are lost to CB_t's plan. Abort is employed if the block is able to guarantee that the aborted action element is not needed within the amended plan. For example if we are discarding an entire physical part in a manufacturing process, then the state of an action element's work on that part is irrelevant, and the ability or inability to rollback is unimportant.

If the amended plan for t still requires that s (or some variant thereof) be completed, then we must *rollforward*. This involves CB_t and CB_s negotiating an *amendment* to a_s (for example of the task parameters, input and output rules, and timing constraints).

If *coord*(a_s) receives a *rollback*(a_s) instruction from *coord*(CB_t), then this requires *coord*(a_s) to undo changes made by a_s todate. *coord*(a_s) should, via negotiation with the lower-level cohorts, be able to ascertain whether or not a proposed rollback is possible. If *rollback*(a_s) is possible, then rollback requires *coord*(a_s) to instruct each of the action elements in CB_s to rollback or abort. Notice that the ability to rollback is not an issue if it occurs prior to invocation. Rollback is not quite the same as in traditional data processing, since the inputs received by a_s may need to be forwarded to some other action element after the rollback. Thus we can (and shall) view rollback as an amendment, in which the underlying task request is being replaced by an empty task request. As indicated below, dynamic negotiation may include the (re)planning of new action elements to compensate for the rollback.

An amended/rolled-back action element continues to appear in the amended plan, whereas an aborted action element does not.

During replanning, rollback, abort and amend could in theory also be applied to action elements that had not reported failure. However in practice we might normally expect that an action element would not be subject to abort or amendment/rollback after invocation if it itself had not failed, and we shall adopt this as a simplifying assumption in later sections.

An action element whose underlying task (i.e. lower-level cooperation block) has committed cannot be rolled-back. It may be necessary to amend the post-conditions/output rules, and abort may be the simplest way of resolving acknowledgement receipt failure. An action element that has terminated cannot be aborted or amended/rolled-back. These issues are illustrated in Figs. 6.7–6.10. Before a cooperation block can commit, all action elements within it must have been aborted or have reported themselves as terminated successfully.

Replanning involves the development of an amended plan. The amended plan must obviously satisfy the current task decomposition. If no such plan can be found, then t might be re-decomposed. Replanning clearly requires rescheduling.

The amended plan must include those action elements from the old plan which had previously terminated, but others may be aborted or amended/rolled-back as necessary, and new action elements may be added. As in Section 6.6, we assume that new action elements and changes to existing action elements are proposed to their respective coordinators. These proposals would be considered by the lower-level cooperation blocks and either accepted or rejected. Note that proposed aborts can always be accepted prior to termination, since they are simply implemented by cascading.

Only if there is unanimous acceptance can the new plan be accepted as a whole – although rejection of only a small number of the plan's action elements might suggest a viable alternative plan. Upon acceptance of the new plan, the proposals regarding action elements are again confirmed.

For failed action elements, amend/rollback proposals are only accepted if they would allow the lower-level block to re-establish consistency, whence confirmation allows such action elements to lower their failed status.

A plan amendment may of course require a particular action element to rollback, and then subsequently schedule a further action element for reprocessing of the required work. Such an action element *compensates* for the inability to process the original action element as intended. Similarly, compensating actions could be scheduled following an abort or amend.

Note that if an agent fails, but the partial work is intended to be completed by some other agent, then formally this is to be viewed as an amendment of the original action element, rather than an abort followed by a compensating action. Note finally that compensating action, or the completion of work following agent failure might in practice require human intervention.

In terms of the cooperation block hierarchy, replanning within CB_t results in

- new child nodes of CB_t, i.e. new action elements;
- the amendment (including rollback) of some of the existing child nodes, subject to the above-mentioned restrictions on rollback; and
- the deletion of those child nodes that have been aborted.

Following replanning, the amended plan again consists of the child nodes of CB_t. In particular, if $a \Rightarrow_P b$ in the plan prior to failure, and a is aborted, then b must either be aborted, or have its input rules amended in order to ensure the internal consistency of the new plan.

6.8 A Computational Model

Figs. 6.7–6.10 illustrate the combined life cycle of an action element a and its underlying cooperation block CB. Figure 6.7 depicts the three phases during the lifetime of a. During the life of CB we can identify four broad states, these being illustrated in Figs. 6.8–6.10.

(a) Internal negotiation, which, as stated in Section 6.7, we assume has the goal of restoring consistency after invocation.
(b) Proposal consideration (a form of internal negotiation) which occurs when a proposal is made (from the parent) for either a new action element or a change to an existing action element.
(c) External negotiation with the parent, in response to a failure of internal negotiation or during the negotiation concerning action element proposals.
(d) Consistent processing – a consistent schedule has been generated and is in progress.

As in Section 6.7 we assume that rollback is a form of amend, and that following invocation, amend/rollback and abort are applied only to failed action elements. Note that if an action element within a block fails, then the block enters internal (and possibly external) negotiation. However, in accordance with our assumption above, existing action elements will continue to execute, unless they themselves fail. Thus, under these circumstances, negotiation and execution of action elements can take place in parallel.

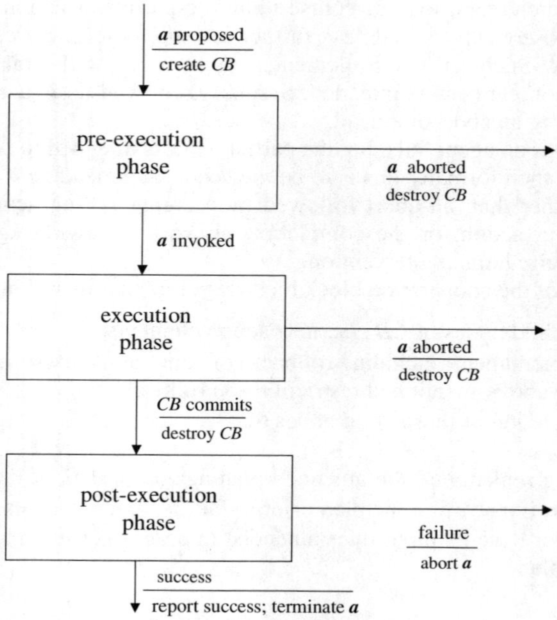

Fig. 6.7. Action element life cycle

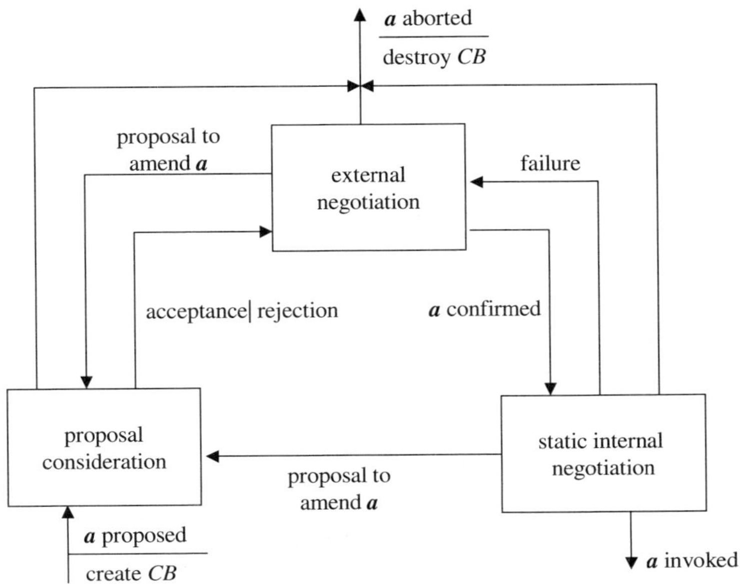

Fig. 6.8. Pre-execution phase

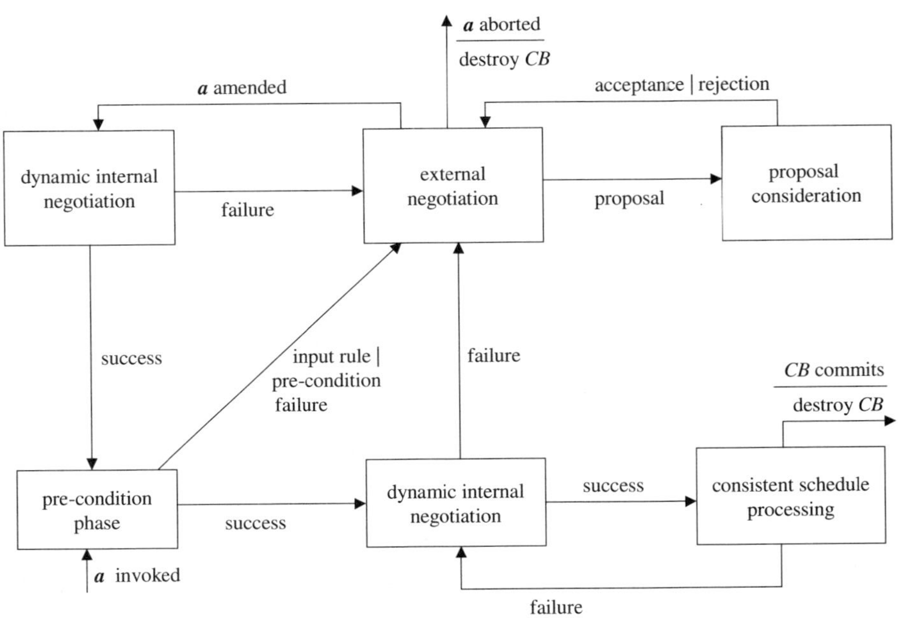

Fig. 6.9. Execution phase

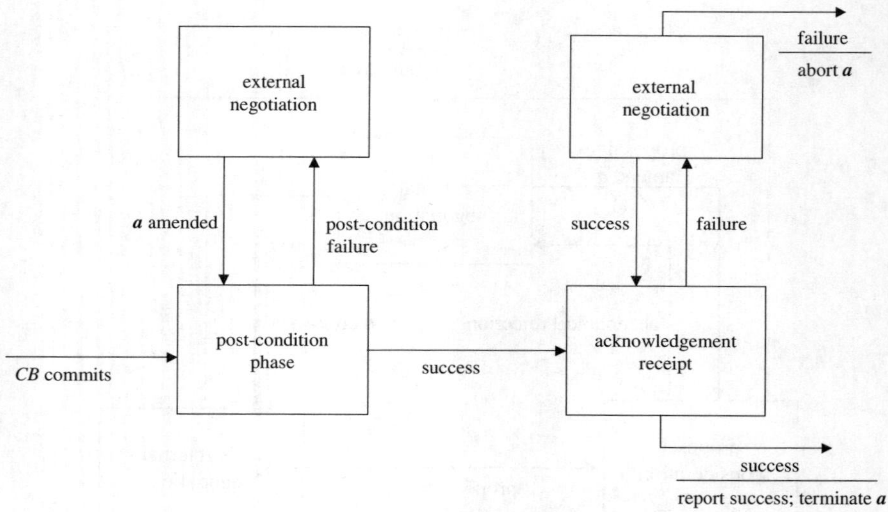

Fig. 6.10. Post-execution phase

6.9 Properties of the Computational Model

There are a number of theoretical properties that we desire of our model, the first being correctness.

Theorem. *An action element is successfully completed if it is reported as being successfully completed. [6.20]*

Ideally, we would also like every task request to terminate successfully. However this is clearly not possible in all cases. For example, if the task request contains unreasonable pre- or post-conditions then, clearly, successful completion cannot be guaranteed. In the case when the task request has a deadline, termination is guaranteed (since all lower-level cooperation blocks must certainly adhere to the deadline or report failure), although, of course, *successful* termination is not guaranteed. In order to ensure termination without such a deadline, we make the following assumptions, which we believe to be reasonable in a practical setting.

(a) *Primitive finiteness.* Clearly, we must insist that the primitive operations within the system are themselves finite. We therefore assume that any negotiation technique(s) that are employed are convergent in that they are guaranteed to either succeed or fail within some finite time. We also assume that the processing of an action element within a trivial block is always accompanied by some time limit after invocation (whose violation leads to failure). Such a time limit would certainly vary from task to task, and possibly from agent to agent.

(b) *Failure finiteness.* We assume that within a given cooperation block, only a finite number of iterations of the execute – fail – recovery cycle are accepted before the block itself reports failure. Again, the number of iterations would be dependent upon the application and on the level of the block, with upper levels allowing more failure than lower-level blocks (since the cost of rollback or abort is much greater). Failures might be weighted according to their type and seriousness. Typically, we might expect that a low-level block would itself fail after two internal failures. As regards the top-level user block, we assume that after the user-determined number of failures is exhausted, the system employs global abort.

(c) *Task decomposition stratification.* We insist that cooperation block hierarchies do not extend indefinitely downwards. The most obvious way in which this could be achieved is by insisting upon a stratification constraint which ensures that a task request at one level is only decomposed into task requests at a lower level.

(d) *Invocation fairness.* As in Section 6.5, we assume that the invocation time of an action element a is specified in terms of (i) an absolute time or the essential inputs, and (ii) *coord*(a)'s internal schedule. Moreover, we need to assume that *coord*(a) adopts a fair internal scheduling policy, in that if condition (i) is satisfied, then a will indeed be invoked eventually.

The following theorem and corollaries make the above-mentioned assumptions concerning primitive and failure finiteness, task decomposition stratification, and invocation fairness. Note from Figs. 6.7–6.10, that every action element either is aborted, terminates successfully, or fails to terminate.

Theorem. *Suppose that at time T the action element* a_t *has been invoked, but has neither terminated nor been aborted. Then at some time > T,* a_t *either terminates successfully or reports failure.* [6.20]

Corollary. *Let* a_t *be an invoked action element; then, in a finite time,* a_t *either terminates successfully or is aborted.* [6.20]

Corollary. *Suppose that the top-level user task is invoked, and that* a_t *is an action element in the corresponding cooperation block hierarchy. Then, in a finite time,* a_t *either terminates successfully or is aborted.* [6.20]

6.10 Dynamic Behaviour

The behaviour within a cooperation block is dynamic for a number of reasons.

(a) An action element's behaviour can be amended as a result of the supplied parameter values,
(b) Dynamic negotiation is allowed, both in developing the initial plan, and in response to failure and amendment,
(c) An action element might be unscheduled, meaning that it does not have an assigned invocation time (see Section 6.5),
(d) Some action elements might be *contingent*, i.e. they may not necessarily be invoked in a successful execution of the block's plan.

Action elements employing dynamic negotiation could still have timing constraints. If so, the deadline also acts as a (somewhat inelegant) limit on the extent to which dynamic interaction can proliferate. We envisage dynamic negotiation as replacing static negotiation when the task request is poorly prescribed (e.g. sparse), whence a decomposition is difficult to choose in advance. Unscheduled and dynamically scheduled action elements are clearly more prone to failure, which in turn may suggest duplicating action elements within the plan (with each of the duplicates being contingent). An unscheduled action element is not necessarily dynamic, since its cooperation block could well have a plan (and even a relative schedule).

There are some fairly obvious ways in which additional dynamic behaviour could be allowed. For example, we might allow an action element's output rules to attach additional input and output rules to other action elements, or indeed add additional action elements to the plan. Such an ability would have to be limited in some way (for example by insisting that all action elements have a fixed deadline) so as to guarantee termination. In our model, the inclusion of (a finite number of) contingent action elements as described above provides a limited variant of this approach. These issues relate to the extent to which we allow continuous rather than failure-driven replanning. We have insisted (somewhat arbitrarily and for the sake of simplicity) that following invocation, consistency requires the construction of a complete plan (at that level within the cooperation block hierarchy). Allowing execution to proceed without a complete plan would be an extreme form of dynamism. In our model, dynamic amendments to a plan could be instigated by a cohort issuing a pseudo-failure message (incorporating a request for an amendment).

6.11 Mapping CKBS to HMS

The fundamental concepts within the HMS abstract model are those of a *holon* and a *cooperation domain*. These concepts are then mapped down onto an implementation model, which employs the concept of a *function block* [6.12] to provide the base-level functionality.

A holon is defined within the HMS architecture as a *cooperative* and *autonomous* functional unit of manufacturing production. Holons cooperate with each other via cooperation domains, which can be regarded as concrete realisations of cooperation blocks.

Autonomy is defined as the independent ability to generate plans and schedules, to monitor task execution, and to respond to and rectify failure. In the current architecture, this form of autonomy is a property of each subtree within a cooperation block hierarchy. Indeed this unit is also cooperative in the sense that it cooperates (via its top-level coordinator) with the cooperation block of which it is a child. It is thus holonic. Note, however, that holons are not necessarily agents. A subtree within a cooperation block hierarchy occupies a specific position within a larger hierarchy, whence its role is relatively fixed during its lifetime, which is determined by how long the alliance needs to persist to process the assigned task.

Individual agents, on the other hand, are persistent, and can decide whether (and how) to participate in any given cooperation block. Indeed, they may well belong to many such blocks simultaneously. Agents themselves may or may not be holonic. Processing agents can certainly generate internal plans and schedules for tasks for which they have the appropriate processing skill. Whether they are able to respond to failure depends partly upon the architecture (in particular, on the extent to which skill-processing components within the agent are replaceable) and also on whether the task assigned to an agent can be amended to a task outside the agent's skill class.

Holons are also said to be *encapsulated*, in that their inner structure is hidden from the outside world. In the CKBS model, this form of encapsulation is represented by insisting that each action element (which corresponds to a particular subtree of the cooperation block hierarchy) is a black box as far as its peer action elements are concerned.

Self-similarity is the property of holons in which they have similar properties to their constituent parts. Again this is a property of cooperation block hierarchies, which are made up entirely of sub-hierarchies, extending down to the base level with a trivial cooperation block (and hence a trivial hierarchy).

6.12 Related Work

Beyond the current text, there are a number of research papers available that address the area of agent-based manufacturing. Norrie, Balasubramanian and Maturana [6.1, 6.22] define a multi-agent architecture for concurrent manufacturing using the concept of a virtual agent cluster (which is similar to a cooperation block) and mediator agents providing both intra- and inter-cluster coordination (similar to coordinators/minders) to meet local and global objectives. A prototype implementation of this architecture has been developed using C++/Smalltalk. Jarvis [6.16] investigated quality control issues (encompassing monitoring, diagnostics and repair) in manufacturing and reviewed the solutions currently available. That paper also includes a strategy for holonic diagnostics. Parunak [6.27, 6.28] discussed the application of agent-based techniques to problems in manufacturing applications, and compared such techniques with traditional approaches.

In contrast to the engineering paradigm, there has been considerable research effort within the mentalistic paradigm, entailing the ascription of mental propositions or intentionality to a computer system. Central to this approach are the concepts of belief, desire and intention (BDI) which were introduced by Bratman and others [6.3, 6.4] as the basis of a single-agent architecture, and which were further developed by Rao and Georgeff [6.29, 6.30]. The BDI scheme defines the internal processing of an agent through a set of mental categories, within a control framework for the rational selection of a course of action, often with the added concepts of goals and plans. BDI concepts, along with communication, are employed as the foundation for the following DAI/multi-agent architectures.

The speech-act theory of Searle [6.31] encouraged multi-agent researchers [6.9, 6.21, 6.24] to develop inter-agent communications protocols, treating communication as a type of action to be incorporated in the planning and reasoning processes. Primitives inspired by the speech-act theory and used in many multi-agent prototypes include "propose", "refuse", "oppose", "respond", "refute" and "inform".

Shoham's Agent Oriented Programming Language (AOPL) [6.32] is a logic-based instrument for multi-agent systems, in which agents are specified with a set of capabilities, beliefs, commitments and commitment rules, the rules determining how an agent acts. However, Shoham's agents are unable to plan or to communicate requests for certain types of actions, and a major drawback of AOPL is its inability to execute the associated logic.

The Cosy architecture [6.13, 6.14] developed by Haddadi and others at Daimler Chrysler has been applied to the simulation of some road traffic problems. It is considered to be a hybrid BDI architecture, with five main components: sensors, actuators, communications, cognition and intentions. The cognition component is made up of a knowledge base (containing beliefs), script execution, protocol execution and reasoning. Its communication protocols are a set of predefined dialogue trees.

Mueller's INTERRAP model [6.23, 6.24] is built on a layered architecture with hierarchical abstractions. The highest layer is the cooperation layer and the lowest is the world interface, with behaviour- and plan-based layers in between. The inter-agent communications are expressed through primitives, such as "inform", "announce", "reject" and "accept" within logical expressions. The effectiveness of the model has been demonstrated in a simulated environment to resolve conflicts amongst forklift robots placing boxes on shelves.

The European ESPRIT programme supported two major projects on multi-agent systems: ARCHON [6.17] for applications in complex systems such as power plants, and IMAGINE [6.15] as a more general approach. The GRATE model [6.19] of Jennings is based on the ARCHON project and has been employed in a number of simulated environments. It contains a cooperation and control layer and an application domain layer. The behaviour of the cooperation layer is determined from the built-in knowledge held in the system, but does not address the issue of how to reconcile reactivity and deliberation. Domain-dependent information includes a self model and acquaintance models.

The MECCA architecture [6.21] is an off shoot of the IMAGINE project, and (as in [6.6]) divides an agent structurally into a head and a body. However, the MECCA architecture is AI-based, supporting four parallel processes in the agent head: goal activation, planning, scheduling and execution. A set of cooperation methods and cooperation primitives are defined as required for specific applications. An extended version of concurrent Prolog is seen as the basis of distributed processing in the model, presently upgraded with Java. The model has been applied to a simulated traffic control environment.

A different approach has been advocated by Singh [6.33], who employs a triplet of intention, know-how (ability plus knowledge) and communication, which he has formalised into a model of action and time. According to this model, agents' intentions, along with their knowledge and know-how, constrain the actions they choose to perform. Agents that know how to achieve what they intend can succeed provided they act on their intentions, exercise their know-how, and persist with their intentions long enough.

More recently, FIPA (Foundation for Intelligent Physical Agents) [6.10] has proposed a detailed agent architecture, which includes an infrastructure, a communications facility, commands and languages. It uses modified BDI concepts, and supports the speech-act -inspired primitives mentioned above.

A formalised logic-based model, centred on the concepts of commitment and convention (a variant of the BDI approach), has been proposed by Woolridge and Jennings [6.35]. The model defines terms and semantics along with some operational primitives for multi-agent systems.

An interesting contrast to the manufacturing domain is to be found in [6.2], which describes the 3T robot architecture. This separates the control problem into three tiers: (i) a set of situated skills, (ii) a sequencing capability that can activate these skills in order to accomplish specific tasks, and (iii) a deliberative planning capability. The planning and sequencing capabilities develop a set of reaction action packages which we can think of as being analogous to the concept of an action plan. Event monitors detect certain world events and notify the sequencing layer. This integration of planning and control is also found in CIRCA [6.25, 6.26], where real-time control is employed to guarantee hard deadlines, whilst an AI subsystem is employed to perform high-level reasoning about goals, capabilities and plans.

Conclusions

We have presented a high-level model for CKBS in terms of agents, skills, cooperation blocks and hierarchies, together with details of how failures and recoveries are handled. The model is presented at a sufficiently high level to allow it to be mapped onto holonic systems without overly constraining such systems. The model is also sufficiently abstract and formal to allow the theoretical properties of correctness and termination to be proven. We have made very few assumptions regarding agents and work is ongoing to identify the requirements on agent behaviour that then allow us to infer the system properties discussed in the current paper.

References

[6.1] S. Balasubramanian and D. Norrie: "A multi-agent architecture for concurrent engineering", Concurrent Engineering: Research and Applications, **4**, 7–16 (1996).

[6.2] R.P. Bonasso, R.J. Firby, E. Gat, D. Kortenkamp, D. Miller and M. Slack: "Experiences with an architecture for intelligent reactive agents", Journal of Experimental and Theoretical Artificial Intelligence, **9**, 237–256 (1997).

[6.3] M.E. Bratman, *Intentions, Plans and Practical Reason*, Harvard University Press, Cambridge, MA (1987).

[6.4] M.E. Bratman, "Plans and resource bounded practical reasoning", Computational Intelligence, **4**, 349–355 (1998).

[6.5] S.M. Deen, ed., "Proceedings of the Second International Working Conference on CKBS", DAKE Centre, Keele University (1994).

[6.6] S.M. Deen, "An architectural framework for CKBS applications", IEEE Transactions on Knowledge and Data Engineering, **8**, 663–671 (1996).

[6.7] S.M. Deen, "A fault tolerant cooperative distributed system", *IEEE DEXA Workshop*, Vienna, ed. R. Wagner (1998), pp. 508-513.

[6.8] S.M.Deen, DREAM: Dake Research into an Engineering Approach to Multi-agent systems, http://www.keele.ac.uk/depts/cs/dake/ (2001).

[6.9] M. Fisher, "Specifying and executing protocols for cooperative action", *Proceedings of the Second International Working Conference on CKBS,* ed. S. M. Deen, Keele, UK (1994), pp. 295–306.

[6.10] FIPA: Foundation for Intelligent Physical Agents, http://www.fipa.org (2001).

[6.11] M. Fletcher, ed., *Holonic Systems Architecture,* HMS Project Deliverable D1.1-1 (1998).

[6.12] Function Blocks for Industrial Process-measurement and Control Systems, Part 1: Architecture, Technical Report TC65 WG6, International Electrotechnical Commission (1997).

[6.13] A. Haddadi, "A hybrid architecture for multi-agent systems", *CKBS-SIG Proceedings*, ed. S.M. Deen, DAKE Centre, Keele University (1993), pp. 13–26.
[6.14] A. Haddadi, Communication and Cooperation in Agent Systems – A Pragmatic Theory, Springer, Berlin, Heidelberg (1996).
[6.15] H. Haugeneder, "IMAGINE: A Framework for building multi-agent systems", *Proceedings of the Second International Working Conference on CKBS,* ed. S.M. Deen, Keele, UK (1994), pp. 31–64.
[6.16] D. Jarvis, Quality Control Holonic Manufacturing Systems, CSIRO, Australia (1998).
[6.17] N.R. Jennings, "The ARCHON project and its applications", *Proceedings of the Second International Working Conference on CKBS,* ed. S.M. Deen, Keele, UK (1994), pp. 13–30.
[6.19] N.R. Jennings, "Controlling cooperative problem solving in industrial multi-agent systems using joint intentions", Artificial Intelligence, **74**, 195–240 (1995).
[6.20] S.M. Deen and C.A. Johnson: "Formalising an engineering approach to cooperating knowledge based systems", IEEE Transactions on Knowledge and Data Engineering, **15**, 103–177 (2003).
[6.21] A. Lux and D.D. Steiner: "Understanding cooperation – an agent's perspectives", in Readings in Agents, eds. M.N. Huhns and M.P. Singh, Morgan Kaufmann (1997), pp. 471–480.
[6.22] F.P. Maturana: *MetaMorph: An Adaptive Multi-Agent Architecture for Advanced Manufacturing Systems*, Doctorial dissertation, University of Calgary, Canada (1997).
[6.23] J.P. Mueller, "A conceptual model for agent interaction", *Proceedings of the Second International Working Conference on CKBS*, ed. S. M. Deen, Keele, UK (1994), pp. 213–234.
[6.24] J.P. Mueller, *The Design of Intelligent Agents*, Springer, Berlin, Heidelberg (1996).
[6.25] D.J. Musliner, E. Durfee and K. Shin: "CIRCA: A cooperative intelligent, real time control architecture", IEEE Transactions on Systems, Man and Cybernetics, **23**, 1561–1574 (1993).
[6.26] D.J. Musliner, J.A. Hendler, A.K. Agrawala, E.H. Durfee, J.K. Strosnider and C.J. Paul: "The challenges of real time AI", IEEE Computer, **28**, 58–66 (1995).
[6.27] H.V.D. Parunak, A.D. Baker and S.J. Clark, "The AARIA agent architecture", Integrated Computer-Aided Engineering, **8**, 45–58 (2001).
[6.28] H.V.D. Parunak, *Practical and industrial application of agent-based systems*, The Industrial Technology Institute (1998).
[6.29] A.S. Rao and M.P. Georgeff: "Modeling rational agents within a BDI architecture", *Proceedings of the 2nd International Conference on Principles of Knowledge Representation and Reasoning*, Cambridge, MA, eds. J. Allen, R. Fikes and E. Sandewall (1991), pp. 473–484.
[6.30] A.S. Rao and M.P. Georgeff: "BDI agents – from theory to practice", *Proceedings of the 1st International Conference on Multi-Agent Systems*, San Francisco, ed. V. Lesser (1995), pp. 312-319.
[6.31] J.R. Searle, *Speech Act*, Cambridge University Press (1969).
[6.32] Y. Shoham, "Agent oriented programming", Artificial Intelligence, **60**, 51–92 (1993).
[6.33] M.P. Singh, Multi-Agent Systems: A Theoretical Framework for Intentions, Know-How and Communications, Springer, Berlin, Heidelberg (1994).
[6.34] R.G. Smith, "The contract net protocol: High level communication and control in a distributed problem solver", IEEE Transactions on Computers, **C-29**, 1104–1113 (1980).
[6.35] M. Woolridge and N.R. Jennings: "Formalising the cooperative problem solving process", in Readings in Agents, eds. M.N. Huhns and M.P. Singh, Morgan Kaufmann, San Francisco (1997), pp. 430–440.

Part III

Systems Operation

7. An Investigation into a Computational Model for HMS

S. M. Deen

Department of Computer Science
University of Keele
England
Email: deen@cs.keele.ac.uk

Abstract. This paper investigates a computational model for an abstract HMS operational scenario with a view to outlining rules that should be implemented in a holonic system to ensure a correct dynamic behaviour, particularly during operational disturbances. The model considers operations at the substructure levels of holons. It is claimed that without such a model the system behaviour will be unreliable, potentially causing not only wasted production but also uncontrolled executions. Three features considered under the computational model are execution consistency, execution termination in a finite time and operational stability. A failed action must be rolled back to a recoverable point for possible redoing, execution must follow a sequence of rules, termination must be guaranteed, and stability must be ensured. The paper argues against the implementation of a holonic system without such behavioural features.

7.1 Introduction

Since the HMS environment is complex and highly distributed, the potential for something going wrong with wasted production is naturally high, and hence it is important to construct a computational model that specifies the correct system behaviour through all its operational states (including failure states) during execution. In constructing such a model we may begin with the following four concepts:

- *Completeness*: A software system is computationally complete if it supports all the required computations. This necessitates the analysis of all the computational requirements and demands that all the necessary procedures have been implemented.
- *Consistency*: This ensures that the final result produced in an error-prone environment is equivalent to that produced in an error-free environment in spite of possible systems malfunctions. We have considered two elements here: Task completion (not to be confused with completeness) and serialisation. As we are dealing with physical objects, automatic restart after a system malfunction may not be feasible in many cases, without some human inspection and decision. This will require the execution to stop at what we call recoverable points. This in turn will mean one of the following: (i) the process will complete successfully or (ii) if it fails, it must be rolled back to a recoverable point. Our second element, serialisation, ensures that the execution is carried out in a valid sequence avoiding conflicts.

- *Termination*: The execution must terminate in a finite time. That is, the execution of a task should not explode exponentially into further tasks and should not lead to cycles, but should converge into a result.
- *Stability*: The operational behaviour of the distributed system must be stable, so that system malfunctions are containable and do not cause a catastrophic outcome.

A system must have structures and procedures that ensure these behaviours. As implied above, a system is computationally complete if all the necessary software for handling all its inputs and outputs has been implemented correctly. This theoretical definition is not a great deal of help in real systems, where the only practical way of ensuring completeness is by rigorous program testing. Therefore, we shall not pursue this topic any further. As regards termination, one can examine algorithms and processes to check if they terminate, but as we shall see below, termination in our limited context is relatively easier to ensure, enforced by time-out if necessary. Consistency and Stability, however, will require a more thorough investigation.

A set of rules in logic is consistent if their successful execution in the same situation does not lead to conflicting conclusions [7.1]. We have adapted this definition of consistency for manufacturing operations as follows: a set of operations must not lead to conflicting results (conclusions) in spite of systems malfunctions; that is, rules (operations) must be specified to ensure this non-conflicting behaviour. Non-conflicting results at different executions under the same set of inputs will imply identical results (as the final result) if the process has been successfully completed at each time. If the process has not been successfully completed, then a non-conflicting result can be a partial result, but such a partial result will be an acceptable step towards (and hence not conflicting with) the final result.

Many of the concepts used here are drawn from our earlier papers [7.2, 7.3] on Cooperating Knowledge Based Systems (CKBS) in the DREAM (Dake Research into an Engineering Approach to Multi-agents) project, in which we have studied agent-based processing for transaction-oriented applications, amalgamating ideas from both Distributed Databases (DDBs) and the multi-agent systems of Distributed Artificial Intelligence (DAI/MAS). In transaction-oriented applications, of which Holonic Manufacturing Systems is an example, a task is decomposed into lower-level tasks for execution. Since distribution is involved, it is important to show that the dynamic behaviour of the system is sound, as we have attempted to do here, using some concepts from the world of databases.

This CKBS model is more general and more suitable than the HMS/FB model for exploring ideas as done in this chapter, as well as in Chapter 6 and 8. The relationship between the CKBS model and HMS/FB model, so far as this chapter is concerned, is explained later. Chapter 6 addresses the issues of correctness, termination, recovery and stability, but largely from the perspective of *inter*-agent cooperation as part of the model verification. In contrast this chapter examines *intra*-agent operations, particularly operational primitives within the functional components (funcoms) of agents as part of the operational behaviour, with a different perspective from the one in Chapter 6. Thus to some extent, these two chapters are complementary, but each has a different viewpoint and each can be read as a stand-alone chapter. An earlier version of this computational model appears in [7.2d].

The plan of this chapter is as follows: in Section 7.2 we present an operational framework, which forms the background of this paper. Section 7.3 forms the core, where we have investigated the behaviour of the holon components and outlined rules that must be followed to ensure correct execution. A conclusion is given in Section 7.4. However, before we proceed any further, we shall briefly review the related research activities.

Related Activities

Concepts of belief, desire and intention (BDI) as the basis of a single-agent architecture were introduced by Bratman et al [7.4, 7.5] and developed further by Rao and Georgeff [7.6, 7.7]. These concepts are also employed in DAI/multi-agent architectures. The BDI scheme defines the internal processing of an agent through a set of mental categories, within a control framework for rational selection of a course of action, often with the added concepts of goals and plans.

The Speech-Act theory of Searle [7.8] encouraged multi-agent researchers [7.9] to develop inter-agent communications protocols, treating communication as a type of action to be incorporated into the planning and reasoning processes. Primitives inspired by the Speech-Act theory and used in many multi-agent prototypes include *propose*, *refuse*, *oppose*, *respond*, *refute* and *inform*. DARPA introduced a Lisp-based environment, with KIF (Knowledge Interchange Format) [7.10] and KQML (Knowledge Query Manipulation Language) [7.11], which has made an impact particularly in the United States. Shoham's [7.12] Agent Oriented Programming Language (AOPL), apparently motivated by Searle's Speech-Act theory, is a logic based instrument for multi-agent systems. Agents are specified with a set of capabilities, a set of beliefs and commitments and a set of commitment rules, the rules determining how an agent acts. However, Shoham's agents are unable to plan or to communicate requests for certain types of actions, and a major drawback of AOPL is its inability to execute the associated logic.

Much of the early DAI/Multi-agent research was founded on deliberative rather than reactive architectures, and latterly on a combination of both, sometimes packaged as hybrid architectures [7.13]. The importance of reactive architectures for real-world applications has been emphasised by Brooks [7.14]. Manufacturing applications are best modelled as reactive, as we have assumed in this chapter. The basic FIPA architecture still uses the BDI concepts to define the internal agent states and communication primitives/protocols to communicate them among agents, the HMS/FIPA architecture presented in Chapter 4 employs a reactive approach.

Van Brussel et al [7.15] have proposed an agent-based architecture called PROSA which relates to manufacturing production but not to the controller issues. Some elements of this architecture appear in Chapter 13 of this book. Norrie et al [7.16] have investigated the area of agent-based manufacturing systems and have published a number of works on the architectures and processing of both concurrent engineering and holonic systems. Some aspects of these ideas in some form can be found in Chapter 2.

A major difference between the above approaches and ours is that our model is highly distributed, paralleling closely the HMS/FB architecture (Chapter 4). However, none of the models we know of includes any computational model to ensure the correct dynamic behaviour of the agent-based system.

7.2 Operational Framework

We shall start here with some architectural features, covered earlier in Chapters 4, and 6, but restated below from the perspective of this chapter alone, with its own slant on the definition of the concepts used. This is intended to make the chapter stand-alone.

7.2.1 Holons and Cooperation

A holon has been defined as an autonomous cooperative unit of production, which has a compulsory software element and optional support for a hardware element. The "support" in this context implies the relevant operational model of the hardware and an interface to the hardware, but not the hardware itself. If such support is included, we shall call the holon a hardware holon as opposed to a purely software holon. If the hardware breaks down, then the (hardware) holon will be deemed to have broken down.

We regard a holon as having a head and a body, connected by a neck which is a communication link. The head is responsible for communications and interactions with other holons via the cooperation block, while the body gives the holon its individuality. Both the head and the body are made up of functional components, *funcoms* for short, but in this paper we have considered only those in the agent body (body funcoms). The division of a holon into funcoms permits the localisation of faults into smaller units and enhances choice and flexibility. For a hardware holon, some of these funcoms may hold the hardware model and some others may hold the interface. There are several categories of holon (as in Chapter 8), of which only two types, coordinator and skill holons are of interest to this chapter.

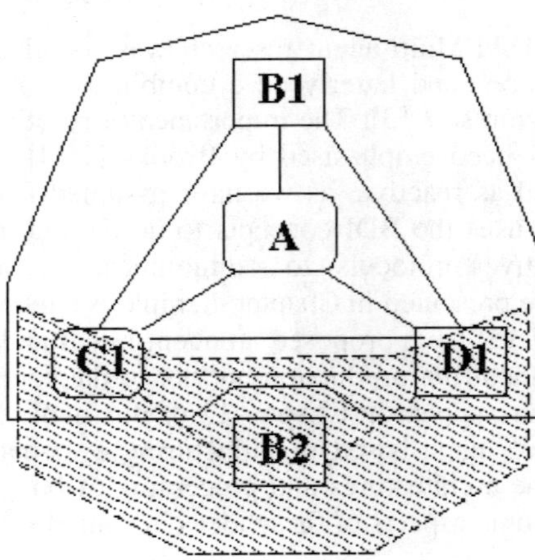

Fig. 7.1. Two Cooperation Blocks (CBs)

To accomplish a joint task (or, equivalently a global task), a coordinator holon (acting as a user holon) decomposes the joint task into (lower-level) tasks and then invites other holons with appropriate skills, to assist it as *cohorts* to carry out those tasks in cooperation. Thus a *Cooperation Block (CB)* is formed, in which the coordinator coordinates and skill holons act as cohorts. A cohort may join or leave a CB as and when required. However, a skill holon in general should be able also to act as the coordinator holon for a lower-level CB, thus supporting a hierarchy of CBs, as shown in Fig 7.1 below with two CBs: holon A acts as the coordinator in the higher block, its cohort C1 acts also as the coordinator of the lower block, and cohort D1 participates in both the blocks. There can be a hierarchy of skills as well, each skill to be offered by an appropriate skill holon. Some skill holons may be capable of offering more than one skill, or even overlapping skills.

The HMS/FB architecture is based on three main concepts: Holon, Cooperation Domain (CD) and Function Block (FB) drawing on the proposed standard of the distributed controller architecture (IEC 61499) described in Chapter 4. Essentially an FB is like a special kind of C++ object (typed) – but in addition to data, events are also public items. Both holons and CDs are made up of function blocks. In this paper, we have used (as stated above) the concept of functional components (*funcoms*) to represent the logical properties of a function block (without any loss of generality), thus avoiding the details of the FB specification of IEC61499 that are redundant to our purpose here. Both FBs and funcoms can be composite, and can form heterarchies with atomic instances as leaves. Additionally we shall use the term *holon* and *agent* synonymously unless otherwise qualified.

While a holon in the HMS/FB architecture provides the skill for task execution, a CD provides the structure needed for inter-holon cooperation, holds blackboards and executes procedures/predicates. The CD thus combines the following three functions:

(a) It acts as a coordinator of the joint task.
(b) It provides a logical framework for inter-holon cooperation.
(c) It supports a physical infrastructure for inter-holon communications.

For clarity of analysis, in this paper, we shall use the concept of a coordinator holon for function (a), and the concept of a cooperation block for (b), and ignore (c), treating it as a blackbox. Thus, the concepts of a cooperation block (CB) and a funcom provide the basic constructs for our abstract CKBS framework. This section concludes the background necessary for a clearer understanding of the remainder of this paper.

7.2.2 Holonic Actions

Given a holon, we can visualise five broad types of actions:

(1) Task execution actions: These are the Do actions for the execution of the basic task – the primary actions (or core activities) of a holon. A Do action will include Schedule and Negotiate, depending on the ability of the holon.

(2) Validation actions: These are the actions to check the truth of validation conditions, such as pre, post and end conditions.

(3) Corrective actions: These are the secondary actions resulting from of delays, failures and recovery procedures, and may include: Rollback, Redo, Quit and Abort.

(4) Information actions: These are information processing and exchange actions, such as: Retrieve, Respond, Insert, Delete, Modify, Confirm, Accept and Reject. They will become instances of the Do action.

(5) Message actions: There are many types of messages, such as send, receive and acknowledge, associated with the other actions, in addition to housekeeping messages.

We have not considered above failures in the execution of information actions, as such failures are not specific to HMS and have standard solutions. We have also assumed (see later) a fail-safe method for the ordered delivery of messages, although there can be delays, which we must take into account. These actions can be implemented differently in different models and different language environments. Each action will have lower-level actions, the lowest-level (leaf) actions being called action primitives, which will either succeed or fail, with nothing in between. For example Negotiate is a composite action made up of other actions, which in the case of the Contract Net protocol can be: Invite, Bid, Evaluate, Accept, Reject and Confirm.

We may define a recovery point as the start point of an action primitive where a Redo can commence. There is a distinction between Quit and Abort. Quit implies a stoppage followed by an automatic rollback to a recoverable point, ready for an appropriate Redo. Abort implies a stoppage followed by an automatic rollback to a recoverable point, but ready for a compensating action, *not* a Redo. A compensating action may involve human intervention, alternative task decomposition and appropriate revised processing to complete the global task. Such a compensating action will be application-dependent and hence has not been pursued here any further. Thus Abort is the termination of the current execution following a breakdown that cannot be automatically rectified by a standard Redo.

7.3 Inter-Funcom Processing

Since funcoms are the constituents of holon bodies, the funcom-level (intra-agent) interactions are more fundamental, and therefore we shall focus only on them. The funcom level interactions during runtime for a funcom serving two parents and two children are shown in Fig. 7.2 below:

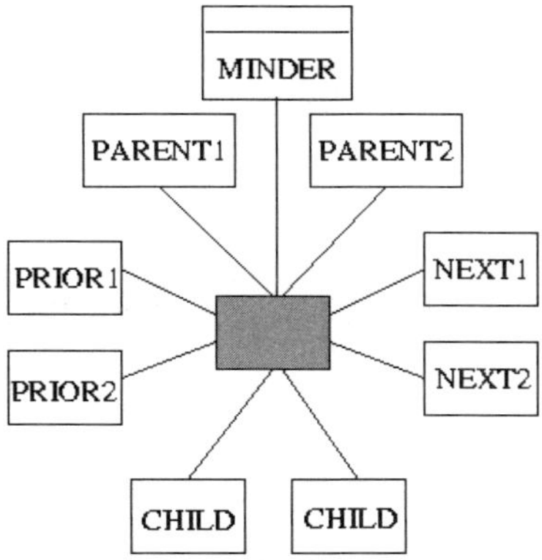

Fig. 7.2. Funcom Interactions. A physical funcom can be shared

The specific interactions are:

(a) parent to child
(b) child to parent
(c) prior (transmission of pre-condition parameters, or information on delays)
(d) next (inquiry on execution status or delays)

For simplicity, we shall not consider (c) and (d) assuming inter-sibling communications to be carried out through the common parents. The funcoms are not scheduled but event-driven; this apart, a funcom execution represents a more general scenario for a computational model as discussed below. Observe that a holon body can consist of a single funcom.

7.3.1 Funcom Primitives

We view each lowest-level (atomic) activity in a funcom as an action primitive. What is a primitive in one funcom can be composite in its lower-level funcoms. During task execution in a funcom, the following activities may take place:

Validation Primitives
Pre-condition check (P)
Post-condition check (P')
End-condition check (E)

Do Primitives
Do (D) – a composite action representing, say *r*, lower-level tasks

Corrective Primitives
Rollback (B)
Redo (R)
Quit (Q) – at a recoverable point
Abort (A) – at a recoverable point

Message Primitives
Processing of an input message (Ii)
Generation of an output message (Io).

These primitives are meant to be backed by an appropriate logging and recovery facility. We assume that all the necessary primitives have been defined and correctly implemented, and therefore the system is *Complete*. We shall now consider Consistency, Termination and Stability.

7.3.2 Consistency

We shall consider below both task completion and serialisation, starting with task completion.

Task Completion

The computationally complete system will have the following characteristics, as required for consistency:

(a) If the execution **E** is successfully completed without any malfunction, then it will have proceeded in the following sequence:
E == <P, D, P', E>
where == means "defined as".

(b) If the execution **E** fails in between two primitives, then it must be rolled back to the end of the last successful primitive (a recoverable point). For two successive primitives in our context, "the end of the previous primitive " means the same as "the beginning of the next primitive". We must assume this will be done using the other defined primitives, along with the logging and recovery facility.

(c) If a failure occurs at the beginning, then the execution **E** must be rolled back to the start, as if nothing has happened. This is a special case of (b).

Thus we can assert that a task ends successfully or ends at a recoverable point.

Serialisation

Serialisation is a database concept to ensure a correct serial execution order of conflicting read and write operations in transactions from different sources. In our case a schedule of activities (primitives) is correct if it forms a valid execution order. A valid execution order will preserve consistency. *Serialisation rules* produce a valid execution order, and if that order is followed, then the execution will preserve consistency. However, in our case the execution order changes dynamically in an error-prone environment, full of failures and interruptions. Therefore we need to determine dynamically the next correct primitive for execution. For this, we require a special logical funcom structure (referred to simply as the *funcom structure*) as shown in Fig 7.3.

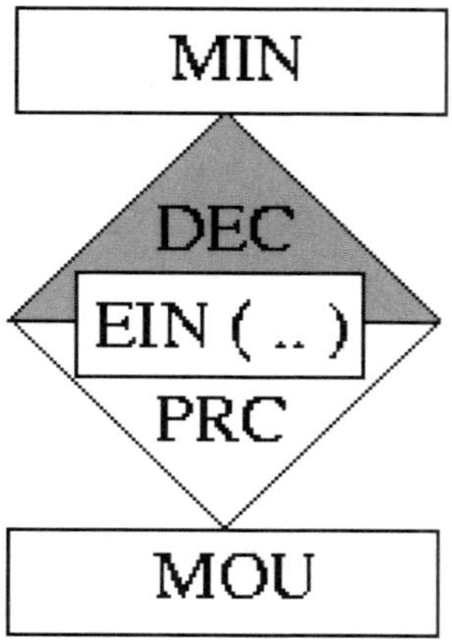

Fig. 7.3. Funcom Structure Diagram

The components are:

EIN (parameters) – Event Indicator
MIN – Input messages
MOU – Output messages
DEC – Decider
PRC – Processor

The successful execution of a primitive is an event, registered as an EIN tuple with appropriate attributes, along with the begin time and end time of each event state. The EIN captures the event states, the MIN holds the input messages to be processed, the MOU holds the output messages generated, and the DEC decides which input message from the MIN is to be processed next by the PRC. All invalid inputs are rejected by the DEC, possibly with appropriate error messages.

We present here a set of rules (to be referred to as the *rule-set*) which serialises activities (primitives) in a correct order. We assume:

- All illegal inputs will be rejected by the DEC.
- Quit and Abort never fail and can follow any primitive.
- Any primitive must be aborted after a specified number of Redo's.
- Any primitive after a specified time-out must be aborted.
- Rollback (B) never fails.
- All Redo's (R) must be preceded by B.
- Ii and Io never fail and can follow any primitive.
- All communications (MOU/MIN) are ordered and failsafe.
- Quit from the MIN may be followed by R or A.
- Redo, Rollback, Quit and Abort are *idempotent*.

The last item implies that if $\varphi = \{R, B, Q, A\}$ and if x is the argument of φ, then $\varphi(\varphi(x)) = \varphi(x)$. That is, the repeated execution of φ over x will be equivalent to a single execution of φ over x.

Consider the execution of the primitive X shown as follows:

X: {Y | Z} ;

if X succeeds, its next primitive is Y,
if X fails, its next primitive is Z.

A sample of serialisation rules is shown below:

P: {D | R' }, where R' = <B, R>
D: {P' | R' }
P': {E | R }
(D + P'): {E | R' }
E: { – | R }
R: {n | R }
B: {R | - }

where n is the next primitive in the correct sequence. Some of these rules can be depicted diagrammatically as shown in Fig. 7.4.

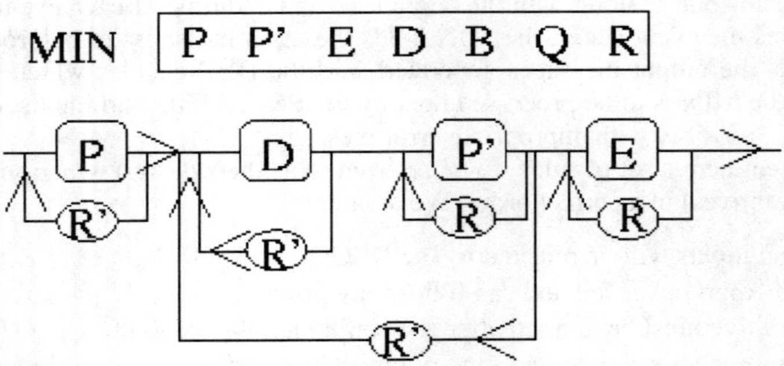

Fig. 7.4. An example of Serialisation Rules

These rules tell us that each functional component must be designed with

- a set of primitives
- a set of serialisation rules
- EIN, MIN, MOU, DEC and PRC
- backup facilities (logging, stable storage and software ...)

to ensure Consistency. The logical structure of a funcom shown in Fig. 7.3 can be implemented differently.

7.3.3 Termination

Termination is correct if the execution terminates in a finite time. The following cases can violate this quality:

- if the execution enters a cycle
- if the task explodes

Cycles can arise from an incorrect specification of the dependencies (including precedent constraints). Decomposition and planning, among others, can lead to explosive alternatives. Finally inadvertent program error can lead to cycles. The general solutions that are available are:

- process time-out
- control of the number of Redo's

Both of these are included under the serialisation rules. Observe that if every child of a parent terminates, the parent does not necessarily terminate. But if a parent terminates, then all its children will terminate by definition. Thus if a parent issues a Quit (or Abort), all its children must also quit (or abort) right down to the leaf level. Therefore, this rule needs to be added to our *rule-set* given earlier.

7.3.4 Operational Stability

A distributed agent system, equivalently a distributed system of funcoms as discussed here, has a parallel with the neural network system in a human brain. Neurons send messages (signals) to one another, and a neuron fires (sends a signal other neurons) when the incoming set of signals satisfies a threshold criterion. As there is no other control, if something goes wrong on the threshold, the firing can become uncontrolled, in turn causing other neurons to fire continuously. This eventually can lead to instability in the neural network, and to seizure (epileptic fit) or other problems. The maintenance of stability is therefore very important in any distributed system where there are a large number of elements present, as in our case. An example of such a distributed system is shown in Fig 7.5 by an inter-connected set of funcoms:

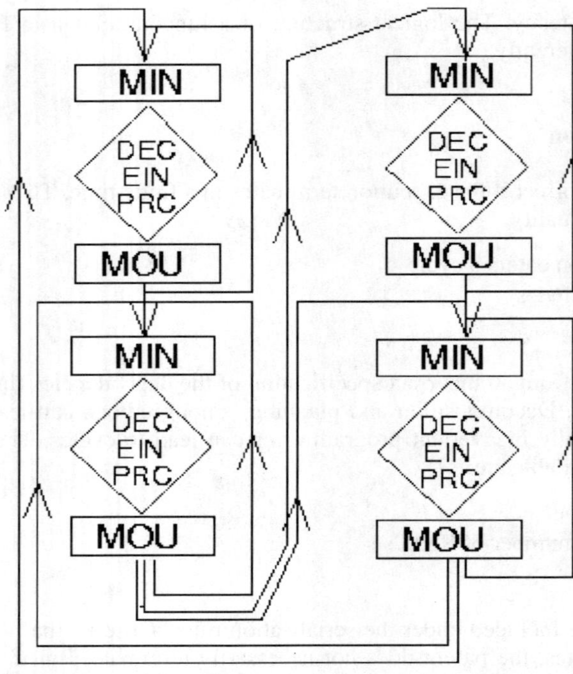

Fig. 7.5. A Funcom Network

Stability implies the absence of oscillations in the system. To control it, one must have bounded inputs and bounded outputs. This requires:

1. Only legal inputs are allowed to enter the MIN.
2. Only relevant inputs are transferred to the EIN.
3. The Decider selects the next correct item for processing.
4. Only valid outputs are generated by the Processor to enter the MOU.

These rules will ensure bounded inputs and bounded outputs in each funcom. A message, to be accepted by a funcom, has to be relevant for the current execution path. All inputs will arrive through the MIN, held in a general message buffer for each agent. The buffer will select the appropriate messages for each funcom. We assume that there is a sufficiently large local funcom message buffer available within the MIN/MOU.

Some issues: Suppose that incoming messages cannot get into the buffer (say, due to buffer overflows or delays). If the message is unexpected, the current task should proceed normally, although this task may have to be abandoned or redone later. If the message is expected, the task must wait (pre-condition) for the message before it starts, although this could make the task fail by time-out. This failure is fine, as it would not corrupt the system. If the target of an input message does not exist any longer (or even never existed), the system must behave correctly, ensured by appropriate rules.

Each message must have a hierarchical path: (i) the funcom must not act on the same message twice (*idempotent*), (ii) the message must not be resent to any funcom on that path, and (iii) a funcom must not act on a message, if the former is included in the hierarchical path of the latter.

7.3.5 Execution Characteristics

Concurrent tasks in an HMS do not lead to lost updates, nor to corrupt knowledge states, since a task is assumed to be an action which does not change the state of global knowledge held. Therefore concurrent activities do not concern us here. What is important is the execution of a task without lost actions, and its correct termination. A task is terminated correctly if it has been executed successfully or if it is left unfinished at a recoverable point. A corollary to this is that the task execution should not lead to cycles. The execution model presented above has the following characteristics:

(1) *Execution of an activity*: During the execution of an activity, the execution may succeed or fail. In case of failure, it can be followed with R' or A. If successive attempts at R fail, then the task can be aborted. A malfunction can cause an Abort. The request for an Abort is propagated from a funcom via successive parents to the root, when a human decision can be made. Thus the system will always remain under control. Since all child activities are linked to the parent, the loss of a child activity will eventually show up at each parent, and thus eventually at the root.

(2) *Next Activity*: The next activity is well-defined, and as long as the specification is correct, the system will behave correctly. If the specification is wrong, then it will have contradictions, which will be detected in some stages of execution leading to an Abort. The activities under the EIN have well-defined execution order and hence have a valid specification; tasks under a parent also have a well-defined execution order enforced by precedent constraints.

(3) *Interrupt from MIN:* The execution process checks information from the MIN and accepts the execution of an activity if it is deemed valid, within the correct sequence. If R is accepted, then it is preceded by B without any side-effect. Therefore the multiplicity of information arriving at the MIN for execution will not cause any uncontrolled behaviour.

(4) *Delays and Malfunctions:* If a required piece of information arrives late there could be several consequences. If it is pre-condition information, then the task can fail but might succeed under a Redo if the information arrives by then. A system of acknowledgements can be instituted to protect against such problems. If a request for R or A arrives late, then there will be wasted activities, but no uncontrolled behaviour. If a funcom malfunctions during the execution, then the task is aborted for a subsequent compensating action. A message for action on a non-existing activity is ignored.

(5) *Dynamic Decomposition of Tasks:* If tasks are dynamically decomposed by a funcom, they are meant to be executed by children or by a lower-level cooperation block (which we have not considered). However, in either case there should not be any problem with correct execution.

7.4 Conclusion

Since an HMS is highly distributed, it is important to demonstrate that its operational behaviour is sound. We have investigated a computational model using ideas derived partly from the transaction concept of databases, and have shown how rules should be used to ensure that its behaviour is sound, particularly during delays and malfunctions, without causing any instability. It is our conclusion that rules such as those discussed in this paper should be implemented in all holonic systems to avoid undesirable results, instability and wasted production.

The operational scenario depicted in this paper is necessarily an abstract version of the reality. A real-world system will have more complex rules which must be investigated, analysed, formalised and implemented to ensure correct operational behaviour.

Acknowledgement

The work presented here forms part of the activities of the HMS project, but the author alone is responsible for the opinions expressed in this paper. The work has been partially funded by the European BRITE-EURAM initiative (Project No. IM 26508, Contract No. N.BRPR-CT97-9000). The author is particularly thankful to James Cole, Martyn Fletcher, Rashid Jayousi and Thomas Neligwa of the Keele DAKE Group for their comments, suggestions and corrections.

References

[7.1] P. Leemans et al, "A Semantic Perspective on Verification of Knowledge", Data & Knowledge Engineering, Vol (40:1), Jan, 2002, p 33–70.

[7.2a] S. M. Deen: "An architectural Framework for CKBS Applications", IEEE Transactions on Knowledge and Data Engineering, Vol (8:4), Aug 1996, pp 663–671.

[7.2b] S. M. Deen: „A Database Perspective to a Cooperation Environment", Proceedings of CIA'97, edited by P. Kanzia and M. Klusch, Springer, 1997, pp 19–41.

[7.2c] S. M. Deen: "A Fault-Tolerant Cooperative Distributed System", Proceedings. of the 9th IEEE DEXA Workshop, Vienna, Aug 1998, Edited by R. Wagner, pp508–513.

[7.2d] S. M. Deen: "Cooperating Agents for Holonic Manufacturing Systems", Multi-agent Systems and Applications II, Eds: V. Marik et al, LNAI, No. 2322, Springer, 2002, pp119–133.

[7.3] S. M. Deen and C. A. Johnson: "Formalising an Engineering Approach to Cooperating Knowledge Based Systems", IEEE Transactions on Knowledge and Data Engineering, **15** (2003), pp. 103–177.

[7.4] M. E. Bratman: "Intentions, Plans and Practical Reason", Harvard University Press, Cambridge, MA 1987.

[7.5] M. E. Bratman, et al: "Plans and Resource Bounded Practical Reasoning", Computational Intelligence, Vol. (4:4), pp349–355, Nov 1988.

[7.6] A. S. Rao and M. P. Georgeff: "Modeling Agents within a BDI architecture", KR'91, edited by R. Fikes and E. Sandewall, pp473–484, published by Morgan Kaufmann.

[7.7] A. S. Rao and M. P. Georgeff: "BDI agents – from theory to practice", Proceedings of the 1st International Conference on Multi-agent Systems, San Francisco, 1995.

[7.8] J. R. Searle: Speech Act, Cambridge University Press, 1969.

[7.9a] A. Sehmi et al: "Support for Distributed Multi-Agent Systems", Proceedings of the Second International Working Conference on CKBS, Keele University, edited by S.M. Deen, published by the DAKE Centre, pp 357–376, June 1994, ISBN 0 9521789 2 3

[7.9b] M. Fisher et al: "Specifying and Executing Protocols for Cooperative Action", Proceedings of the Second International Working Conference on CKBS, June, 1994 pp 295–306.

[7.10] M.R. Genesereth and R.E. Fikes: "Knowledge Interchange Format Reference Manual", Computer Science Department, Stanford University, June 92.

[7.11] T. Finin and R. Fritszon: "KQML – A language for protocol and information exchange", Proceedings of the 13th DAI workshop, pp 127–136, Seattle, WA.

[7.12] Y. Shoham: "Agent Oriented Programming", Artificial Intelligence, Vol: 60, pp51–92, 1993.

[7.13] M. Wooldridge: "Intelligent Agents – A modern Approach to DAI", Multi-agent Systems, edited by G. Weiss, MIT Press, 1999, ISBN: 0-262-23203-0, pp1–27.

[7.14] R. A Brooks, et al: "Alternative Essence of Intelligence", American Association for Artificial Intelligence, 1998; brooks@ai.mit.edu

[7.15a] H. Van Brussel et al: "Reference Architecture for Holonic Manufacturing Systems: PROSA", Computers In Industry, Special Issue in Intelligent Manufacturing Systems, Vol (37:3), pp 255–276).

[7.15b] J. Wyns et all: "Design Pattern for Integrating Çentralised Scheduling in distributed holonic manufacturing control systems", 75–84. Proceedings of the 2nd International Workshop on Intelligent Manufacturing Systems, 1999, edited by H. Van Brussel and P. Valckenaers, SEPT 22–24, 1999, Leuven, Belgium, published by KU Leuven, ISBN 90-73802-69-5.

[7.16a] S. Balasubramanian and D. Norrie: "A Multi-agent Architecture for Concurrent Engineering" Concurrent Engineering: Research and Applications, Vol. (4:1), March 96, pp7–16.

[7.16b] R. W. Brennan and D. H. Norrie: "The Performance of Partial Dynamic Hierarchies", pp 05–13. Proceedings of the 2nd International Workshop on Intelligent Manufacturing Systems, 1999, edited by H. Van Brussel and P. Valckenaers, SEPT 22–24, 1999, Leuven, Belgium, published by KU Leuven, ISBN 90-73802-69-5.

[7.16c] W. Shen and D.H. Norrie: "Developing Intelligent Manufacturing Systems Using Collaborating Agents", pp157–184. Proceedings of the 2nd International Workshop on Intelligent Manufacturing Systems, 1999, edited by H. Van Brussel and P. Valckenaers, SEPT 22–24, 1999, Leuven, Belgium, published by KU Leuven, ISBN 90-73802-69-5.

8. An HMS Operational Model

Thomas Neligwa and Martyn Fletcher[1]

Department of Computer Science, University of Keele
Staffordshire, ST5 5BG, United Kingdom
Email: neligwa@cs.keele.ac.uk

Abstract. This chapter investigates the function block-based Holonic Manufacturing System (HMS/FB) architecture and specifies the HMS operational model by blending concepts from the HMS/FB architecture and Co-operating Knowledge Based Systems (CKBS). The HMS operational model addresses holonic interaction issues viz. negotiation and co-operation, during task distribution, scheduling, execution and execution control. We envisage two levels of sophistication of the HMS operational model: (1) a basic operational model, supporting only the minimum functionality of the HMS, and (2) a fault-tolerant operational model, supporting advanced operational strategies such as fault tolerance. We have used motor assembly testbed requirements, provided by the Japanese HMS consortium to illustrate the basic operational model, and we have generalised two holonic shopfloor layouts, namely, the DaimlerChrysler's engine assembly testbed and Rockwell Automation's proposed simulation testbed for elaborating the fault-tolerant operational model. Although the manufacturing requirements used for this study are necessarily abstract, the ideas discussed in this chapter can provide a useful abstract foundation for many HMS implementations.

8.1 Background and Introduction

The increasing number of low-volume high-variety orders, shorter delivery deadlines, and competitive pricing are compelling manufacturing enterprises to investigate new manufacturing philosophies. This is because conventional manufacturing systems were purposefully designed for mass production and are therefore unsuitable for the new requirements. Such systems are characterised by monolithic devices and centralised control, which must be reconfigured from scratch each time a new batch is introduced or whenever a production schedule is disturbed. With make-to-order requirements, batch sizes are smaller (with a worst-case batch size of 1), and hence production systems are reconfigured as often, which is obviously uneconomical. Furthermore, the rigidity of the conventional systems makes them highly vulnerable to adverse events such as breakdowns, delays, etc. resulting in huge delays, production waste and increased production cost. All these factors are critical in determining the product cost and business competitiveness.

There are many research projects striving to devise the "next generation of manufacturing systems" suited for low-volume high-variety manufacturing requirements, and one of those projects (in which the authors are participants) is entitled "Holonic

[1] Present address: Agent Oriented Software Ltd, Institute for Manufacturing, Mill Lane, Cambridge CB2 1RX, United Kingdom. Email: martyn.fletcher@agent-software.co.uk

Manufacturing Systems" [8.1]. This is a large, industry-driven international manufacturing consortium working on a new manufacturing approach, bearing the title of the project, "Holonic Manufacturing Systems". The holonic approach to manufacturing promises flexibility, adaptability and robustness demanded by the new manufacturing requirements. It has been argued that these operational features can only be realised through *modularisation*, *standardisation* and *distribution* of manufacturing processes and devices.

A Holonic Manufacturing System (HMS) consists of a number of distributed co-operating autonomous production units called *holons* [8.2], which model production tasks, equipment and control functions in a modern manufacturing shopfloor. A holon can be defined as an autonomous co-operative software component that may also include optional hardware [8.3]. Holons are thought to be capable of acting intelligently and functioning independently, while co-operating to solve distributed manufacturing problems.

One of the envisaged characteristic operational features of an HMS is the ability to adapt dynamically but seamlessly to changes in either production requirements, e.g. new tasks, cancellations, order modifications, etc., or to changes within the production system, e.g. faults, power fluctuations, etc. In this context, HMS operational strategies are required to encourage scheduling, execution control, fault diagnosis and monitoring mechanisms to react dynamically to unforeseen events and undesirable effects in order to guarantee robustness of the system. This may involve, for example, automatic rescheduling of the tasks or reconfiguration of the production machinery. Holonic Manufacturing Systems are also expected to ensure efficient use of available resources by marshalling their autonomy and co-operation strategies. These should be deployed with a view of minimising bottlenecks, compensating failures, balancing the workload amongst holons and so on, in order to maintain or improve a predefined performance criterion e.g. throughput maximisation.

A lot of research work has been conducted within the HMS consortium and very good results have been published on the HMS architecture, e.g. the PROSA architecture [8.23]; case studies, e.g. DaimlerChrysler's engine assembly testbed [8.29] and the Japanese motor assembly testbed [8.6]; prototype implementations, e.g. HMS shell[2]; and many more. However, there are two major deficiencies in the current contributions from the consortium partners: (i) the proposed HMS models are either too simplistic or too specialised, and (ii) most HMS test cases are implemented on diverse platforms (programming languages, operating systems and simulation packages). In this view, it is difficult to capture the general architecture and operations for standardisation and formalisation purposes. While a standardised HMS model will facilitate portability of HMS applications, a formalised model can be used to verify the operational behaviour of HMS applications.

A significant amount of work has been done in the HMS project to standardise the architecture [8.7] using the IEC 61499 (International Electrotechnical Commission) Function Block Standard [8.5], but this architecture has not been fully exploited by the consortium partners. Apparently, this is due to the fact that the specification of this architecture and other project activities were carried out simultaneously. However, this architecture remains the most significant attempt to standardise and generalise the HMS architecture. The function block-based HMS architecture specifies how

[2] HMS Shell is an HMS-Application Programming Interface developed by Toshiba Corporation, Japan.

holons can be constructed in the context of the function block standard and how the co-operation and communications infrastructure can be established. However, holonic interaction issues, namely, negotiation and co-operation during task distribution, scheduling of dependent activities, execution, and execution control strategies, are not addressed. We shall collectively refer to these operational issues as the *operational model* in this chapter. Other aspects of the operational model include task notation schemes and decomposition strategies.

In this chapter we shall study the function block-based HMS architecture and use it as a basis for specifying the operational model. We envisage two levels of sophistication of the HMS operational model, (1) a basic operational model, which supports just the minimum HMS functionality, and (2) a fault-tolerant model, which extends the basic model to include the advanced operational strategies, e.g. fault tolerance.

We have illustrated the basic operational model using a motor assembly scenario, abstracted from a case study of the HMS requirements of an induction motor production testbed provided by the Japanese HMS consortium [8.6]. And, for the fault-tolerant model, we have generalised two holonic shopfloor layouts, namely, DaimlerChrysler's engine assembly testbed and Rockwell Automation's proposed simulation testbed and applied our advanced operational strategies. (See also [8.29] for a description of DaimlerChrysler's testbed). Although the manufacturing requirements used for this study are necessarily abstract, the ideas discussed in this chapter can provide a useful abstract foundation for many HMS implementations.

The rest of this chapter is organised as follows: Section 8.2 discusses the foundations of the HMS architecture and operations, Section 8.3 presents an HMS architecture based on the function block standard, and Sections 8.4 and 8.5 present the basic and fault-tolerant operational models respectively, for HMS applications, extending the ideas presented in Section 8.3. Concluding remarks are presented in Section 8.6.

8.2 Foundations of the HMS Architecture and Operations

Within the scope of the HMS requirements, the function block standard [8.5] and some models of Multi-Agent Systems [8.3, 8.8, 8.11, 8.12, 8.13, 8.14–20, 8.26] have been generally accepted as the most relevant approaches to Holonic Manufacturing Systems. We briefly discuss the function block standard in Section 8.2.1 and Multi-Agent Systems in Section 8.2.2, and compare the benefits of adopting either the function block standard or Multi-Agent Systems for implementing the HMS architecture in Section 8.2.3.

8.2.1 Function Block Standard

According to Chapter 4 of this book, the function block (FB) standard is preferred because it has the following features:

- it supports distribution – Holonic Manufacturing Systems are inherently distributed;

- it facilitates straightforward mapping to physical devices – the HMS is directly involved in the control of physical devices and machines;
- it is portable between both software tools and runtime platforms – a useful feature for distributed systems;
- it is component-based – in order to support encapsulation and protection of intellectual property.

In addition, the FB approach can have the following benefits:

- a system (e.g. a holon) can exhibit multiple skills by sharing components over various dynamic configurations depending on the task requirements, and
- faults can be localised into smaller units, which can be easily replaced without causing major disruptions to the global system.

The function block standard was originally developed for implementing industrial real-time control applications. It defines a generic architecture and presents guidelines for the use of function blocks in distributed industrial-process measurement and control systems (IPMCSs).

The function block is a fundamental building block of control systems implemented in this standard. Function block *types* are declared as data structures (cf. data and methods in object-oriented paradigms), while *instances* specify their behaviour or the way event responses are handled. A function block instance is a functional unit of software comprising an individual, named copy of the data structure specified by a function block type, which persists from one invocation of the function block to the next. The characteristic features of a function block are depicted in Fig. 8.1.

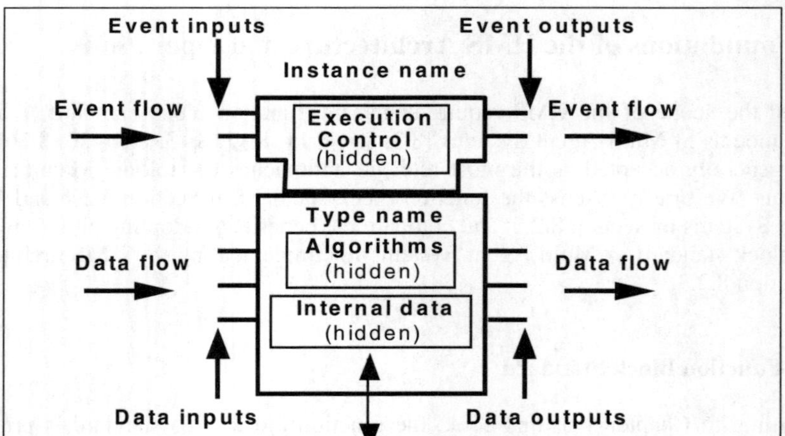

Fig. 8.1. Characteristics of function blocks

It has been argued that the function block environment can support both the low-level control and high-level control requirements. It is also anticipated that HMS applications modelled and implemented using the function block standard will directly benefit from the existing libraries of standardised components, documentation, etc.

An HMS architecture based on the function block standard is discussed in Section 8.3 of this chapter but the implementation details of function blocks are not included because these are extensively covered in Chapter 4 of this book (HMS/FB Architecture and its Implementation).

8.2.2 Multi-Agent Systems and Co-operating Knowledge Based Systems

Multi-agent systems (MAS) have been considered as a prime approach for providing the higher-level functionality (i.e. autonomy and co-operation) of Holonic Manufacturing Systems. Ideally, agents can contribute towards "intelligent" information processing and sharing in the shopfloors, which is necessary for the holons to behave "intelligently". In addition, multi-agent systems provide a potential opportunity for an extensive use of information technologies in manufacturing. It is envisaged that agent-oriented Holonic Manufacturing Systems will provide the necessary agility in manufacturing activities such as planning, co-operation, execution control, diagnostics, etc.

Multi-agent systems is a sub-discipline of Distributed Artificial Intelligence (DAI) [8.14, 8.27], whereby autonomous intelligent units called agents co-operate or compete to solve distributed computational problems that are beyond the scope of any monolithic system. Ideal agents possess an evolving knowledge base that helps them to act intelligently in their environment, and have skills to perform particular tasks or work towards achieving predetermined goals. Agents are also social entities, and hence can interact with each other for communication purposes, participate in solving joint tasks, negotiate over conflicting goals, and so on.

Most of the DAI/MAS models use psychological approaches to problem solving by imitating socio-biological (i.e. those of humans, insects, animals and biological systems) interactions and behaviours. For example, BDI (Belief, Desire and Intention) was introduced as a basis for research into multi-agent architectures [8.16, 8.17]. This model defines the intra-agent behaviour in terms of mental faculties for the rational selection of action plans to satisfy goals. Another example is the Speech-Act Theory [8.15], which treats agent interaction and communication as a type of action to be incorporated into planning and reasoning processes. Primitives inspired by speech-act theory include *propose*, *refuse*, *respond*, *inform*, etc.

There are several languages and standards that have been proposed for specifying and implementing multi-agent systems. During the 1990s, DARPA introduced a LISP-based environment that integrates a Knowledge Interchange Format (KIF) [8.18] and a Knowledge Querying and Manipulation Language (KQML) [8.19]. These formats are becoming de facto standards for exchanging knowledge between agents. Recently, FIPA (Foundation for Intelligent Physical Agents) [8.20] defined a multi-agent framework and an agent communication language (ACL) based on the Speech-Act Theory and KQML. Messages between agents are "actions or communicative acts as they are intended to perform a specific action by virtue of being sent"

[8.28]. FIPA is one the potential platforms for implementing Holonic Manufacturing Systems. For a detailed discussion of HMS and FIPA, see Chapter 5 of this book.

Building a multi-agent system in which agents "do what they should do" is a difficult task [8.27]. Actually, despite the fact that multi-agent systems use some general features such as blackboards, shared knowledge bases, etc, each MAS architecture is unique and each system is always designed from scratch [8.26]. Furthermore, the mental faculties, i.e. BDI and Speech-Act Theory, upon which DAI/MAS is founded, are inherently complex and their operational behaviour can be unpredictable. DAI-based multi-agent systems may therefore be unsuitable for some real-world applications, e.g. manufacturing where the computational behaviour (completeness, correctness, predictability and termination) is of paramount importance.

In order to have a computationally robust model, which could be applied to real-world applications, the DREAM[3] project developed Co-operating Knowledge Based Systems (CKBS)[8.8], which can be regarded as an engineering approach to DAI/MAS. CKBS blends ideas from DAI/MAS and distributed databases [8.25, 8.26], endowing its Knowledge Based Systems or agents with exact skills so that they can demonstrate correct computational behaviour rather than emulating human behavioural attributes. In the CKBS model, multiple autonomous agents co-operate in a decentralised environment to solve problems of real-world distributed applications. Each agent is a high-grain autonomous entity, with a compulsory software component and an optional hardware component.

Problem solving in CKBS focuses on efficiency, effectiveness, fault tolerance, reliability, user-friendliness, multi-user facilities and computational concepts such as consistency, completeness, and correctness [8.13]. Details and formalisation of these operational features are discussed in Chapter 7 of this book, "A Computational Model for Holonic Manufacturing Systems" and Chapter 6 of this book "Towards a Formalised HMS Model", respectively.

The CKBS model has been successfully applied in applications such as Air-Traffic Control Systems [8.21], Telecommunications Network Management [8.22], and has been generally accepted an abstract model for HMS applications. We shall therefore use CKBS ideas to address the high-level operational issues of the HMS architecture in Sections 8.4 and 8.5 of this chapter. The next Section will compare the anticipated benefits of deploying either the function block standard or Multi-Agent Systems for the HMS architecture.

8.2.3 Choosing between Function Block and Agent-Based HMS Architectures

Despite the fact that agent models and the supporting technology provide robust distributed architectures supporting both autonomy and co-operation, multi-agent systems cannot be directly applied to control manufacturing operations, because they lack the low-level (i.e. physical) integration. For example, the CKBS model assumes the presence of an optional hardware component, but no mapping between hardware and software has been defined to date and it is not obvious how the low-level control functions can be achieved. In this position, current MAS models can only be used to control the operations offline.

---

[3] DREAM: **D**AKE **R**esearch into an **E**ngineering **A**pproach to **M**ulti-Agents, University of Keele, United Kingdom.

There have been several independent attempts within the HMS consortium to bridge the gap between agent technologies and Holonic Manufacturing Systems. For example, the PROSA (Product Resource Order Staff Architecture) framework [8.23] is based on object-oriented concepts, viz. specialisation and aggregation, and consists of three basic types of holons; viz. product, resource and order holons, functioning alongside staff holons. Staff holons assist basic holons by providing expert knowledge that is beyond the capability of basic holons. This model is, however, too simplistic to generalise Holonic Manufacturing Systems. Another example is the MetaMorph multi-agent architecture [8.24], but this model does not fully incorporate the control elements needed by Holonic Manufacturing Systems.

On the other hand, the function block-based HMS architecture is still exploratory and there is no straightforward mapping between function blocks and advanced holonic concepts such as autonomy and co-operation. Apparently, neither MAS nor function blocks can accomplish all of the operational features desired for Holonic Manufacturing Systems, and hence blending ideas from the function block standard and from multi-agent systems is expected to deliver the best results.

The HMS architecture discussed in Section 8.3 blends ideas from the function block standard and multi-agent systems, particularly, Co-operating Knowledge Based Systems. The low-level control architecture is exclusively a subject of the HMS/FB architecture but the high-level architecture can benefit from the integration of the agent technology (see Table 8.1). Clearly, the success of FB/MAS interoperability is largely dependent on how cleanly the interfaces between the two technologies can be defined.

Table 8.1. FB/MAS integration opportunities

HMS	HMS/FB	HMS/MAS
HLC	✔	✔
LLC	✔	✖

Table 8.2 summarises the advantages and disadvantages of agent-based and function block-based HMS architectures and lists the benefits of integrating the two approaches.

Table 8.2. Advantages and disadvantages of possible HMS architectures

Architecture	Advantages	Disadvantages
Agent-based (HMS/MAS)	High flexibility	Demands high software maintenance
	Focuses on intelligent behaviour	System managed without real-time control
Function blocks (HMS/FB)	Has open interfaces with modularity	Enhancement needed in a holonic kernel
	Developed on the basis of an industrial standard	Limited user interfaces
	Focuses on distributed hard real-time control	
Integrated (FB/MAS)	Versatile	Interfaces need to be cleanly defined
	Enhanced cooperation functionality	Requires effort to amalgamate

8.3 HMS Architecture based on the Function Block Standard

For clarity and ease of discussion, and to be consistent with Chapter 4 of this book, we have distinguished the architectural concepts into the following two levels:

(i) Low-level control (LLC) architecture, which encapsulates non-holonic control and automation functions. Such functions may include PLC-based control logic and other interfaces to physical devices as specified in the function block standard.
(ii) High-level control (HLC) architecture, which provides a facility for supporting inter-holon interactions, i.e. communication, co-operation and information exchange. Nevertheless, the HLC architecture operates as a higher level of the LLC architecture (see also Fig. 8.2).

Within the function block context, holons consist of components responsible for both the LLC functionality, i.e. those which provide individualistic skills and interfaces to physical equipment, and components responsible for the HLC functionality i.e. autonomy and co-operation. Holons are built from a collection of function blocks distributed across multiple devices, e.g. computers and resources, e.g. processors. Function blocks are used for defining the individualistic skills of a holon (including interfaces to physical equipment), its autonomy, and services and data to support communications and co-operation with other holons.

The LLC and HLC architectures are discussed further in Sections 8.3.1 and 8.3.2, respectively.

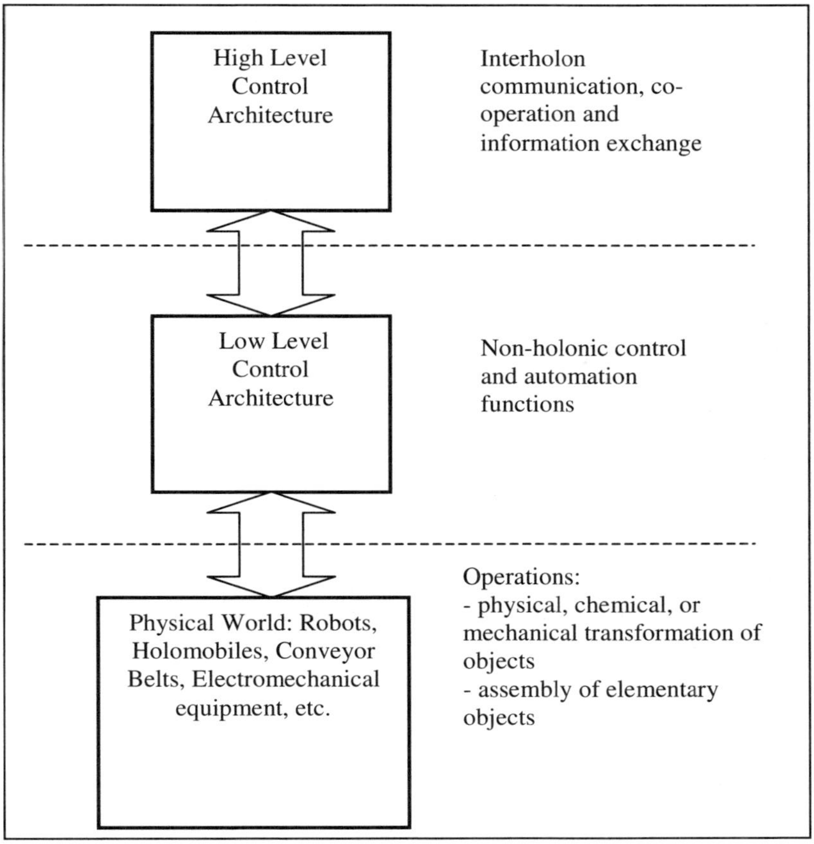

Fig. 8.2. HMS architectural layers

8.3.1 Low-Level Control Architecture

The low-level control architecture ([8.7] and Chapter 4) is composed of parts of the holon structures (e.g. H1, H2 and H3 in Fig. 8.3) that are responsible for the non-holonic real-time control and automation functions of the HMS, including:

- The COMM function block – real-time communications.
- The CTL function block – control and automation of physical equipment, task execution, fault detection and recovery and task co-ordination (jointly with the HLC).
- The I/O function block – input/output associated with industrial processes, machines and their human operators, designers, installers, and maintainers.

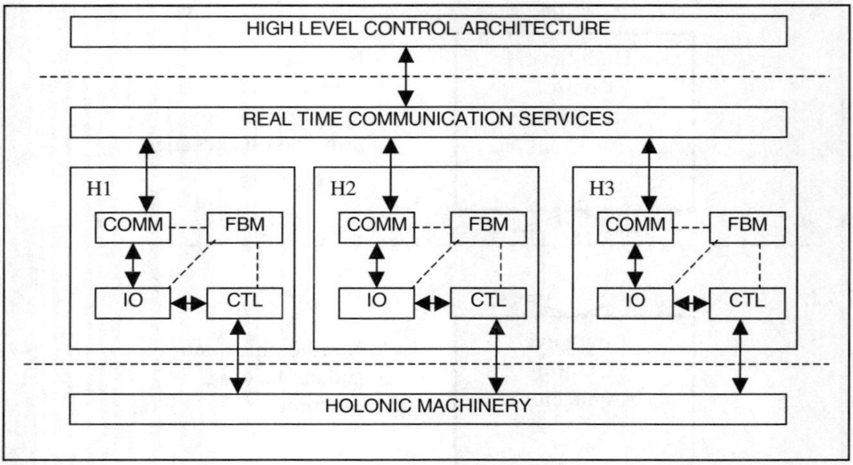

Fig. 8.3. Low-level control architecture

- The FBM (Function Block Management) function block is responsible for creation, execution and destruction of other function blocks.
- The real-time communication services (RTCS) interface provides real-time message exchange services between function blocks, of the same or different holons.

8.3.2 High-Level Control Architecture

The purpose of the HLC architecture is to provide an interface between the holon and its corresponding *Co-operation Domain* (CD). The CD is a core structure that facilitates transparent holonic interactions and information exchange during co-operation, and in effect it decouples inter-holon interactions, thus allowing individual holons to be autonomous. Co-operation between holons is only possible through the CD infrastructure, and membership of the co-operation domain is by subscription only via the CDI (see below). However, not all subscribers need be involved in solving a joint problem. Only a subset of CD members might be required for some tasks.

There are three main function blocks identified for the high-level functionality of every holon (Fig. 8.4):

- The CDI (Co-operation Domain Interface) acts as a low-level library to interface with co-operation domains and enables holons to exchange task data, schedules, ontologies, etc. The CDI function block provides an interface to the communication services required to support cooperation domains. These services include the following. Firstly, message broadcasting to all members of a cooperation domain, where the message (which we model as a combination of data and events flowing between function blocks) includes attributes such as the originating holon's name, a unique message identifier, reference to preceding message identifiers, and the message content. Secondly, rejection of an attempt to send a message under fault conditions such as the holon not belonging to the specified cooperation domain, the holon identifier being unknown, or invalid message contents. Thirdly, receipt of messages and their associated attributes from all cooperation domains to which the holon belongs.

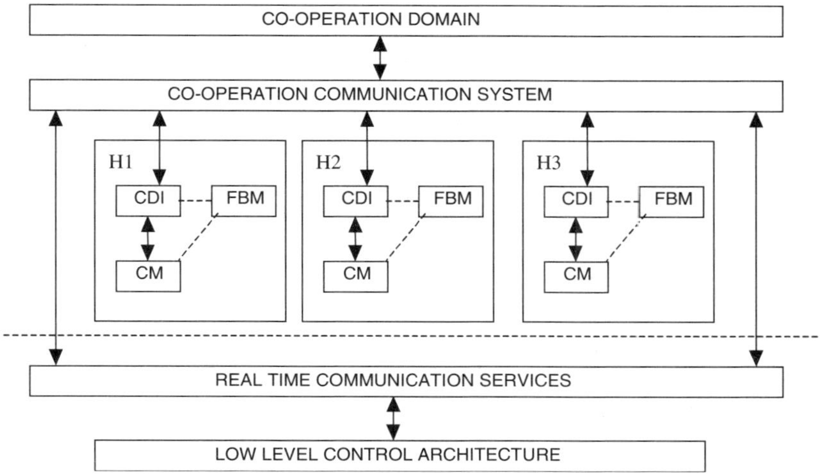

Fig. 8.4. High-level control architecture

- The CM (Co-ordination Management) is a high-level library managing co-ordination by executing task decomposition, conflict resolution strategies, negotiation, etc. This function block interacts with another function block called the KBI (Knowledge Base Interface) to read/write rules, relationships, etc. to a repository.
- The FBM (Function Block Management) is for creation, execution, and destruction of function blocks. It also determines whether a Co-operation Domain is active or inactive.

The holon is only a logical amalgamation of the independent (LLC and HLC) function blocks, which may also include distributed or sharable physical equipment or interfaces to physical equipment. The integrated LLC and HLC architectures are shown in Fig. 8.5. We can also interpret this view as the complete holon structures of H1, H2 and H3 interacting via the RTCS and the CD as part of the HMS.

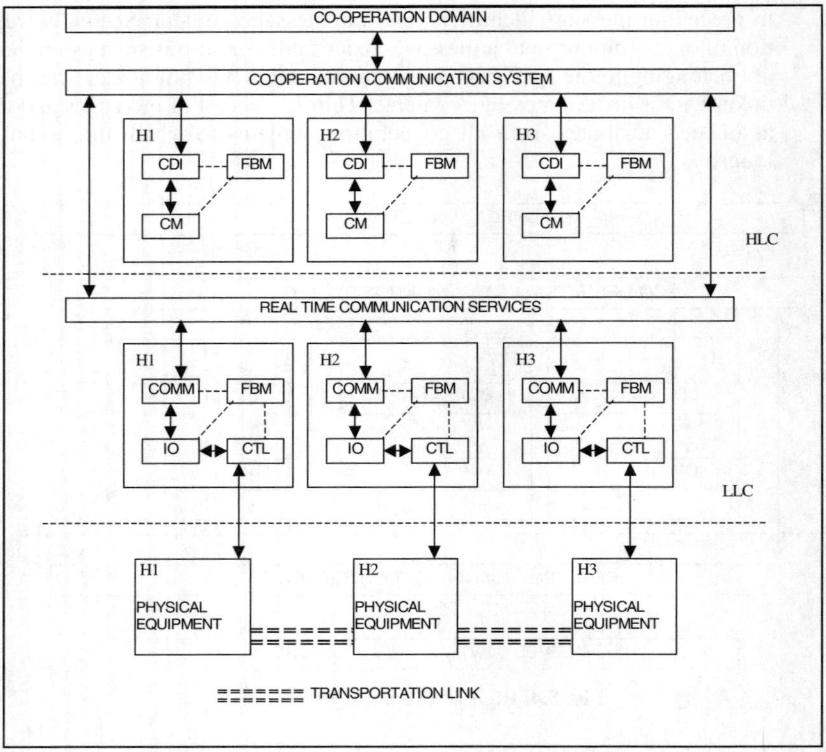

Fig. 8.5. Integrated LLC and HLC architectures

The LLC and HLC architectures presented above are concerned only with the internal structure of the holon and how it can interface to the CD. Although the CDI-CD interface has a high potential for facilitating peer-to-peer interactions in a highly distributed HMS, high-level inter-holon activities, e.g. task decomposition, scheduling, execution, execution control, etc., that require dynamic information exchange, negotiation and co-operation between holons are not yet covered. As mentioned earlier, we shall collectively refer to these operational issues as an ‘‘operational model” discussed in the following section. Furthermore, for operational reasons such as efficiency, effectiveness, cost and operational strategy, a pragmatic HMS architecture must allow some holons to undertake specialised roles. One role, for example, could be co-ordination, which involves task decomposition, conflict resolution and decision-making at the HMS level.

For the reasons discussed earlier in Section 8.2.2, we have addressed these operational issues using agent concepts borrowed from the CKBS model [8.3, 8.4, 8.10, 8.11], thus effectively extending the HMS/FB architecture to cater for a complete range of HMS operational features. The next two sections will discuss abstract operational models for supporting inter-holon interactions, communications and operational strategies. As mentioned earlier, we have divided the operational requirements into two categories: the basic operational model and a fault-tolerant operational model. These are discussed in Sections 8.4 and 8.5, respectively.

8.4 Basic Operational Model for HMS Applications

Co-operation is a process through which holons interacting via the co-operation domain execute dependent activities of a joint task. We pointed out in the previous section that not all subscribers of the CD need be involved in solving a joint task and only a subset of CD members might be required for some tasks. This subset of CD members (from this point forward referred to as cohorts) is often determined dynamically during either the decomposition of the task or during the allocation of the subtasks to holons, as we shall see later in this and the following sections. This section presents a basic operational model that supports only the minimum functionality of the HMS. We have illustrated this model using requirements of a motor assembly testbed provided by the Japanese HMS consortium [8.6].

8.4.1 Co-operation Blocks

To refer to the dynamic association of co-operating holons, i.e. cohorts, we shall use the term *Co-operation Block* (CB) [8.7]. The CB is a logical structure denoting a dynamic alliance between holons interacting through a CD (which is a physical infrastructure). In Fig. 8.6, H1 acts as a co-ordinator and H2, H4 and H5 as cohorts. H3 and H6 are part of the CD but not the CB. The CB is similar to Arthur Koestler's *holarchy* [8.2], but the latter is a hierarchical structure while the former is often a heterarchy.

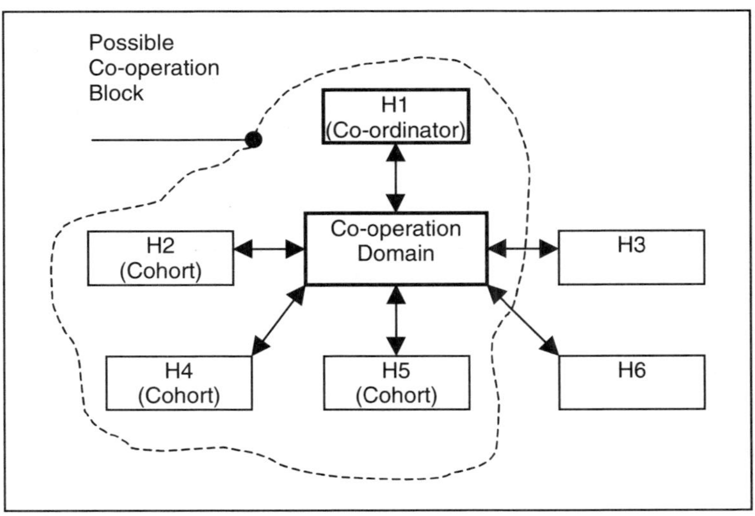

Fig. 8.6. Basic Co-operation Block

Every co-operation block has a *co-ordinator* holon and at least one participant or *cohort*. Each basic co-operation block is a heterarchy of holons working together to solve a joint task subject to constraints, conditions and preferences (see Section 8.4.3). Several CBs may be formed dynamically, each for one joint task and have a finite life equal to the duration of the task for which they were formed, i.e. a CB is a dynamic entity, created for the execution of that task and destroyed when the task is completed.

The co-ordinator of each CB is responsible for receiving tasks from external users or other holons and decomposing them into appropriate subtasks, initiating negotiations for scheduling or rescheduling, and monitoring the execution by handling the dynamic task constraints and disruptive events (in co-operation with cohorts).

These concepts are discussed further in Sections 8.4.2, 8.4.3 and 8.4.4 respectively.

8.4.2 Task Representation and Decomposition

Each manufacturing task has a task scheme, which describes the task *procedure*, *constraints* and *preferences* [8.8, 8.13]. While the procedure comprises the task execution logic, constraints represent (i) a dependency scheme (static constraints) and (ii) a dependency satisfaction rule (dynamic constraints and alternative decompositions).

The dependency scheme describes the static constraints arising from decompositional dependencies. These are called *pre-*, *post-* and *end-*conditions. Consider, for example, a global task $T = GT$, where G represents the global constraints applied to T in order to complete T. Task T may be decomposed into a set of lower tasks as $T = \{T_0, T_1, \ldots, T_n\}$, where each $T_i = g_i\, t_i$, and g_i is a constraint on a subtask t_i at level i (Fig. 8.7). Further decompositions into lower levels are possible but these are dependent on the cohort, which is capable of acting as a lower-level co-ordinator, thus leading to a *hierarchy of Co-operation Blocks* (see Section 8.4.3).

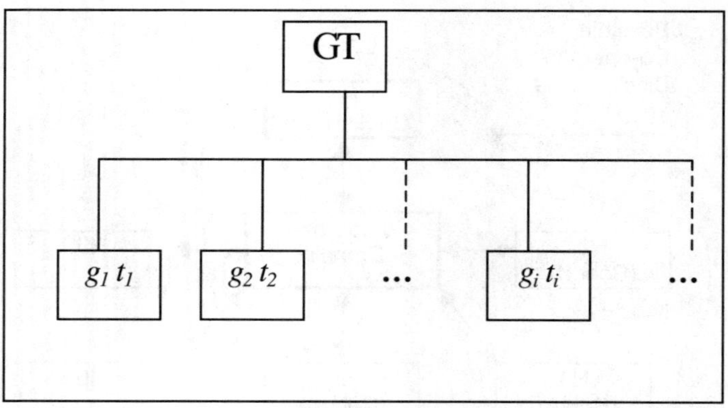

Fig. 8.7. Task decomposition hierarchy

The three types of conditions may be described as follows:

(i) For any task T_i, there are four types of *pre-conditions*:
  - *System-dependent*: the system must be available in a fault-free state before the task can start.
  - *Task-dependent*: specific requirements must be met in order to guarantee a successful completion of the given task, otherwise a task cannot begin or can be delayed to start.
  - *Precedent-dependent*: these are precedence constraints, some of which may be derived dynamically by the system from the dependency scheme. If a required input is not available it might lead to delays or aborts.
  - *Child-dependent*: these constraints will apply to T_{i0}, which may also have other types of *pre-*, *post-* and *end*-conditions.

(ii) The *post-condition* refers to a situation where a completed task must satisfy a validity criterion, without which a result will be invalid. If a post-condition is not satisfied, the task will require redoing or aborting, which may consequently result in delays in dependent tasks.

(iii) The *end-condition* applies to a successfully completed task and includes messages to be sent out announcing the completion of a task. End-conditions can be either obligatory or non-obligatory. Failure of an obligatory end-condition will trigger a re-send of the message while a non-obligatory end-condition can simply be ignored.

The task decomposition knowledge is often held by the co-ordinator but in other cases, it might be necessary to negotiate with all potential cohorts before the criteria for decomposition are established. The next section will discuss how a decomposed task can be scheduled via negotiation, consequently establishing a Co-operation Block.

8.4.3 Negotiation and Scheduling

Negotiation and scheduling are explicit steps in the formation of a Co-operation Block. As in CKBS, holons interact using the Contract-Net Protocol (CNP) [8.9], which is arguably the most popular interaction protocol for MAS applications. The CNP is a market-based negotiation strategy that allows participant agents (henceforth referred to as holons) to exhibit full autonomy since they can choose to decline or accept the terms being negotiated on, purely on the basis of their own decision-making mechanisms. The exact steps of the CNP are as follows:

- the "co-ordinator" broadcasts an announcement to invite potential bidders (i.e. potential cohorts) for a particular task.
- bidders respond by sending their bids to the co-ordinator.
- The co-ordinator evaluates the bids and decides to accept some, and informs all bidders of the outcome.
- Successful bidders confirm the agreements.
- The work is done and the protocol ends.

Cohorts may decompose their subtasks even further and distribute them to their own lower-level cohorts, thus becoming co-ordinators of a lower-level echelon of tasks and cohorts of a higher level co-ordinator. This phenomenon leads to a hierarchy of co-operation blocks (Fig. 8.8). Multi-skilled holons can take part in multiple co-operation blocks, until their capacity is exhausted or no more tasks are available in the system.

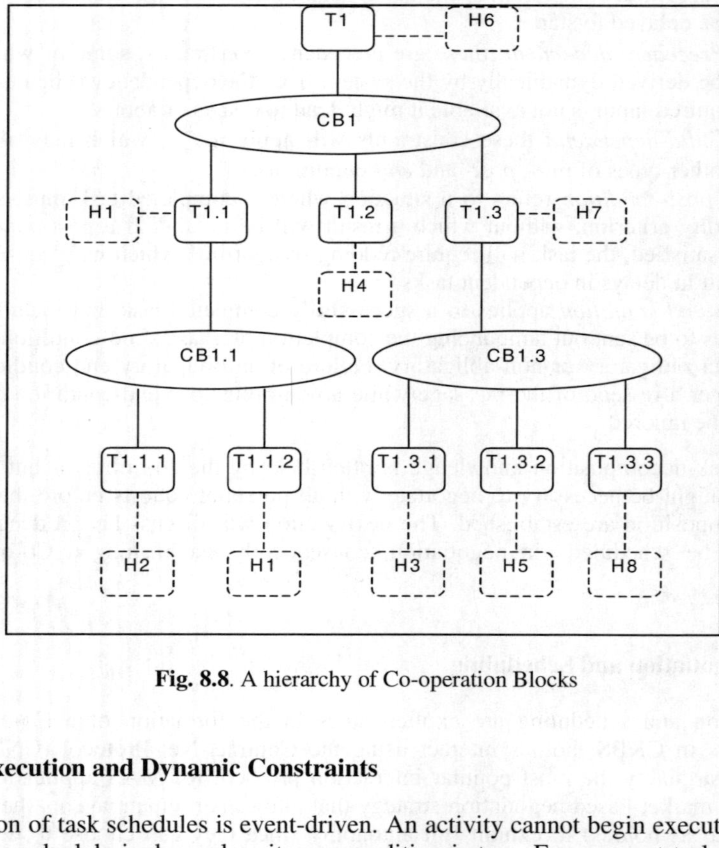

Fig. 8.8. A hierarchy of Co-operation Blocks

8.4.4 Execution and Dynamic Constraints

Execution of task schedules is event-driven. An activity cannot begin execution, even when the schedule is due unless its pre-condition is true. For a parent task T_f at any level f decomposed into child tasks { T_{f1}, T_{f2}, …. T_{fn} }, there is a parent-child relationship implying that the parent cannot complete its task until all the children have completed theirs. Furthermore, child tasks can come up with local solutions, which might limit solutions from other children in the task hierarchy. These partial solutions are referred to as dynamic constraints. It is likely that dynamic constraints may result into unresolved cyclic dependencies such as $T_{f1} \rightarrow T_{f2} \rightarrow \ldots\ T_{fm} \rightarrow T_{f1}$ that must be broken.

To break such a cycle, all m children should send their raw solutions to their respective parent T_{f0}, which can consequently decide how to resolve the dynamic constraints. This strategy does increase computational overheads but it guarantees that a feasible solution can be found. Alternatively and perhaps more efficiently, sub-tasks $T_{f1} \ldots T_{fm}$ may be reduced to a single task that produces only one solution rather than all possible solutions (called a *solution space*) at that time. Holons must negotiate and try to relax their *preferences* until a satisfactory solution is found, and this solution can then be sent to T_{f0} for constraint validation. However, this is beyond the scope of this chapter.

8.4.5 Applying the Basic Model in a Motor Assembly Scenario

This application scenario has been abstracted from the requirements of a motor assembly testbed (J1) provided by the Japanese HMS consortium. This testbed was constructed to fulfil the HMS requirements as viewed across the whole Japanese consortium of the HMS project.

The central element of the basic operational model is the Co-operation Block. We shall therefore discuss only how the motor assembly scenario can be transformed up to this level of abstraction, assuming that the rest of the details of the model can be applied accordingly.

Motor Assembly Requirements. Induction motor assembly processes described in [8.6] involve machining, fixturing and assembly of motor elementary parts, namely the stator, rotor and bearing brackets. The rotor and the stator are also individually assembled at lower levels from a set of other component parts. Sub-processes include cutting, lathing, slotting, folding, coiling, mounting, insertion, die-casting and assembly of elementary parts to form higher-level parts as necessary. Non-assembly activities include testing, packing and dispatching complete induction motors.

Based on the existing motor production equipment, total machining time is about 1.33 hours, total assembly time is about 1.00 hours and machining set-up time is 0.50 hours per product. However, it takes about 20 to 25 days of lead time to manufacture a complete induction motor, for two major reasons. First, specialised machines are not responsive enough to customised orders, i.e. make-to-order requirements, and hence factory management is complex and inefficient. The complexity of the system increases because the assembly plant is required to accommodate an unprecedented daily order mix of about 200 different types of motors with an average lot size of 5, and thus produce up to 1000 motors on a daily basis. Second, production planning and scheduling are carried out offline, and dynamic rescheduling of jobs and reconfiguration of production machinery are not supported, thus disruptive events such as delays or changes to schedules cannot be tolerated. This leads to high maintenance overheads, long lead times and waste production.

Although it is apparent that machining and assembly times may not be much improved, holonic shopfloors are expected to reduce set-up and waiting times of the assembly machines. It is estimated that if resources were fully utilised, an average motor could be manufactured within 5 days instead of the present lead time of 20 to 25 days. Three of the most significant operational requirements are discussed below.

(1) Minimising the number of inflexible special-purpose machines and developing intelligent production machinery (*holonic machines*) that are capable of exhibiting multiple manufacturing skills. It should also be possible to reconfigure these skills dynamically in order to satisfy the make-to-order requirements in the shortest possible set-up time and with the least overhead. Constructing re-

configurable production hardware is beyond the scope of our work but such hardware will be assumed to exist.

(2) Integrating production planning and scheduling with real-time production to enhance the robustness of the system in the face of disturbances. It should therefore be possible to reschedule the production activities dynamically and reconfigure the production hardware as required. There are three possible levels of process integration:

- *Low-level integration*: allows automated planners and schedulers to update information related to dynamically varying manufacturing quantities (i.e. orders specification, skills availability and execution, and production constraints) without interfering with the system operations.
- *Intermediate-level integration*: allows the system to be partly centralised. Designated co-ordinators supervise production based on their own static strategies.
- *High-level integration*: allows production entities (holons) to co-operate and negotiate to determine, during run time, the best course of action as influenced by their adaptive strategies. This is a purely distributed system of autonomous entities.

(3) Developing robust task execution and control strategies that would guarantee effectiveness, efficiency, flexibility, adaptability, scalability and fault tolerance of the system.

Basic Operational Model for the Motor Assembly Scenario.

From the HMS point of view, the motor production processes form a hierarchy, which can be modelled as a holarchy or more precisely, a co-operation block. On top of the hierarchy is the final motor assembly process, which receives parts from three lower-level buffers/warehouses (brackets, stators and rotors). The rotor and stator assembly processes are also independent assembly processes at a lower level, receiving parts from lower-level processes and so on, until the lowest level, which produces elementary parts, is reached (Fig. 8.9).

Notice how buffers/warehouses decouple the assembly processes from each other. Because the assembly processes can be independent of each other, each can be replaced by a holon, in order to manifest their autonomy and co-operation in the joint assembly of the motor. Obviously, buffers are autonomous with respect to each other and with respect to the assemblers, thus they are holons of another kind (Fig. 8.10).

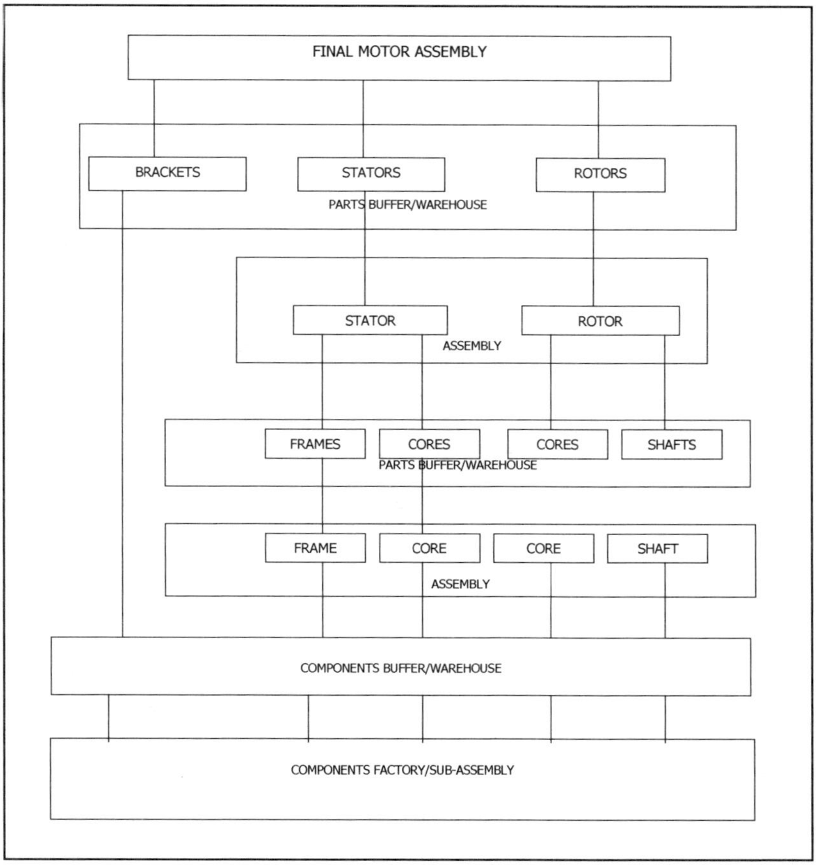

Fig. 8.9. Motor assembly process hierarchy

The Contract-Net protocol can be applied directly in facilitating holonic interactions within the motor assembly shop. For example:

1. The motor assembly holon can act as a co-ordinator with a user interface to help define and receive different tasks (i.e. motor production orders) and decompose them into the corresponding stator types (S), rotor types (R) and bearing brackets (B) for the next lower-level subassembly process.
2. The co-ordinator will then invite bids from its cohorts, namely, the buffers/warehouses for stators, rotors and brackets, and for the task scheduling proposals. In effect, this process forms a Co-operation Block of S, R and B, while S, R and B can themselves replicate this scheduling process for their lower level assembly activities specified in Fig. 8.9, consequently constructing a hierarchy of co-operation blocks similar to that shown in Fig. 8.11.

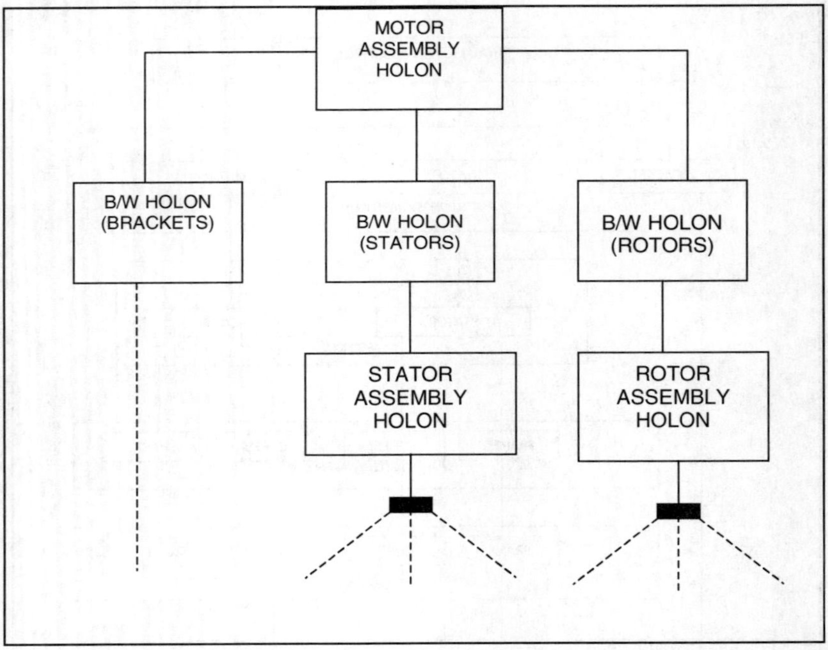

Fig. 8.10. A hierarchy of assemblers and warehouse holons

3. Based on the available information, the co-ordinator will construct an execution plan and initiate execution once the schedule is due.
4. In each case, S, R and B will decompose their problems into a lower echelon of assembly activities, involving the stator, rotor and bracket holons, respectively, and so on until the lowest production level is invoked.

During co-operation, advanced HMS execution strategies could be applied to achieve robustness, effectiveness and fault tolerance. For example, in order to enhance fault tolerance, replacement classes [8.10, 8.11] could be defined such that if a parent holon, e.g. motor holon, breaks down its twin can take over without disrupting the overall system performance. Furthermore, operational delays due to unforeseen random events could be absorbed and effectively compensated by sharing the workload among the twins, as prescribed in the temperature model [8.4]. These advanced operational strategies and the corresponding shopfloor layout are discussed in Section 8.5.

8.5 Fault-Tolerant Operational Model for HMS Applications

Owing to unforeseen events e.g. machine breakdowns, communication delays, etc., in manufacturing processes, jobs take longer than originally scheduled. In addition, the sophistication introduced by the HMS architecture increases the system complexity because distributed tasks are highly dependent on each other. Rescheduling cascades are more significant in HMS than in the traditional manufacturing systems. In addition, physical transformation on workpieces cannot simply be redone or rolled back, but a compensation action must be carefully planned.

Holonic Manufacturing Systems should be able to anticipate and react "intelligently" to disruptive real-time events that are likely to degrade the overall system effectiveness and efficiency. In the interest of fault tolerance, graceful degradation and high throughput, we have proposed to use *replacement classes* [8.10, 8,11] and what we have called the *temperature model* [8.4]. A replacement class contains holons of the same skill, called *twins*, which can be used towards replacement of faulty cohorts, load balancing or parallel processing of tasks. The temperature model is a dynamic rescheduling strategy used to schedule manufacturing tasks on demand and reschedule dynamically those tasks affected by run-time events such as breakdowns or overloads.

Detailed discussions of replacement classes and the temperature model are presented below in the Sections 8.5.1 and 8.5.3, respectively.

8.5.1 Replacement Classes

With ideas borrowed from the fault-tolerant CKBS model [8.11], we categorise holons into four functional classes: co-ordinator class, skill class, provider class and support class. Such classes are called *replacement classes* because their members, called *twins*, have the same skill and thus can replace each other in the case of malfunctions or when extra processing capacity is required. There is a *minder* holon for each skill class (but we have not considered a minder of minders or a minder of co-ordinators).

The minder holon is responsible for both the welfare of its twins and brokering their services. It provides the necessary backup and replacement if one of the twins breaks down. The minder is also responsible for scheduling its twins for efficiency and effectiveness, initiating action when the schedule is due, monitoring and rescheduling its twins, informing its higher minder when the task is complete or otherwise, and keeping links with other skill classes. Twins can be multi-skilled, and hence might belong to more than one skill class. Further description of the replacement classes is given below.

Co-ordinator Holons. The nature of Holonic Manufacturing Systems allows them to accept multiple (make-to-order) tasks with each task (type) T_i having its specialised co-ordinator holon, C_i. Multiple instances of T_i would therefore require an equal number of instances of C_i. Each co-ordinator instance would decompose its task into

$$T_i = \{t_1^i,\ t_2^i, \mathrm{K}\ , t_j^i, \mathrm{K}\ t_N^i\},$$

where t_j^i denote the dependent sub-tasks or activities of T_i.

Each co-ordinator maintains the skill classes required for completing its tasks along with the hierarchical structure. Co-ordinators will negotiate with skill classes using a suitable protocol such as the contract net protocol described earlier, in order to schedule the activities and construct an execution plan. Scheduling assigns t^i_j to a skill holon, H_{ij} which is capable of completing the subtask and is available.

The task *schedule* S_i, will therefore reflect both precedence and temporal constraints among the subtasks and corresponding skill holons. When the activity becomes due, the minder will initiate action and each twin will confirm the start of the action.

Skill Holons. We assume that there are multiple skill classes in the HMS shopfloor. Each skill class supplies skill holons or twins, which can further subdivided into lower-level skill classes. The lowest subclass represents an atomic skill capable of solving atomic tasks. The twin (also called a cohort) has the following roles:

- Bid for tasks, negotiate, and schedule its operations.
- Confirm the start of the scheduled action when a task becomes due.
- Maintain the values of its execution status with states such as normal, late, abnormal, dead, etc and inform the minder (see below for the definition of a minder) if anything has gone wrong. Faults may originate from the twin itself, an auxiliary service or its lower-level twins. If a twin fails during negotiation, its transactions can be rolled back to a recoverable point as in databases. If a twin fails during execution, its minder will allocate the orphaned workload to a new twin. If a physical workpiece is involved, this has to be inspected by a human and reintroduced at an appropriate point with a compensation action. However, this is beyond the scope of our model.
- Confirm the completion of the task to the minder and co-ordinator, and send a pre-condition to the child activities.

Provider and Support Classes. The provider class provides the HMS with services such as the directory, blackboards, etc. The directory service contains both descriptive information and physical addresses of holons, listed according to either functionality or locality to help holons find each other. The support class provides physical devices on which holons can exist and execute their operations, e.g. host computers, communication networks, etc.

Members of these two classes are non-holon but provide important auxiliary services to holons, and are very significant in the functioning of the HMS because they are potential sources of faults that can have a severe effect on the HMS. Nevertheless, their discussion is beyond the scope of this work. We shall therefore focus our attention on the services provided by the co-ordinator and skill classes only.

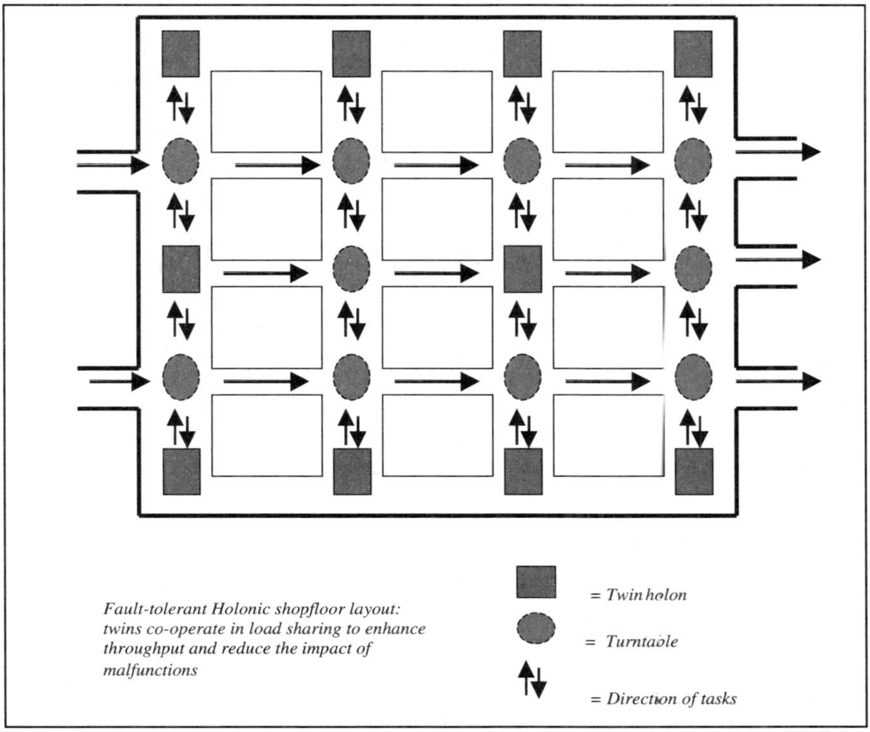

Fig. 8.11. Fault-tolerant HMS shopfloor layout

8.5.2 Fault-Tolerant HMS Shopfloor Layout

Each major production entity in the HMS shopfloor, e.g. an individual machine or a group of machines, warehouses, turntables, conveyor belts, automated guided vehicles, humans, etc., can be modelled as a skill holon. These holons are expected to co-operate with each other in order to contribute to the manufacture of workpieces. A workpiece is defined as a tangible object that is produced by means of either (i) chemical, physical or mechanical transformation, or (ii) assembly of elementary workpieces. In our model, workpieces are represented as tasks, as described above in Section 8.5.1.

The ideal HMS shopfloor is strategically organised such that all holons are connected by a suitable transportation link, e.g. a conveyor belt or an automated guided vehicle, so that they can transfer jobs between each other in order to minimise the impact of failure or overload on throughput. Fig. 8.11 illustrates how jobs would be flowing in such a shopfloor, into and out of manufacturing workstations or warehouses. Co-ordinator and minders are orthogonal to the diagram, and hence are not shown. The layout shown in Fig. 8.11 is a generalisation of two holonic shopfloor layouts, viz. DaimlerChrysler's engine assembly testbed and Rockwell Automation's proposed simulation testbed.

In this HMS model, twins are expected to be entirely responsible for deciding on the best course of action affecting their set of activities, with little or no human intervention. This can be achieved via peer negotiations among them and with the help of co-ordinators and minders. There is no global controller of inter-holon interactions and in most cases, therefore, the real-time HMS behaviour is difficult to predict because due to non-linearity of the interactions, the state of the HMS can be computationally intractable. This is, however, a trade-off between robustness and predictability of the system behaviour.

Nevertheless, the HMS should always ensure that co-operating holons are generating a correct operational behaviour. For instance, the system should be able to settle down automatically whenever unexpected events such as breakdowns, overloads, delays, etc., occur, while guaranteeing that jobs are being satisfactorily completed as scheduled. In this view, holons should be able to adapt to ongoing changes both internal to themselves and in the external environment.

We have developed a co-operation strategy called the temperature model for modelling the behaviour of individual holons and effectively gauging the overall system behaviour. This model is discussed in the next section.

8.5.3 Temperature Model

The temperature model is an individualistic strategy that operates in every holon as a primary behaviour with a view to achieving robust system behaviour as an emergent behaviour. As highlighted earlier, the temperature model is a dynamic scheduling and rescheduling strategy for distributed manufacturing tasks. We have used temperature as a metaphor to exemplify the load levels of individual holons as their primary behaviour. Overloaded holons are regarded as hot (or infinitely hot for broken ones), while underloaded holons are regarded as cold.

According to the principles of thermodynamics, heat flows from hot to cold. Hot holons will therefore seek to dissipate their heat towards cooler holons having the same skill, i.e. twins, thus sharing their workload and leading the system into equilibrium or a uniform workload distribution. As a side effect, distributing temperature evenly helps the system to tolerate disruptive run-time events, e.g. faults, delays, etc., and to exhibit high levels of robustness, as desired for HMS operations.

There are two ways to achieve this result, (i) sharing among holons of the same skill class, called twins. These are managed by a class holon, called the minder (ii) if no twin can take a task, the affected twin or its minder informs the relevant co-ordinator, which can decide to find an alternative decomposition (and hence an alternative set of holons) for the task. Priorities and constraints can be applied to favour some transfers and to disallow others. But the decision whether to offload or accept tasks to or from twins or the co-ordinator is entirely based upon the individual holon's autonomy and its operational state.

The ultimate objective of the temperature model is to improve the system effectiveness by minimising waiting times for tasks, reducing idle times of skill holons and minimising waste in an error-prone production environment. The temperature model is an individualistic strategy, that operates in every holon as a primary behaviour, with a view of achieving a robust system behaviour, i.e. global coherence in co-operation, as an emergent behaviour (cf. the synergy of co-operation), hence controlling the HMS operations without relying on a central control facility.

The temperature model is based on three basic concepts of thermodynamics: *relative heat, temperature* and *latent heat.*

Relative Heat. The *relative heat* $H_t(A_i^k)$ at a time t represents the total duration of the expected activities in agent A_i of skill k. This quantity can be expressed as

$$H_t(A_i^k) = D_b + D_e - D_p$$

D_b is the total duration of activities already waiting in the holon's input buffer, and can be calculated as

$$D_b = \sum_{x=0}^{b} d_x / q_x$$

Here x is the activity number, ranging from $x = 0$ to $x = b$ (buffer size), d_x is the standard duration of activity x, and q_x is the quality of service offered for activity x. q_x can vary from $q = 0$ (poorest quality, i.e. the holon is broken down) to $q = 1$ (the best quality). Note that it is also possible to use costs to measure the relative heat, instead of using the duration.

D_e is the total duration of the new arrivals, obtained from the schedule of A_i^k or statistically from a Poissonian rate, λ:

$$D_e = \sum_{x=1}^{\lambda t} d_x / q_x$$

Here λt is the number of activities expected in time t. D_p is given by

$$D_p = t/\mu,$$

where μ is the average job completion time.

Temperature. The *temperature* of the holon at time t can be defined as

$$T_t(A_i^k) = H_t(A_i^k)/t$$

Therefore, at time t,

$$T_t(A_i^k) = 0 \text{ if } \mathrm{D_p} = \mathrm{D_b} + \mathrm{D_e}$$

$$T_t(A_i^k) < 0 \text{ if } \mathrm{D_p} > \mathrm{D_b} + \mathrm{D_e}$$

$$T_t(A_i^k) > 0 \text{ if } \mathrm{D_p} < \mathrm{D_b} + \mathrm{D_e}$$

Latent Heat. The price paid to transfer activity x from holon A_i^k to holon A_j^k is called the *latent heat* $L_x(A_i^k, A_j^k)$. The latent heat is calculated as the time needed for a workpiece to be available at the new location, i.e. the time taken to unload the workpiece, transport it, set it up and make it ready. Latent heat does not change the temperature and its role is similar to that of latent heat in changes of state in matter. We assume that latent heat tables will be available within each minder class to facilitate the evaluation of cost effective transfers.

The transfer of activity x from A_i^k, with $H_t(A_i^k) > 0$, to A_j^k, with $H_t(A_j^k) < 0$, is beneficial *iff* $H_t(A_j^k) + d_x \leq 0$. This condition cools A_i^k and simultaneously heats A_j^k, and hence the transfer is feasible.

Note that it is also possible to use costs to measure the relative heat instead of using duration. According to the principles of thermodynamics, heat flows from hot to cold. Therefore workload should be transferred from hot holons to cool holons in order to achieve a uniform temperature distribution. Overloaded holons are hot while underloaded holons are cool. Hot holons will attempt to dissipate their heat to cool holons, thus keeping the overall temperature (load distribution) uniform. Priorities and constraints can be applied to favour some transfers and disallow some others.

Systems Effectiveness as an Emergent Behaviour. To examine *systems effecttiveness* using the temperature model, let us consider m holons processing n activities in a manufacturing session with a begin time B_{time} and an end time E_{time} for each activity. The combined operational duration D_{op} of all n activities is

$$D_{op} = \sum_{x=1}^{n} d_x$$

which is the same as the total relative heat. If the operational cost of activity x is C_x, then the total cost is

$$C_{tot} = \sum_{x=1}^{n} C_x d_x$$

If the session duration is S_d, the duration available is

$$D_{tot} = mS_d,$$

Therefore the relative time efficiency is D_{op}/D_{tot}

Alternatively, if we assume the same cost C_O for per unit duration for each holon, then the total wasted cost is

$$W = (D_{tot} - D_{op})\, C_o$$

If during this processing session, z activities have been transferred, then the transfer cost C_{tr} is

$$C_{tr} = \sum_{j=1}^{z} c_{av} \times L_x(A_i, A_j)$$

where C_{av} is the average cost of transfer per unit duration.

To maintain the equilibrium, the following general strategy is applied:

(i) At time t, each holon evaluates its temperature.
(ii) If the temperature is negative, the holon will seek transfer from hot holons.
(iii) If the temperature is high, the holon will not accept any new work but will seek to transfer some of its current work to cooler holons.

(iv) A transfer will take place from the hot holon, subject to the feasibility criterion explained earlier and any other relevant constraints.
(v) In the case of contention, priorities can be used in selecting which one to transfer.

Observe that redistribution should be carried out as late as possible to reduce the probability of further redistributions, i.e. *rescheduling cascades*. Equally, a redistribution that is too late can be a waste of time.

The interactions in this model are highly non-linear, and we believe it is easier to study the system behaviour in a simulation. We are currently implementing a Java-based simulation of this model in order to study the overall system behaviour and verify the temperature model. The main objective of the simulation is to show that the HMS environment presented in this chapter, can tolerate faults, reduce delays and idle times, by rescheduling activities orphaned by broken down twins and those affected by other sources of delays e.g. precedence, internal due operational factors, etc. It is envisaged that replacement classes and the temperature model will not only enhance the robustness of the HMS but also expedite completion of some tasks and therefore a higher throughput can be expected, i.e. some tasks will be completed earlier than prescheduled. Results from this simulation will be published in our future papers.

8.6 Conclusion and Remarks

We have devised an operational model for a highly distributed HMS architecture by blending concepts from the function block-based HMS (HMS/FB) architecture and Co-operating Knowledge Based Systems (CKBS). This operational model addresses holonic interaction issues during task decomposition, scheduling, execution and execution control. We have also introduced two levels of the HMS operational model: (1) a basic operational model that supports only the basic interaction requirements of the HMS, and (2) an advanced or fault-tolerant model that supports fault tolerance via replacement classes and the temperature model.

While replacement classes enable co-operating holons of the same skill, i.e. twins, to transfer their workload to one another in order to minimise the impact of faults and delays, the temperature model ensures that the transfer criteria expressed by the latent heat between the twins are satisfied. In addition, the temperature model encourages individual holons to distribute their workload until the entire system is in equilibrium. This has two merits. (1) It can be argued that the HMS becomes more resilient and tolerant of random faults in a complex and error-prone environment, and hence production waste is minimised. (2) Idle time for holons is minimised and some tasks can be completed earlier than prescheduled. In both cases (1) and (2), throughput is continually improved. We are currently implementing a simulation of the fault-tolerant HMS model so that we can verify these expected results quantitatively. Both operational models have been illustrated using pragmatic HMS requirements. The basic model has been applied to an induction motor assembly testbed, which represents HMS requirements as viewed by the Japanese HMS consortium, and the advanced model has been elaborated using a generalisation of two holonic shopfloor layouts, namely, DaimlerChrysler's engine assembly testbed and Rockwell Automation's simulation testbed.

Acknowledgements

The work presented here was supported by EU grant BRPR-CT97-9000 for the IMS/HMS project. The authors would also like to extend their gratitude to members of Keele's DAKE group for providing a thorough review of the early versions of this chapter, and to the Japanese HMS partners, DaimlerChrysler and Rockwell Automation for allowing us to use their testbeds for our study.

References

[8.1] HMS Consortium, *Holonic Manufacturing Systems Programme Overview*, Technical Report, http://hms.ifw.uni-hannover.de/ (1998).

[8.2] A. Koestler, *The Ghost in the Machine*, Pan, London (1967).

[8.3] S.M. Deen, "A Co-operation Framework for Holonic Interactions in Manufacturing", *Proceedings of the Second International Conference on CKBS*, ed. S.M. Deen, DAKE Centre, University of Keele, England (1994).

[8.4] S.M. Deen and M. Fletcher: "Temperature Equilibrium in Multi-Agent Manufacturing Systems", *Proceedings of the 11th International Conference on Database and Expert Systems Applications (DEXA'00)*, London (2000).

[8.5] International Electrotechnical Commission, *Function Blocks Architecture for Industrial Process Measurement and Control Systems*, ftp://ftp.ab.com/stds/iec/tc65wg6/document/pt2acd1.doc (1997).

[8.6] S. Tamura, *Description of J1 Testbed*, HMS Japan (1998).

[8.7] M. Fletcher, *Holon and Holonic System Architectures: Deliverables D1.1-1 & D1.1-2*, HMS Consortium, http://hms.ifw.uni-hannover.de/ (1998).

[8.8] S.M. Deen, "An Architectural Framework for CKBS Applications", IEEE Transactions on Knowledge and Data Engineering, **8**(4), 663–671 (1996).

[8.9] R. G. Smith, "The Contract-Net Protocol: high-level communication and control in a distributed problem solver", IEEE Transactions on Computers, **29**(12), 1104–1113 (1980).

[8.10] S. M. Deen and M. Fletcher: "Fault Tolerance in Holonic Manufacturing Systems", *IEEE International Workshop on Embedded Fault-Tolerant Systems,* Boston,USA, (1998).

[8.11] S. M. Deen, "A Fault-Tolerant Co-operative Distributed System", *Proceedings of the 9th IEEE DEXA Workshop: Database and Expert Systems Applications*, ed. R. Wagner, Vienna (1998).

[8.12] C. A. Johnson and S. M. Deen, "Formalising an Engineering Approach to Co-operating Knowledge Based Systems", Technical Report, Department of Computer Science, University of Keele (1996).

[8.13] S. M. Deen: "A Computational Model for a Cooperating Agent System", in *Cooperative Information Agents III*, eds. M. Klusch, O. Shehory and G. Weiss, Lecture Notes in Artificial Intelligence, No. 1652, Springer, Berlin, Heidelberg (1999), pp. 175–195.

[8.14] C. Hewitt, "Open information systems semantics for Distributed Artificial Intelligence", Artificial Intelligence Journal, **47**(2), 79–106 (1991).

[8.15] J. R. Searle, *Speech Acts*, Cambridge University Press, Cambridge (1969).

[8.16] M.E. Bratman, D.J. Israel and M.E. Pollack: "Plans and Resource Bounded Practical Reasoning", Computational Intelligence Journal, **4**(4), 349–355 (1988).

[8.17] A.S. Rao and M.P. Georgeff: "BDI Agents: From Theory to Practice", Proceedings of the 1st International Conference on Multi Agent Systems, 1995.

[8.18] M.R. Genesereth and R.E. Fikes: *Knowledge Interchange Format Reference Manual*, Stanford University, Stanford, CA (1992).

[8.19] T. Finin and R. Fitzson: "KQML: A Language for Protocol and Information Exchange", *Proceedings of the 13th International Conference on Distributed Artificial Intelligence*, ed. M. Singh, University of Maryland, USA (1993).

[8.20] FIPA (Foundation for Intelligent Physical Agents), *Agent Communication Language and Specification of Agent Architecture*, FIPA'97 Specification, Parts 2 and 3, http://www.nortelnetworks.com/fipa-os.html (1997).

[8.21] S. M. Deen, R. Jayousi, B. Ndovie and B. Taha: *An agent-Based Approach to Dynamic Scheduling in Traffic Control*, Technical Report, Department of Computer Science, University of Keele, England (1997).

[8.22] S. M. Deen and M. Fletcher: "Agent-Based Service Brokerage in Telecommunications Management Networks", *Proceedings of Intelligent Agents in Communications*, London (1998).

[8.23] H. van Brussel, J. Wyns, P. Valckenaers, L. Bongaerts, and P. Peeters: "Reference Architecture for Holonic Manufacturing Systems: PROSA", Computers in Industry, Special Issue on Intelligent Manufacturing, 37(3), 255–276 (1998).

[8.24] F. Maturana, W. Shen and D. H. Norrie: "MetaMorph: An Adaptive Agent-Based Architecture for Intelligent Manufacturing", International Journal of Production Research, **37**(10), 2159–2173 (1999).

[8.25] S. M Deen, *Practical Database Techniques*, Pitman, London (1990).

[8.26] S. M. Deen, *A Relational Approach to a Co-operation Environment*, Technical Report, Department of Computer Science, University of Keele (1996).

[8.27] G. Weiss, ed., Multiagent Systems: A Modern Approach to Distributed Artificial Intelligence, MIT Press, Cambridge, MA (1999).

[8.28] L. Chiariglione, *FIPA: Agent Technologies Achieve Maturity*, Agent Link Newsletter, http://www.agentlink.org (1998).

[8.29] S. Bussmann and J. Sieverding, *Specification of Holonic Agent Control Concepts in Manufacturing Logistics*, HMS deliverable D7.5.2, http://hms.ifw.uni-hannover.de/ (1999).

9. Holonic Diagnosis for an Automotive Assembly Line

D.H. Jarvis and J.H. Jarvis

Agent Oriented Software Pty. Ltd.
Australia
Email: {dennis.jarvis | jacquie.jarvis}@agent-software.com

Abstract. Diagnosis is an important function of a holonic manufacturing system if the desired levels of stability, adaptability and flexibility are to be achieved. Our research agenda is to study holonic behaviours (such as diagnosis and control) through the incorporation of these behaviours into operational industrial systems. Given the lack of fielded holonic solutions in industry, we are currently constrained to use conventional systems in our work. In this paper we describe the development of a holonic diagnostic capability for a PLC-controlled vehicle assembly line. A novel model-based strategy is used for diagnosis. Because of the constraints imposed on model formation in this environment, a two-phase approach consisting of off-line fault space generation and online fault space analysis is used. The fault space analysis utilises heuristics to achieve the desired performance levels (diagnosis in less than 60 seconds and success rates of greater than 90%). Areas for further research in holonic diagnosis are identified.

9.1 Introduction

In [9.1], the Hungarian author and philosopher Arthur Koestler proposed the word "holon" to describe a basic unit of organisation in biological and social systems. "Holon" is a combination of the Greek word *holos*, meaning whole, and the suffix *on*, meaning particle or part. Koestler observed that in living organisms and in social organisations, entirely self-supporting non-interacting entitities did not exist. Every identifiable unit of organisation, such as a single cell in an animal or a family unit in a society, consists of more basic units (plasma and nucleus, parents and siblings) while at the same time forming a part of a larger unit of organisation (a muscle tissue or community). A holon, as Koestler devised the term, is an identifiable part of a system that has a unique identity yet is made up of subordinate parts and in turn is part of a larger whole. The strength of holonic organisation, or holarchy, is that it enables the construction of very complex systems that are nonetheless efficient in the use of resources, highly resilient to disturbances (both internal and external) and adaptable to changes in the environment in which they exist. All these characteristics can be observed in biological and social systems. The task of the Holonic Manufacturing Systems consortium is to translate the concepts that Koestler developed for social organisations and organisms into a set of appropriate concepts for manufacturing industries. The goal of this work is to attain in manufacturing the benefits that holonic organisation provides to living organisms and societies, e.g. stability in the face of disturbances, adaptability and flexibility in the face of change and efficient use of available resources.

Viewed in the above context, diagnosis is clearly an important function of a holonic manufacturing system if the desired levels of stability, adaptability and flexibility are to be achieved. Our research agenda is to study holonic behaviours (such as diagnosis) through the incorporation of those behaviours into operational industrial systems. Given the lack of fielded holonic solutions in industry, we are currently constrained to use conventional systems in our work. In this paper we describe the development of a holonic diagnostic capability for a PLC-controlled vehicle assembly line.

9.2 The Testbed

The testbed was a low-volume vehicle assembly line for an Australian automotive manufacturer. It consisted of three assembly stations (Stations 10, 20 and 30) linked by a transfer line. It had a well-defined hierarchical structure, making it ideal for an exploration of holonic diagnosis. Two types of bodies were assembled on the line (Style A and Style B). The line was controlled by a single PLC with approximately 700 I/O points. The control logic was similar to that of the larger systems within the company.

The operations performed at each station were an alternating sequence of automatic steps (known as stages) and manual steps. Automatic operation at a station was initiated by an operator (or operators) depressing one or more sets of palm buttons. Those buttons remained depressed for the duration of that step. The activities performed during automatic operation typically involved the opening and closing of clamps. On completion of an automatic step (indicated by a lamp being illuminated on the station control panel), the operator removed his or her hands from the palm buttons and initiated a sequence of manual activities. These typically involved the loading of panels and/or welding. When the assembly operations had been completed at a station, the operator needed to wait until assembly has been completed in the other stations. At that point, the transfer line was activated (by the operators at each station simultaneously depressing their palm buttons) and the station cycle began again with a new body. Note that the number of stages in a cycle depended on the model that was being built.

When a fault occurs because of a malfunction of a PLC-controlled I/O point, (for example, a limit switch not making contact), it is often very difficult to determine the cause of the problem. The reason for this is that the complexity of the control pathways means that the manufacturing process might not stop until some time after the actual fault has occurred (and further along in the manufacturing sequence). To compound this problem, it is also difficult to establish valid conditions for resuming production on such systems – the downtime between fault diagnosis and production restart is often significant and costly.

One of the major obstacles that has prevented the development of effective diagnostic systems in such environments is that conventional rule-based approaches are unable to produce comprehensive systems in reasonable timeframes [9.2]. The underlying reason for this is that rule-based diagnostic systems are constructed by associating a set of states with a set of observed faults. This association is normally performed by people who have had extensive experience with the underlying manufacturing system. Unfortunately, when dealing with complex, one-off manufacturing systems, it becomes impossible for a single person (or even a group of people) to experience a sufficiently wide range of fault states to generate a comprehensive diagnostic system. Furthermore, the knowledge acquisition task involved in the development of such a system is extremely time-consuming (and therefore expensive), thus providing more disincentive.

It is our belief that a model-based approach offers a much more attractive path for the development of comprehensive and timely diagnostic systems. In a model-based approach, one would first construct a model of the manufacturing system. Reasoning strategies would then be developed that enabled an association to be made dynamically between an observed fault state and an underlying cause. This approach is particularly attractive for holonic diagnosis, as each holon in the system would have its own model. For example, in the case of our testbed, if one formed a holarchy consisting of the assembly line and the three assembly stations, then we would develop models for each of the four holons.

In the case of PLC-controlled manufacturing systems, the bulk of the knowledge required to construct the model is already available in the form of the PLC program. However, in manufacturing systems, it is common for a PLC output to indirectly affect a PLC input. For example, a PLC output may cause a clamp to close. The closure of that clamp might then be detected by a limit switch, which then drives a PLC input. Such associations are not specified in a PLC program, and in order to construct an operational model of the system, the control logic embodied in the PLC program needs to be augmented with information about the state behaviours of the devices that are connected to the PLC and about how those devices affect each other. This additional information can normally be extracted from wiring diagrams and the pneumatic and hydraulic circuit diagrams. Note that our reliance on experts is greatly reduced, and that if efficient reasoning strategies can be devised, our fault coverage will be comprehensive.

9.3 Model Based Diagnosis

In model-based diagnosis, one reasons directly about the system, using an appropriate model. In the case of PLC controlled discrete manufacturing systems, such a model needs to be able to simulate the behaviour of the system at a causal level (i.e. to generate the sequence of manufacturing events that occur). In the previous section, we have shown that such models can be constructed given the control logic and the finite state behaviours of the components of the manufacturing system.

Discrete manufacturing systems are most naturally represented by component-oriented models. Such models contain objects which correspond to the system's components on a one-to-one basis. These components will normally correspond to physical entitities (e.g. adders and multipliers in a digital circuit, or limit switches or clamps in a manufacturing system). Component-oriented models explicitly state how a system's components are connected together and how their outputs are derived from their inputs. They directly represent the structure and function of the system and can simulate its operation. Model-based reasoning has been successfully applied to models other than component-oriented; the interested reader is referred to [9.3] for further details. A fundamental issue in the design of diagnostic systems that use a component-oriented approach is whether to build faulty behaviours into component descriptions or to dispense with explicit fault models and dynamically infer which components are faulty. The latter approach was pioneered by Davis [9.4] (constraint suspension) and de Kleer and Williams [9.5] (GDE); the former approach was pioneered by Bratko et al. [9.6] (Kardio). These three approaches are discussed below.

1. *Constraint suspension*: Davis modelled the intended behaviour of a system as a network of interconnected constraints, where each constraint captures the behaviour of one of the components (or connections) [9.4]. If the device is exhibiting abnormal behaviour, then the observed outputs will not correspond with those predicted from a consideration of the constraint network. Normally, contradictions in constraint networks are resolved by the retraction of one or more input values. Davis, however, retracted constraints (i.e. component behaviours) in order to generate a consistent network. He called this process constraint suspension. Diagnosis then becomes a process of finding a set of constraints which, when suspended, accounts for the observed behaviour.
2. *GDE*: De Kleer and Williams used an assumption-based truth maintenance system (ATMS) as the cornerstone of the diagnostic process in their system, called GDE [9.5]. An ATMS [9.7, 9.8] is a mechanism for maintaining alternative states of a world which are consistent with some theory, but in which different assumptions hold. In a diagnostic application, this theory is the device description, and the alternative states of the world are characterised by all possible combinations of failed components. These combinations are known as candidate sets. Diagnosis in GDE can be viewed as the search of a candidate set space for the best candidate set. In performing this task, GDE will propose further measurements to help distinguish between candidate sets. The input to GDE consists of the structure of the system, the component behaviour descriptions, the a priori probabilities of component failure and a set of observations.
3. *Kardio*: Dynamically inferring the presence of faults in a system from a knowledge of the correct behaviour of that system is a technique which has been little used outside the realm of digital circuits. This is perhaps no coincidence, as the technique presupposes that the domain in question is governed by a finite (and preferably small) set of laws that are capable of being precisely stated. The alternative approach is to embed fault behaviours into component descriptions. An example of this method can be found in the Kardio project [9.6]. Kardio is a qualitative model of the electrical control system of the heart. The model was constructed as a network of components such as impulse generators, conduction pathways, impulse summators and ECG generators. Normal and abnormal behaviours were associated with components and a dictionary of simple arrhythmias was incorporated into the model. A simulation algorithm was then implemented which enabled an ECG to be generated for a given combination of arrhythmias. Fault diagnosis with such a system could be achieved by running the simulation backwards, but the number of alternatives that would need to be considered made such an approach unattractive. Instead, what was done was to generate all legal arrhythmia/ECG pairs by exhaustive simulation. An inductive learning program was then used to compact this rule set. These rules could then be used for either diagnosis or prediction. A similar approach was used by Pearce in the development of a diagnostic system for the electrical power sub-system of a satellite [9.9].

Recent work in model-based diagnosis has focused on the use of richer behaviour models which incorporate both normal and faulty behaviours, the development of more sophisticated reasoning strategies to dynamically determine faulty components, and deployment in practical applications. This work is exemplified by projects undertaken at NASA [9.10] and Xerox PARC [9.11]. As part of its New Millennium Program, NASA is exploring the issues of monitoring, diagnosis and repair in the context of autonomous spacecraft. An embedded real-time execution kernel (known as Livingstone) has been developed that uses a compositional, component-based model of the underlying system to perform appropriate actions automatically [9.10]. Each component is modelled as a transition system that specifies the behaviours of operating and failure modes of the component, nominal and failure transitions between modes, and the costs and likelihoods of transitions. The transition system model is a composition of its component transition systems. Livingstone uses its transition system model both to identify the current configuration of the system and to move the system into a new configuration that achieves the desired configuration goals. To achieve the stringent demands of real-time performance, Livingstone reduces each function to a deductive search problem on a propositional database. This search must be completed before the system moves to the next state, with required response times of the order of hundreds of milliseconds. Hence the success of Livingstone's model-based execution paradigm hinges critically upon the efficiency of the propositional deductive database. Livingstone uses a logic-based truth maintenance system (LTMS) [9.12] to achieve the desired performance.

The Model Based Computing project at Xerox PARC [9.11] arose from the observation that large amounts of software are developed for electromechanical machines, such as software for design, control, scheduling, simulation, diagnosis and productivity analysis. Software like this has to contain a lot of information about the machines, and about their components, particular configurations, normal and faulty behaviours, interaction and timing issues, etc. There is an emerging view that not only should this machine information not be hard-wired into software, but also that it should be prepared separately in the form of declarative, constraint-based models. These models need to capture the structure and behaviour of the machine and they should also be compositional.

Concurrent constraint-based programming provides a natural framework for developing such models [9.11]. A particularly hard part in developing machine software is analysing how machine components or operations interact. This analysis should be done automatically and is facilitated if we can start with models of simple machine elements and then connect them to form larger models. Overall behaviour is then derived automatically from the behaviour of the simpler elements. Besides making modelling easier and less error-prone, this also allows one to automatically and dynamically put together and customise configurations. If such models can developed, the intuition is that they can be combined with software systems corresponding to specific tasks (e.g. diagnosis) via customised reasoning engines. These engines will take as input the specification of a system configuration, task-specific but machine-independent architectures and algorithms, and the model of the components, to generate the target application. Initial efforts in this field have focused on the development of appropriate modelling and constraint languages [9.13] and kernels for supporting model-based autonomous activity [9.10].

We are unaware of any model-based diagnostic strategies that have been developed specifically for PLC-controlled manufacturing systems. As we shall see, the separation that exists in these systems between the control logic for the total system (as embodied in the PLC program) and the finite state behaviour of the physical components imposes significant constraints on the determination of a suitable strategy.

9.4 The Modelling Process

A model of a PLC-controlled manufacturing system, if it is to be used for fault diagnosis, will need to simulate the following activities:

- *PLC operation*: A PLC typically executes the following algorithm:

```
for (ever)
{
  Read all inputs;
  Evaluate the PLC program;
  Set all outputs;
  Service remote i/o requests;
}
```

  One loop is known as a scan. Our objective was not to emulate the detailed operation of the PLC so that accurate scan times could be determined, but rather to simulate its input/output behaviour. That is, having read all the PLC inputs, we need to determine what PLC outputs should be set by evaluating a representation of the PLC program.

- *Manufacturing system operation*: The PLC controls a manufacturing system. We can view the manufacturing system as consisting of two different entity types:

  1. *Actuated devices*: An actuated device is a collection of electromechanical devices that can exist in one of several states. State selection is controlled by one or more PLC outputs. One or more events in the manufacturing process are associated with each state. As an example, consider the actuated device SAV23 in the testbed. It consists of two solenoids, an air valve, a piston and three clamps (12K2-K2, 12S-S and 12O-O). It has two states, designated SAV23A and SAV23B. In the first state, the clamps are open; in the second state, the clamps are closed.
  2. *Sensors*: Sensors enable us to determine whether a particular manufacturing event has occurred. Examples of sensors include limit switches, proximity switches and palm buttons.

These entities typically interact in the following manner:

```
Set a PLC output;
The device changes state;
A manufacturing event occurs (e.g. clamp closure);
if (there is a sensor for this event / state)
{
   The new state is detected (e.g. by a limit switch);
   The associated PLC input is set;
}
```

Ideally we would like to model each actuated device as a holon, but in general the control logic for a single device is difficult to extract from the PLC code because its state transitions are invariably dependent on the state of other devices. The control logic can be extracted in terms of finite-state machines [9.14] but one then has a representation that is different from that of the underlying system (the PLC code) and diagnosis becomes problematic. Consequently, we decided to model each station (and the overall line) as holons. This was a pragmatic decision based on the fact that the control logic for these entities was easily identifiable in the PLC code. The holon models were thus partitioned into two sub-models, one dealing with the control logic (represented as PLC code) and the other dealing with the state behaviour of the holon's actuated devices and sensors We refer to these as the control model and physical model, respectively. Note that the interaction between the models is via PLC inputs and outputs.

The physical model was implemented as a collection of finite-state machines; devices change states when the appropriate PLC outputs are fired. When a device undergoes a state transition, sensors associated with the previous state are deactivated, and those for the current state are activated. The PLC code was represented as a sequence of AND/OR graphs (one per rung); rung evaluation was initially viewed as a single-source shortest-path problem for a graph with unweighted edges.[1] Some sample output from a simulation trace is shown below:

```
Stage 1 Ready. Hands On - Clamp Body

Y527 = 1 -> <2,1>
Previous State
ST20 7A-A,7B-B LOCATOR UNCLAMP BOTH SIDES
X346 X345 X348 X347 X438 X437 -> OFF

Current State
ST20 7A-A,7B-B LOCATOR CLAMP BOTH SIDES
```

[1] Our objective is to determine whether a suitable path between the start node and the finish node exists, not to calculate the shortest such path. Consequently, we can use a shortest-path algorithm, but stop as soon as a suitable path is found.

```
X403 X436 X435 -> ON

Y529 = 1 -> <2,1>
Previous State
ST20 7D-D,7F-F,7J-J LOCATOR UNCLAMP BOTH SIDES
X351 X350 X360 X359 X353 X352 -> OFF

Current State
ST20 7D-D,7F-F,7J-J LOCATOR CLAMP BOTH SIDES
X404 -> ON

Y548 = 1 -> <1,2>
Previous State
ST20 OVERHEAD FRAME CLAMP
X482 X481 X486 X485 -> OFF

Current State
ST20 OVERHEAD FRAME UNCLAMP
X484 X483 X488 X487 -> ON
```

The simulation approach is described in detail elsewhere [9.15, 9.16].

9.5 The Diagnosis Process

9.5.1 Assumptions

The assumptions that we made in developing the diagnostic system are the following:

1. There will be a single point of failure.
2. The point of failure will be a PLC input (typically a limit switch).

The justification for the first assumption is that independent multiple faults which all contribute to a line stoppage are unlikely to occur. Certainly, multiple independent faults can occur in an assembly system, but our observation has been that these faults are benign – that is, limit switches had failed in such a way that the station was still completing its cycle (and consequently maintenance personnel were unaware of the problems). This was possible because of the way in which the system was designed. If one inspects the pneumatic circuit diagrams, one will see that in many cases, only one device state (e.g. clamp open) is explicitly checked for with a limit switch. If the PLC program then only checks to see whether that event has happened, and if the limit switch associated with that event fails in the on position, PLC execution will continue as normal and the station will cycle normally. (However, note that if the clamp fails to operate correctly for some reason, we have the potential for major damage to occur.)

This poses a challenge to our single-fault assumption; even if we were to regularly check for benign faults (e.g. at the end of each shift) we would have no guarantee that, when a line stoppage occurred, a benign fault had not occurred since the last check.

The second assumption can be relaxed if required; if it proves necessary, we can induce faults in PLC outputs. It was not relaxed in this project, because maintenance personnel experienced no difficulties in determining the cause of the stoppage in these situations – their major difficulty was in finding faults associated with a single PLC input. Note that this would allow us to diagnose multiple dependent faults; e.g. if a cylinder failed to advance, then all limit switches associated with all clamp actions associated with cylinder advancement would be in error.

9.5.2 Strategy

The conceptual strategy that we used for diagnosis within a holon is presented below:

```
capture the PLC state when an automatic step fails to
complete
determine the stage during which failure occurred
for (all possible faults in the holon)
{
  simulate the behaviour of the holon with that fault
  induced
  if (the simulation fails in the same stage as the
      physical system)
  {
    compare observed state and simulated state
    if (observed state == simulated state)
    {
       report cause of fault
       exit
    }
  }
}
report failure to identify cause of fault
```

In implementing such a strategy, there are two major issues that need to be considered (assuming that have a working simulation of each holon, that we have the necessary infrastructure to capture PLC states and that efficiency is not a concern):

1. How do we determine the fault space?
2. How do we compare observed and simulated states?

The fault space for a particular station is the set of all possible faults that can occur. As we are dealing with a system which changes with time, a faulty limit switch will be characterised not only by its state (on or off) but also by the point in the cycle when the failure occurred. Our initial belief was that limit switches would fail only when they underwent a change of state from on to off, or vice versa. However, reality proved otherwise: switches get knocked by welding guns, they can get covered in glue, they can accumulate grime or they can just gradually become misaligned or fall apart. In short, a fault can occur at any point in the cycle.

At first glance, the size of the fault space would suggest that the diagnostic strategy described above would be infeasible, as for Station 20 (the most complex of the holons), there are approximately 17,000 points in the full fault space. Note that this is equal to the number of input points (210 for Station 20) multiplied by the number of logical scans in the cycle (75 for Style A, 88 for Style B). While the size of the fault space is large, the size is linearly dependent on these two parameters. Thus if we double the number of inputs or double the number of scans, we double the size of the fault space. Also, note that the number of points in the fault space that need to be examined can be significantly reduced if we realise that:

- *The fault space can be partitioned according to the stage in which failure will occur for a given fault point*: When we perform diagnosis, we shall be focusing on a subset of the full fault space, namely those faults which, when induced, would cause termination to occur in the step at which failure was observed. Thus we can ignore all fault points which would cause termination in other stages. However, one needs to be aware that induced faults can cause a stage incompletion in the step at which they were induced, in any subsequent stage in the current cycle or even during subsequent cycles. Thus when we perform diagnosis, we need to consider all faults in preceding steps, in both the current and the previous cycle. (If we are inducing style-specific faults, then the previous cycle means the last time that the current style was processed in the station.)
- *At each point in the fault space, there is only one (of two) possible faults that needs to be induced*: If we assume that limit switches fail only in one of two modes (on and off) then we need only induce one of those failures at each point in the fault space. (At a given point of the simulation, suppose that a limit switch is on. If it fails in the on position, then it cannot impact on cycle completion until we reach the point in the cycle where it is supposed to go off.) Thus, at each point in our fault space, we induce a fault in a limit switch by determining its current state in the simulation (on or off) and then setting it to the opposite state.
- *Benign faults can be ignored*: As was mentioned earlier, if the PLC program checks only to see if a limit switch is in one state (typically on) and the switch fails in the on position, the line will continue to operate normally. Thus, if we can identify these faults, they can be eliminated (or pruned) from the fault space. We were able to identify the bulk of the faults that exhibited this behaviour through an off-line analysis of the PLC program; approximately half the limit switches in the line are checked for one state only.

Comparison of observed and fault states is not as complicated. The issue here is "What is the minimum amount of information that will enable us to identify a simulated state as being the *same* as the observed fault state?" We have chosen in the current implementation to do the comparison on the basis of selected inputs. Owing to production pressures, maintenance personnel were unable to prepare a system which was free of benign faults, so we chose to ignore those inputs which corresponded to benign faults. This has worked well in practice, although in general a PLC state will be characterised by all its inputs, latched outputs and non-volatile memory locations.

9.5.3 Optimisations

Even with the pruning of the fault space as described in the previous section, the amount of computation required by our strategy is excessive. Consequently, we implemented the following two-stage strategy:

```
//Stage 1 (Off-line)
for (each point in the pruned fault space)
{
  induce the fault
  simulate holon behaviour with the fault induced
  record the fault state
  record the stage at which failure occurred
}

//Stage 2 (On-line)
capture the PLC state when a stage completion fails to
occur
determine the stage during which failure occurred
for (each point in the pruned fault space)
{
   if (recorded failure stage == observed failure
stage)
   {
      if (recorded failure state == observed failure
state)
      {
         report cause of fault
         exit
      }
   }
}
```

Note that in Stage 1, if fault points are examined in scan sequence, the computation required can be significantly reduced by judicious saving and restoring of simulation states. However, the amount of computation that is required is still excessive if the AND/OR representation of the PLC program is used and rung evaluation is treated as a shortest-path problem. We therefore transformed the PLC program into compilable C code. This was achieved by writing a program which, for each rung, generated all possible paths between the start and end of the rung. These paths then formed the logical-expression component of an `if` statement (the nodes in each path were combined with `&&` operators; the paths for each rung were combined with `||` operators); the consequents of the `if` statement become assignment statements involving the rung output node.

Program evaluation is now achieved by embedding the C code that we generate in a function (called `scan()`). This routine is then compiled along with the remainder of the source code for the system. Using the compilable representation reduced our computation time by a factor of 10; stage 1 of our strategy took about 20 minutes on a SPARCstation 10.

9.5.4 Verification

The diagnostic system consisted of four sub-systems, corresponding to the four holons in the system. The diagnostic system was verified by inducing faults during normal production and then attempting to ascertain the cause of the resultant stoppage by using the diagnostic system. If a station had stopped mid-cycle then the diagnosis was initially performed using the model for that station. If that proved unsuccessful, then the assembly line model was used, and failing that, the downstream stations were used. If the stoppage occurred during transfer, then the assembly line model was used first, followed by the models for each of the stations in turn. Three trials were conducted and 55 faults were induced. On completion of the trials three faults remained unresolved, giving a success rate of ~95%.

9.6 Discussion

The system level strategy that we described in the previous section was not implemented as an automated process. At the request of the operators, the individual diagnosis holons were made available to them, and in performing diagnosis, they used the strategy described above. Implementation of this strategy for the testbed would be straightforward, but the development of general system-level strategies is an area for future research. We would expect that any general strategy would need to incorporate expert knowledge in addition to structural and behavioural models of the system level. This issue becomes even more interesting if you have more than two levels in the hierarchy. Our testbed has a well-defined hierarchical structure – the line consists of stations – stations consist of frames and hoists – frames and hoists consist of clamps, motors and sensors. Unfortunately, the structure below stations is not readily discernible in the PLC code, so we were unable to explore this issue in this particular project.

While a model-based approach to holonic diagnosis is attractive in that it can reduce the reliance on experiential knowledge and that the models are available for other uses, it is by no means a prerequisite for building a holonic diagnostic system. Each application needs to be treated on its merits, and depending on the nature of the system, a conventional rule-based approach or perhaps a neural network may be more appropriate. Having said that, in this particular instance we would argue very strongly that a model-based approach was the most appropriate approach.

Diagnostic systems are typically evaluated against fault coverage and execution time. Execution time for our system was acceptable (20–40 seconds per fault); note that this time is fault-independent. Coverage is more difficult to evaluate. We would expect that if our model was correct, then all single faults would be detected. However, it would appear unrealistic to expect that we can replicate the exact behaviour of such complex systems when limit switches fail on a regular basis, clamps fail to completely close because of variations in panel quality and positioning, and the environment is conducive to intermittent faults (such as grime and glue on limit switches, and limit switches and wires being knocked by welding guns). There may well be some inaccuracies remaining in our model that can be rectified and we shall continue to strive for 100% coverage. However, we consider a 95% success rate to be quite satisfactory, particularly considering that in the trials that we have conducted, we produced no false positives (i.e. whenever the system identified a fault, it was correct), and that the effort involved in generating the diagnosis by the operator is minimal (no input is required, and the elapsed time for a diagnostic run is only 20–40 seconds).

We anticipate that fault states will exist for which we cannot assign a unique cause; in such cases, all potential candidates will be reported. Note that the likelihood of this happening is increased because of the limited amount of information that we use in state comparisons. As mentioned earlier, this work has assumed that we shall have a single point of failure, and that it will be a limit switch. If we were to allow devices to fail as well, we could diagnose situations where cylinders had failed to advance or retract. In the current system, these situations would appear as multiple points of failure (all the limit switches associated with a cylinder position would be in error). The system also has the interesting capability of being able to detect intermittent faults, provided that they cause a line stoppage. In this situation, the diagnostic system will indicate that a particular limit switch is faulty, but a casual inspection may indicate that the switch is behaving normally.

Our approach to diagnosis has similarities to that used by Bratko et al. [9.6] in their diagnosis of cardiac arrhythmias. They first developed a qualitative model of the electrical system of the heart. Faults were then induced in this model and the fault states (heart arrhythmias) were recorded. An inductive learning program was then used to generate compressed diagnostic rules from knowledge of the induced faults and their associated fault states. Unlike us, Bratko et al. did not initially prune the search space; the removal of degenerate fault states occurs during the inductive learning phase. Note that we could have used inductive learning to generate compressed diagnostic rules on our pruned fault space, but diagnostic performance was satisfactory using uncompressed rules. This situation may of course change if we apply our approach either to larger systems or to systems where significant pruning of the fault space cannot be achieved.

9.7 Conclusion

A model-based diagnostic system for a PLC-controlled vehicle assembly station has been constructed and tested online. Fault coverage on the test data was 95%, and execution time was 20–40 seconds/fault. Experience with the system has identified areas for further research in the area of holonic diagnosis.

References

[9.1] A. Koestler: *The Ghost in the Machine*, Arkana, London (1967).

[9.2] W.B. Day and M.J. Rostosky: "Diagnostic Expert Systems for PLC Controlled Manufacturing Equipment", International Journal of Computer Integrated Manufacturing, **7**, 116–122 (1994).

[9.3] A. Cohn: "Qualitative Reasoning", in *Lecture Notes in Artificial Intelligence*, No. 345, Springer, Berlin, Heidelberg (1988), pp. 60–95.

[9.4] R. Davis: "Diagnostic Reasoning Based on Structure and Behaviour", Artificial Intelligence, **24**, 347–410 (1984).

[9.5] J. de Kleer and B. Williams: "Diagnosing Multiple Faults", Artificial Intelligence, **32**, 97–130 (1987).

[9.6] I. Bratko, I. Mozetic and N. Lavrac: *Kardio: A Study in Deep and Qualitative Knowledge for Expert Systems*, MIT Press, Cambridge MA (1989).

[9.7] J. de Kleer: "An Assumption-based TMS", Artificial Intelligence **28**, 127–162 (1986).

[9.8] J. de Kleer: "Problem Solving with the ATMS", Artificial Intelligence, **28**, 197–224 (1986).

[9.9] D. Pearce: "The Induction of Fault Diagnosis Systems from Qualitative Models", in *Proceedings of AAAI-88*, AAAI Press, Menlo Park CA (1988), pp. 353–357.

[9.10] B. Williams and P. Nayak: "Immobile Robots: AI in the New Millennium", AI Magazine, **17**, 16–35 (1996).

[9.11] V. Saraswat, D. Bobrow and J. de Kleer: *Infrastructure for Model-Based Computing*, Xerox PARC (1993).

[9.12] K.D. Forbus and J. de Kleer: *Building Problem Solvers*, MIT Press, Cambridge MA (1993).

[9.13] H. Wong, M. Fromherz, V. Gupta and V. Saraswat: "Control-Based Programming of Electro-Mechanical Controllers", in *Proceedings of IJCAI Workshop on Executable Temporal Logics*, Montreal (1995).

[9.14] J.H. Jarvis and D.H. Jarvis: "Design Recovery for PLC Controlled Manufacturing Systems", in *Manufacturing Systems: Modelling, Management and Control*, ed. P. Kopacek, Elsevier (1998).

[9.15] J.H. Jarvis and D.H. Jarvis: "Life Cycle Support for PLC Controlled Manufacturing Systems", in *Software Engineering for Manufacturing Systems*, eds. A. Storr and D. Jarvis, Chapman & Hall, London (1996).

[9.16] J.H. Jarvis and D.H. Jarvis: "Simulation of a PLC Controlled Assembly Line", in *Simulation in Industry: 9th European Simulation Symposium*, eds. W. Hahn and A. Lehmann, SCS (1997).

Part IV

Application Environments and Issues

10. HMS Development and Implementation Environments

Shinsuke Tamura[1], Toshibumi Seki[2], and Tetsuo Hasegawa[2]

[1]Department of Information Science, Faculty of Engineering, Fukui University
3-9-1, Bunkyo, Fukui 910-8507, Japan
Email: tamura@ fuis.fuis.fukui-u.ac.jp

[2]Corporate Research and Development Center, Toshiba Corporation
1, Komukai Toshiba-cho, saiwai-ku, Kawasaki 212-8582, Japan
Email:{toshibumi.seki | tesuo3.hasegawa}@toshiba.co.jp

Abstract. In this chapter, desirable features of Holonic Manufacturing System (HMS) development and implementation environments are extracted, and HMS-Shell and JDPS are explained, as examples that provide these features. HMS-Shell is a GUI-based design tool for HMSs based on the Cooperation Domain (CD) that encapsulates cooperation partners and cooperation protocols. JDPS is a HMS implementation and execution environment based on reliable broadcast messaging that achieves complete location and multiplication transparency of entities. These features of HMS-Shell and JDPS make application programs independent of cooperation mechanisms, and enable easy dynamic reconfiguration and incremental development of HMSs while systems are in operation.

10.1 Introduction

Holonic Manufacturing Systems consist of sets of mutually independent elements, i.e. *autonomous* and *cooperative* entities named *holons*, allocated over distributed computers, and can achieve given goals through cooperation among individual holons without centralized mechanisms. Because of the distributed structure of a HMS, it can be reconfigured dynamically to adapt to various environments. Holons can be added, deleted and/or migrated easily while the system is in operation and, moreover, individual holons find their roles in the given conditions so that they can contribute to varying objectives with their full ability, i.e. without being restricted by the ability of centralized mechanisms. However at the same time, this reconfigurable and adaptive structure makes HMSs difficult to develop and operate. Unlike conventional manufacturing systems, in which the roles of elements and of the information and the control flows among them are determined at design stage, in HMSs, they are determined during the operation of the system. To achieve efficient, reliable development and operation of HMSs with these non-deterministic properties, several new features are required for the support tools also.

In this chapter, desirable features of HMS development and implementation environments are extracted, and *HMS-Shell* and *JDPS* are explained as examples that provide these features. Firstly, the objectives of a HMS are explained together with its architecture in Sect. 10.2, followed by Sect. 10.3, which introduces strategies for supporting HMS development and implementation. Then, in Sects. 10.4 and 10.5, features of HMS-Shell and JDPS are described as comprehensive examples of support tools for HMS development and implementation. In Sect. 10.4, some desirable features of system simulation are also summarized. Simulation capability is more important in the development and operation of HMSs than in conventional manufacturing systems, since it is difficult to estimate nondeterministic behaviors of distributed holons.

10.2 HMS Objectives and Architecture

10.2.1 Objectives

The goal of a HMS is to eliminate various bottlenecks that exist in conventional manufacturing systems. The roles of the elements in many conventional manufacturing systems are managed by centralized mechanisms. Therefore management mechanisms must maintain information about various, numerous elements and control their behavior; consequently these mechanisms show their limits, as they are potential system-wide bottlenecks. Regarding *system development*, their complicated functions are tightly connected to large numbers of elements, making it difficult to extend systems to deal with continuously changing manufacturing environments, i.e. even small changes in an element require modification of many relevant parts of the system. These tight connections to various elements also cause problems in *system reliability*, i.e. whole systems become unavailable because of malfunctions in the centralized mechanism. Also, the capabilities of centralized mechanisms, which are determined and fixed at the design stage, restrict the abilities of individual elements. A task given to the system cannot be accomplished, even if there are elements that have the ability to execute it, when the centralized mechanism does not have enough ability. Consequently, the *adaptability* of the system to changes in the manufacturing environment is very limited.

Holonic Manufacturing Systems are *reconfigurable* and *adaptive* manufacturing systems that enhance the following capabilities of individual elements to reduce system-wide bottlenecks caused by centralized mechanisms.

- *Highly modularized elements through their autonomy*. Autonomous functioning of individual manufacturing elements in a HMS increases their modularity by hiding their inner structure, as objects in the object-model do. The behaviors of HMS elements are further encapsulated, i.e. the *holarchy* (their hierarchical structure) hides the group behaviors of elements. This increased modularity of elements enables continuous, incremental development of manufacturing systems, so that easy addition, modification, deletion and migration of elements become possible to deal with growing manufacturing requirements and rapidly advancing technologies, while the system is in operation.

- *Adaptive elements through their cooperation ability.* Distributed decision-making based on cooperation among HMS elements enables the development of highly adaptive manufacturing systems, in which a set of elements dynamically combine with one another to reconfigure the manufacturing system when the environment changes. Because individual elements know their ability most properly, they can show their full ability to achieve given objectives in varying conditions through cooperation among themselves, without being limited by the abilities of centralized mechanisms.

10.2.2 HMS Architecture

The key concepts in the HMS field for achieving the above features of individual elements are holons and holarchies. A HMS is a manufacturing system that includes holarchies of holons in it [10.1, 10.4]. Here, a *holon* is a manufacturing entity which possesses both autonomy and cooperativeness. The autonomy of a holon, i.e. the ability of a holon to make decisions and control itself autonomously, encapsulates its behaviors and improves the modularity of the holon. For example, a conveyor holon knows that it should start movement when workpieces are placed in its input buffer, that it should continue movement while workpieces are on it, that it cannot convey a workpiece when the weight is over some limit, etc., and it has also mechanisms to control its behaviors within itself. Therefore the conveyor holon can be modified, added to the system and deleted from the system without changing other parts, since it achieves its objectives consistently, independent of others. Cooperativeness, through which holons combine with each other, achieves highly adaptive behavior of HMSs, in which holons share roles with other holons to execute dynamically defined jobs while resolving conflicts among them. Let us consider a manufacturing objective of handling workpieces located at a remote place with rough information about the sizes of the workpieces. This objective can be achieved through role-sharing among robot, vision sensor and conveyor holons, i.e. the vision sensor compensates for the object-handling accuracy of the robot, the conveyor compensates for the ability of the robot to move, and the robot handles workpieces placed around it and also enlarges the visible areas of the vision sensor. Adaptive manufacturing systems must change this role-sharing structure along with the varying objectives. Cooperation among holons achieves a variety of role-sharing structures, because individual holons decide their roles so that they can contribute to given objectives without being restricted by centralized mechanisms.

A *holarchy* is a hierarchical structure of holons. In a holarchy, holon A is called a parent holon of holons B and C and, conversely, holons B and C are called child holons of holon A, when the objectives of holon A are achieved through cooperation among holons B, C and A itself. That is, a parent holon represents cooperative activities, and child holons are the members involved in the cooperation. In this sense, a child holon can be considered as a part of the parent holon. The important feature of the holarchy is that a parent holon represents cooperation activities while encapsulating the combined functions generated by the child holons. By means of this feature, HMSs can achieve complicated tasks through cooperation among holons, while keeping the complexity of the cooperation mechanisms within certain limits.

It is a common approach in many agent systems to configure a system with autonomous, cooperative elements, in order to maximize its adaptability. Compared with conventional centralized management systems, in which the abilities of individual elements are restricted by the ability of the centralized mechanism, decision-making based on cooperation among elements achieves very adaptive behavior because individual elements themselves know their capability and limits best. However, highly sophisticated mechanisms are necessary to achieve complicated behaviors through cooperation among agents, and these mechanisms will generate new bottlenecks. A holarchy is one solution to coping with this paradox; namely, in a HMS, complicated cooperation behaviors are implemented as iterations of simple cooperation in individual layers of the holarchy. A set of mutually cooperating child holons configure a parent holon so that the combined functions of the child holons can be treated as abstract functions of the parent holon, by encapsulating individual functions of child holons into group functions. Also, parent holons configure further abstract holons in the same way to achieve more complicated tasks while maintaining the same level of complexity.

Figure 10.1 shows an example of a holarchy. In the figure, holons A, B and C are parent holons of holons C, D and E, holons E and F, and holons C_1 and C_2, respectively. Holons C (C_1, C_2) and D are cooperating to achieve an objective X and, in the same way, holons D and E and holons E and F are cooperating with each other to achieve objectives Y and Z, respectively. Here, in the cooperation process, holons E and F determine their behaviors as individuals, and holon B determines its behaviors as those of a set of holons, i.e. holons E and F, for example. Holons can have *multiple holarchies*, i.e. can be involved in multiple holarchies simultaneously, according to the given objectives, as shown in the figure. Holons C_1, C_2 and D correspond to the robot, vision sensor and conveyor holons mentioned above, respectively, i.e. the robot holon and the vision sensor holon temporarily constitute a high-performance robot holon with visual processing capability, and this high-performance robot holon constitutes a robot station holon with the conveyor to achieve objective X, i.e. to handle workpieces located at a different place with rough information about the sizes of them. This robot for remote workpiece handling can also accomplish another objective Y with the aid of the other holon E. A manufacturing entity without autonomy or cooperativeness is called a *non-holon*, and by definition, a HMS also includes nonholons.

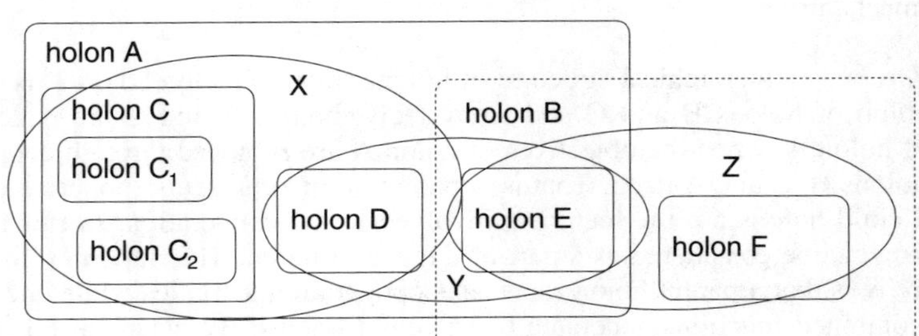

Fig. 10.1. Holon and holarchy

To make a holarchy dynamically reconfigurable, individual holons have *Cooperation Domains* (CDs) in them [10.6], and it is this reconfigurability, supported by CDs, that makes a holarchy different from a simple hierarchy. A CD is a communication channel where cooperating holons are connected, and represents a cooperation activity to achieve a specific objective. A holon can have multiple CDs in it, and can be connected to multiple CDs. A holon that possesses a CD is called the *owner holon* of that CD. A parent holon is the owner of a CD to which child holon(s) are connected, and parent and child holons cooperate with one another through this CD. Figure 10.2 shows the implementation of the holarchy shown in Fig. 10.1 using CDs. Holon A has cooperation domain CD-X and CD-Y to achieve objectives X and Y; it cooperates with holons C and D through CD-X, and cooperates with holons D and E through CD-Y. This relation can be interpreted as follows in the case of the example of the robot, vision sensor and conveyor: the robot station (holon A) owns CD-X, and the high-performance robot (holon C) and the conveyor (holon D) cooperate through this CD to handle workpieces located at a different place. Also, the high-performance robot owns CD-V_1, which handles workpieces with rough information about the sizes of them. The robot (holon C_1) and the vision sensor (holon C_2) cooperate through CD- V_1.

Cooperation Domains enable the comprehensive and efficient manipulation of cooperation in manufacturing environments where many various cooperation groups exist simultaneously and vary with time. Firstly, CDs provide the HMS designers with a clear understanding of complicated, dynamically varying cooperation structures, since they correspond to individual cooperation objectives. Secondly, more importantly, CDs make holon cooperation transparent, i.e. designers can develop holons without knowing the exact cooperation mechanisms. Consequently, it becomes possible to reconfigure holarchies dynamically, and to improve HMSs incrementally by adding new holons or modifying existing ones. In the case of the example of the robot, vision sensor and conveyor, the robot holon can be developed through considerations only of its intrinsic functions, for example workpiece manipulation, since the cooperation mechanisms among the holons are encapsulated in CD-X and CD-V_1, and the robot holon can be incorporated into various kinds of cooperation without changing other parts provided that relevant CDs support those cooperations.

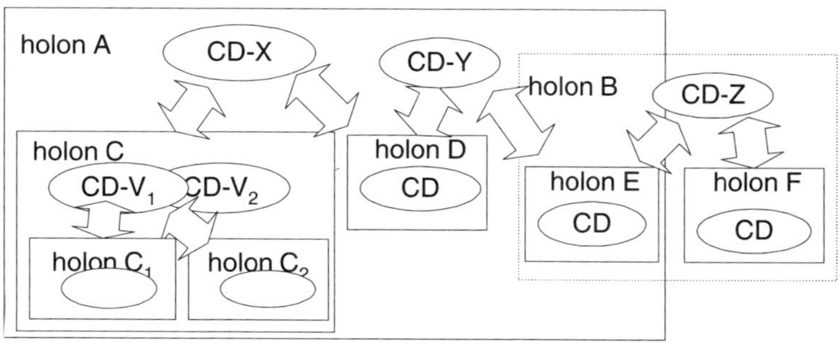

Fig. 10.2. Cooperation domain and holarchy

The introduction of the CD achieves *cooperation-transparent* holons in the following respects, and enables incremental development of HMSs and the dynamic reconfiguration of holarchies.

- *Cooperation partners.* A CD encapsulates holons that are involved in cooperation, i.e. holons can communicate with all holons connected to the CD without knowledge of the individual holons. Therefore HMS designers can develop holons independently of the actual cooperation partners; at the same time, cooperation partners can be changed dynamically during the operation by connecting and/or disconnecting holons to and/or from the CD.
- *Cooperation protocols.* A CD also encapsulates its cooperation protocol. Therefore designers can develop holons independently of the actual cooperation protocols. Holons can use different types of cooperation protocols that are adequate for individual applications (objectives), by changing the CDs that the holon is connected to or that the holon owns.
- *Implementation platforms.* A CD encapsulates computation and communication platforms, and therefore HMS designers can develop holons independently of the platform. For example, designers can reuse holons that utilize contract-net protocols on a simple broadcast messaging platform to create a system based on publish subscribe or other types of communication platforms, by changing CDs (this feature is actually a part of the transparent cooperation protocols).

10.3 Strategies to Support Development and Implementation of HMSs

Development and implementation tools for HMSs must address the following issues, if they are to support in an efficient way the development processes of HMSs that consist of independently developed holons and if centralized management mechanisms as described above are to be avoided.

- *Cooperation-transparent holons.* HMS development and implementation supports must provide facilities that make holons independent of cooperation mechanisms, i.e. cooperation partners and protocols. Individual holons cannot determine how to cooperate with other holons in advance, because a HMS is developed incrementally by adding holons when they become necessary and also because holons dynamically combine with each other to adapt to varying conditions. Incremental development, a reconfigurable holarchy and separate development and maintenance of individual holons can be achieved only through such cooperation-transparent holons.
- *Understanding of holon behavior.* Various tasks coexist simultaneously in a HMS, and the roles of holons are assigned dynamically according to these tasks and the conditions of individual holons. Therefore the information and control flows among holons are complicated and vary with time, and they make HMS development, maintenance and operation inefficient and confusing. In order to develop and operate HMSs efficiently and reliably, HMS development and implementation support must provide HMS designers and operators with facilities to capture and

maintain the complicated, dynamically changing cooperation behavior of holons comprehensively.

- *Platform-independent implementation.* Incremental development and dynamically reconfigurable features inevitably imply the integration of manufacturing management and control systems on heterogeneous computer platforms. Therefore HMS development and implementation support must cope with platform-independent manufacturing management and control systems.

In the previous subsection, it was mentioned that CDs in holons are dedicated to providing HMSs with the above features. Namely, CDs encapsulate cooperation mechanisms within them, and thus designers can deveop holons without knowing the exact cooperation partners and protocols and their implementation methods. Also, CDs assigned to individual objectives provide a comprehensive and quick understanding of the roles of holons that are shared by multiple cooperation objectives, e.g. designers can identify dynamically generated cooperation relations by tracing corresponding CDs individually. HMS development and implementation environments enhance the above features of CDs. They provide HMS designers and operators with separate development and maintenance capabilities for individual holons and a more comprehensive understanding of holon behavior.

Figure 10.3 shows the configuration of HMS development and implementation supports. JDPS [10.3] is a distributed object-oriented program execution environment developed on JACK (a Java-based agent-oriented software tool developed by Agent Oriented Software Pty. Ltd.) [10.5]. JDPS serves as an execution environment for holons located on distributed computers connected by a LAN. This enables the development of location- and multiplication-transparent holons through its reliable broadcast messaging capability. Holons can also utilize the intelligent- agent functions supported by JACK. HMS-Shell [10.2] is a set of tools for developing and evaluating HMSs, and consists of a HMS editor, a HMS translator, and a HMS simulator. The HMS editor provides designers with a holon and holarchy definition facility through a graphical user interface. The HMS translator generates holon programs executable on JDPS from the outcomes of the HMS editor, and the HMS simulator provides a simulation facility for evaluating the behavior and performance of HMSs. The main feature of HMS-Shell is its CD implementation structure, which increases the cooperation transparency of holons. Although HMS-Shell generates programs that are executable only on JDPS in the current implementation, it is also possible to generate programs that are executable on other platforms by replacing translators.

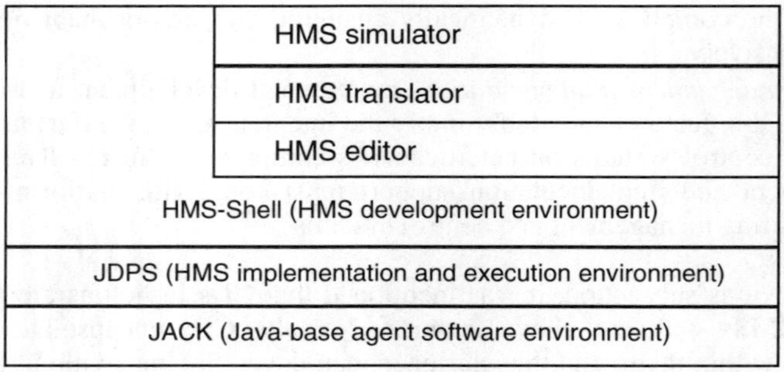

Fig. 10.3. JDPS and HMS-Shell

The strategies utilized by HMS-Shell and JDPS to enhance the capabilities and features of CDs are summarized below:

- *Cooperation Gate and Cooperation Component of CDs*. HMS-Shell further encapsulates the functions of a CD into a set of *Cooperation Gates* (CGs) and *Cooperation Components* (CCs). This makes a CD more adaptive to changes in cooperation partners and protocols. CGs and CCs make holons independent of their cooperation mechanisms with other holons as follows. A CG is a mechanism that connects holons to a CD. Individual holons come to be partners of the cooperation corresponding to the CD by being connected to the CD through CGs; messages sent to the CD are delivered to all holons connected to the CD through CGs. Therefore a holarchy can be reconfigured dynamically by connecting or disconnecting CGs to or from the CD. A CC encapsulates cooperation protocols, and makes holons independent of cooperation protocols. Namely, holons can be developed without knowing the exact cooperation protocols, which differ between different implementations. It is also possible to modify cooperation algorithms easily, by replacing the CC attached to the CD.
- *Visualization of holarchy and behavior of holons*. HMS-Shell improves the HMS behavior-capturing processes. The HMS editor enables graphical editing of holons and the holarchy, to provide designers with a correct and quick understanding of the roles of individual holons and complicated holarchies, in an environment where a lot of holons, and information and the control flows among them, coexist according to the variety of objectives that need to be achieved. The HMS simulator visualizes the dynamic cooperative behavior and performance of holons. Usually, simulation systems and actual systems are developed as different systems. Therefore, in many cases, simulator development is considered as an additional process in system development, and resources for developing simulators are reduced as much as possible. Even if simulators are developed, it is very rare to continuously modify the simulator in accordance with the progress of the design. The *self-simulation* facility of the HMS simulator addresses this issue, i.e. HMS implementations can be used both for simulations and for actual operations by simply changing their running modes; this enables a reduction of the extra effort required to develop simulators.
- *Broadcast messaging*. JDPS adopts *broadcast messaging* as the primary communication mechanism among JDPS objects (holons) allocated over distributed com-

puters connected by a LAN, to achieve their complete location and multiplication transparency. Because broadcast messages arrive at all objects, the objects can communicate with one another without knowing their locations and the existence of replicated objects. Location and multiplication transparency of holons implemented as JDPS objects enables a remarkable improvement in the dynamic reconfigurablility of HMSs. Holons can be added, migrated and replicated at any computer without changing any part of the system, including the communication mechanisms. Broadcast messaging also simplifies the implementation of communication in dynamically reconfigured systems, e.g. the sending of a single message is enough to deliver it to multiple holons, and even mutually unknown holons can communicate with each other.

- *Platform-independent holons based on Java implementation.* JDPS adopts JACK as its implementation base, to make itself platform-independent. Because JACK is Java-based, JDPS objects can be executed on various platforms, and objects implemented on different platforms can communicate with one another correctly. To connect JDPS programs with *legacy programs*, communication mechanisms between JDPS objects and C/C++ programs have also been developed. In this environment, JDPS and legacy programs can send messages to both JDPS and legacy programs without identifying which platform the destination holons are implemented on. HMS-Shell also takes account of the requirements for platform-independent holons. Holon programs executable on a desired platform can be generated from the outcome of the HMS editor, by replacing the HMS translator.

With the above strategies, HMS-Shell and JDPS enable rapid and correct HMS design and implementation. Designers can develop HMSs while capturing their complicated and varying structure adequately, and can allocate functions over distributed computers connected by a LAN with the minimum of constraints. Once developed, HMSs can be improved continuously with growing requirements and technologies. Holons can be added or modified with the minimum of changes to existing systems to satisfy new requirements and/or to introduce new technologies. Addition of computers to execute a growing volume of manufacturing requirements is a purely local operation. Finally, the cooperation-transparent features of holons enable dynamic reconfiguration of holarchies, i.e. holons can cooperate with even mutually unknown holons to achieve dynamically generated objectives. Details of JDPS and HMS-Shell are given in the following sections.

10.4 HMS-Shell

HMS-Shell is a set of tools for designing and maintaining CD-based HMSs. This section describes an implementation structure for CDs that enables cooperation transparent to holons, and the features of individual tools in HMS-Shell.

Figure 10.4 shows the relations between the tools in HMS-Shell and the HMS development processes. In the figure, rectangles and ovals represent tools and the outcomes of tools, respectively. Designers first define the structure of individual holons and holarchies and their cooperation procedures by using the HMS editor through a graphical user interface. The products of the HMS editor are the configuration structure of the holons/holarchies and the cooperation procedures of the holons, and they can be stored in *holon bases* and *Cooperation Component ba*ses, respectively, to be reused in other systems. Therefore the HMS editor provides designers with functions to retrieve stored results from these bases. The HMS translator generates *skeletons* of HMS programs from the outcomes of the editor. Here, a skeleton program is one that consists only of configurations of holons and holarchies, and the cooperation procedures used in the holons, i.e. without the exact application procedures of the holons. Complete source programs are developed by adding application procedures to skeletons by using ordinary program editors, and are compiled to executable modules by ordinary compilers. The current implementation of the HMS translator generates skeletons that are compiled to modules executable only on JDPS.

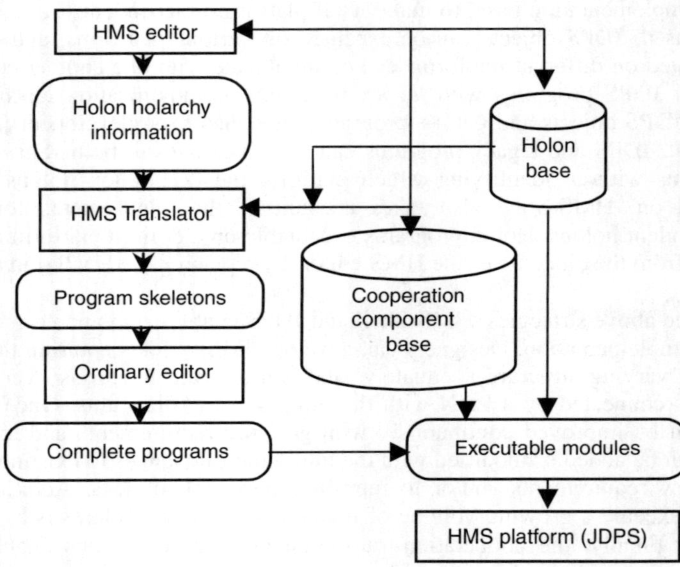

Fig. 10.4. Relations between HMS development processes and HMS-Shell

Elaborate examples of source codes generated by HMS-Shell are shown in Figs. 10.5 and 10.6; they are shown in pseudo-code for clarity.

In this example, the shop holon accomplishes TASK1 through finding a suitable cell holon, which executes the task by using the contract-net protocol. Skeletons of the shop and cell holons generated by the shell create CDs, connect relevant holons to them (i.e. create a holarchy) and implement the contract-net protocol in the CDs (a CD consists of CGs and CCs, which are divided into three parts CC/Cs, CC/Rs and CC/Ss, and the contract-net protocol is distributed over them as described in the following subsection). The intrinsic functions of the shop and cell holons, developed by ordinary programming editors, are attached to the skeletons, as the "shop main" and "select_resource" methods of the shop holons, and the "bid" and "handler" methods of the cell holons. The contract-net protocol, itself implemented on the CC, is retrieved from the CC base. Before we explain individual features of these tools, the approaches of HMS-Shell to implementing and enhancing CDs are described in Sect. 10.4.1.

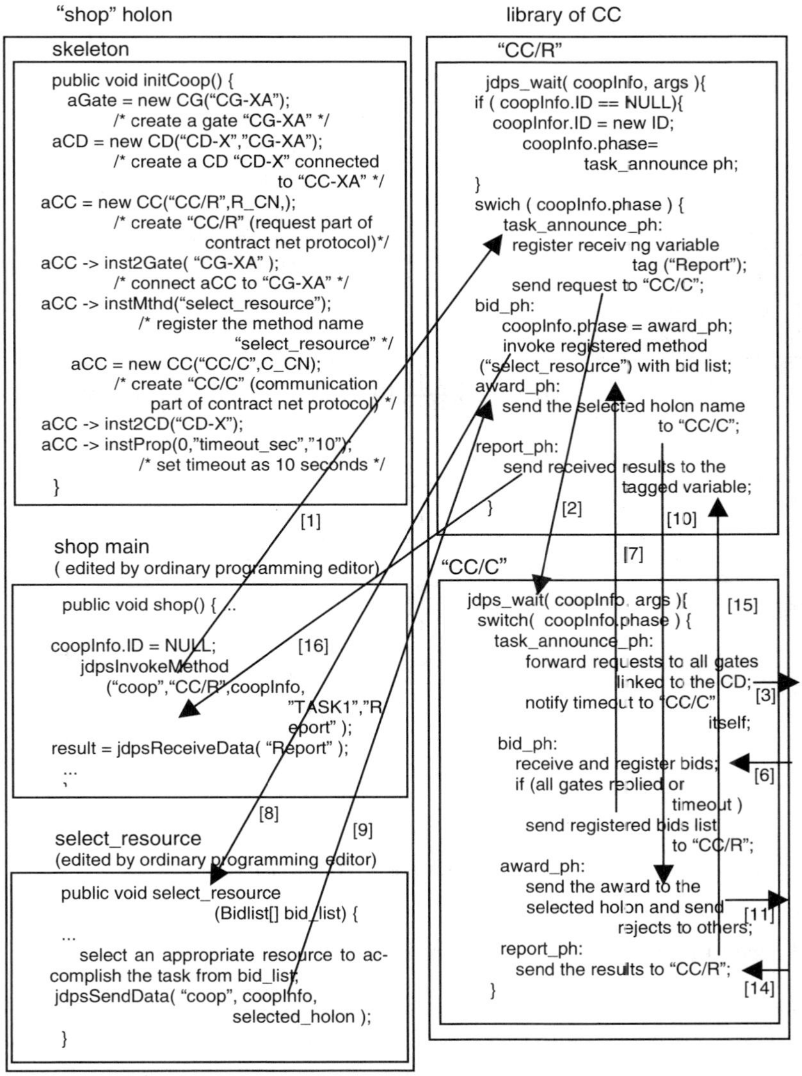

Fig. 10.5. Codes generated by HMS-Shell (shop holon)

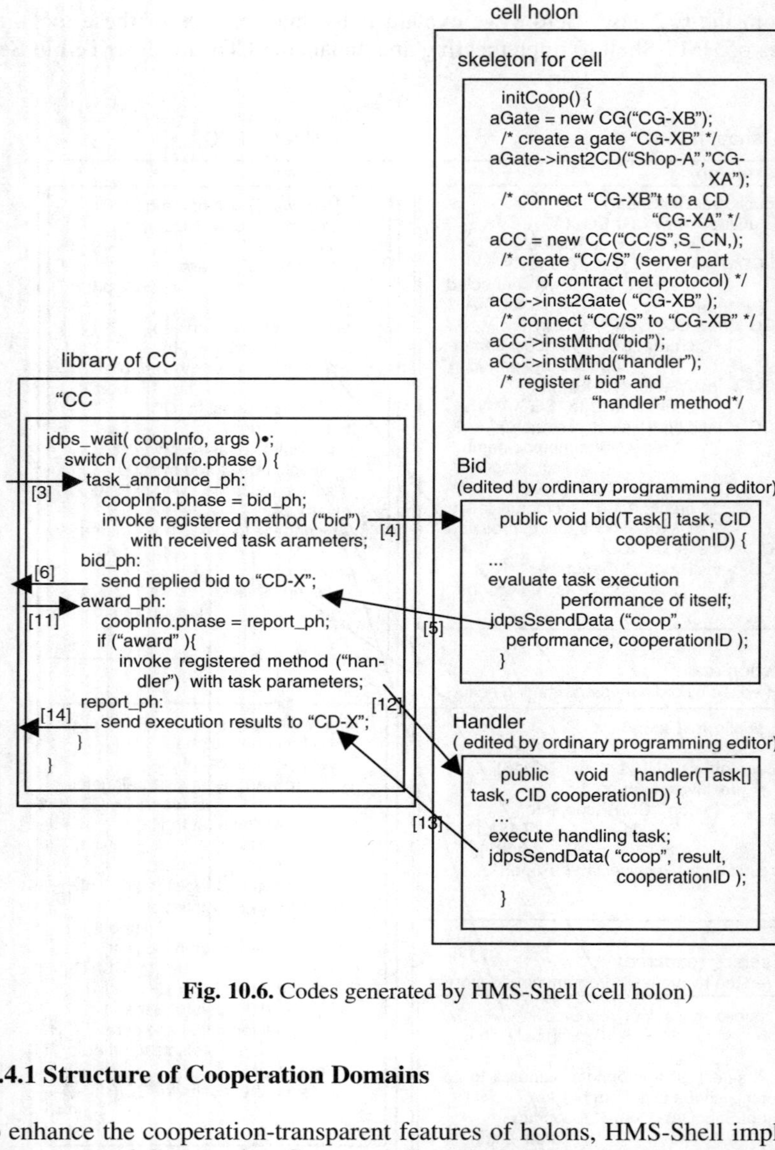

Fig. 10.6. Codes generated by HMS-Shell (cell holon)

10.4.1 Structure of Cooperation Domains

To enhance the cooperation-transparent features of holons, HMS-Shell implements a CD as a set of Cooperation Gates and Cooperation Components, as shown in Fig. 10.7. CGs and CCs encapsulate cooperation partners and protocols, respectively. A CG is a part that connects a holon to the CD, i.e. a holarchy is configured by defining CGs corresponding to individual holons involved in the holarchy (including the owner holon of the CD) and connecting the holons to the CD through these CGs. In the figure, cooperation domain CD-X, owned by holon A, is connected to holon A through cooperation gate CG-XA. CD-X is also connected to holons B and C through CG-XB and CG-XC, respectively, to configure a holarchy that consists of the parent holon A and its child holons B and C.

The behavior of a holon, including cooperation with other holons, are defined as methods of the holon, and cooperation among holons is implemented by message exchanges between methods of holons. Here messages are sent not directly to the destination methods but through CGs. In Fig. 10.7, method-a_1 of holon A, method-b_2 of holon B and method-c_1 of holon C exchange messages with each other by sending and receiving messages to/from CD-X through CG-XA, CG-XB and CG-XC, respectively. By avoiding designating actual holon names as message destinations in this way, individual holons become independent of their cooperation partners, i.e. methods of holons can be designed without knowing the exact names of the cooperation partner holons. Also, cooperation partners can be changed during program execution without changing the methods themselves, by changing the method to CG and CG to CD connections, and by defining new CGs if necessary.

Another important feature of CGs connected to a CD is that messages sent to a CG arrive at all methods connected to the CD through CGs, and by this feature designers can avoid many mistakes caused by complexity. For example, messages inviting partner holons to reconfigure a new holarchy that achieves a given objective can be delivered to relevant holons by the sending of a single message to an appropriate CD. Usually, CDs used for recruiting cooperation partners in holarchy reconfiguration processes are prepared in advance by gathering potential partner holons, and these CDs exist permanently.

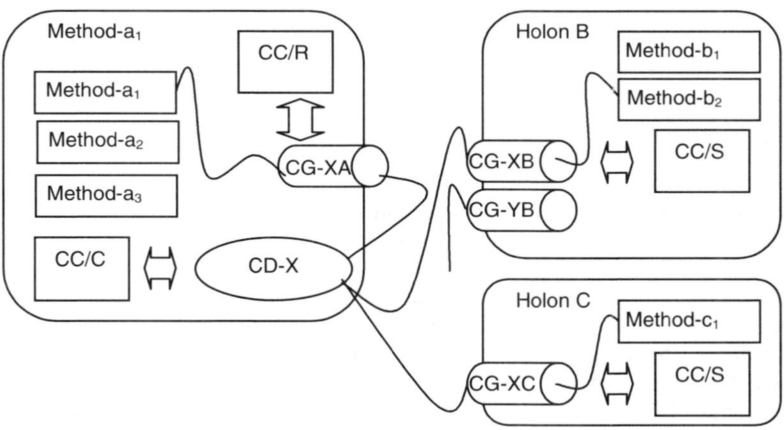

Fig. 10.7. Implementation of a CD

The functions of a CC are divided among three parts, the *CC/R* (requester part), *CC/C* (communicator part) and *CC/S* (server part), within the CD. These parts share roles to hide the details of cooperation protocols, and individual holons become independent of cooperation protocols. Namely, holons can be developed free from cooperation protocols, and they can utilize different protocols properly by changing CCs. Cooperation protocol encapsulation in CCs also enables easy reuse of protocols in many applications. In the following, a *client* means a holon that requests contributions from other holons, and a *cohort* means a holon that is involved in the cooperation and contributes to satisfy the requests of client holons. The roles of the individual parts are as follows.

- *CC/R (requester part).* A CC/R implemented in a CG encapsulates the cooperation protocol used by a client holon. It receives messages from both the client and the CD, to which the CG is connected. Methods in the client holon initiate the cooperation by sending messages to the CC/R, and then the CC/R constructs the necessary messages for initiating the cooperation and transfers them to the CD, to be sent to the cohort holons connected to it, according to the cooperation protocol. When the CC/R receives messages from the cohort holons through the CD, it translates them to forms readable by the client, and invokes methods of the client holon defined by the cooperation protocol to return the results of the contributions of the cohorts to the client. Because the cooperation protocol is hidden from the client as described above, the client holons can initiate cooperation without knowing the detailed cooperation protocols.
- *CC/C (communicator part).* A CC/C simply delivers messages that are received from holons connected to the CD on which the CC/C is implemented to all holons connected to the CD, through CGs.
- *CC/S (server part).* A CC/S encapsulates the cooperation protocol used by a cohort holon. The CC/S also receives messages from both a holon and a CD in the same way as a CC/R. It receives cooperation requests from the client holon through the CD, then translates them to forms readable by the cohort, and invokes the methods of the cohort holons defined by the cooperation protocol to transfer requests to cohort holons connected to the CD through the CG. When the CC/S receives messages from a cohort holon, it transfers them to the CD to be passed to the client holon through the CG. Cohort holons can participate in the cooperation without knowing the detailed cooperation protocols, because they are hidden in the CC/S.

The cooperation among shop and cell holons shown in Figs. 10.5 and 10.6 explains the behavior of CGs, CC/Cs, CC/Rs and CC/Ss. As described before, the shop accomplishes TASK1 by using an appropriate cell existing in the system. To do this, first "initCoop" method of the shop creates a gate "CG-XA" and a CD "CD-X", and connects itself to CD-X through CG-XA. Also the method informs the CC/R of the method name "select_resource" (the shop selects the most preferable cell by this method). Then the shop requests to accomplish TASK1 by sending a message to the CC/R through CG-XA (line [1]), and waits for the results from the CC/R (line [16]). A tag "Report" that defines the variable where the results of TASK1 are stored is designated as an argument of "jdpsInvokeMethod". The message requesting execution of TASK1 is transferred to cell holons that are connected to CD-X, i.e. the CC/R transfers it to the CC/C (line [2]), and the CC/C transfers it via the CC/S (line [3]) to the cell holons (line [4]). Here, the CC/S is informed of the method name "bid" from the "initCoop" method of the cell in advance. Each cell holon evaluates its performance in accomplishing TASK1, and returns the result to the CC/S (line [5]), so that the CC/S transfers the results to the CC/C (line [6]), the CC/R (line [7]) and, finally, "select_resource" method of the shop holon (line [8]). The algorithm to select an appropriate cell from holons that return their performance is implemented in the "select_resource" method, and it informs the CC/R of the selection results (line [9]).

Then the CC/R transfers the message via the CC/C (line [10]) to the CC/S (line [11]), to invoke the "handler" method of the selected cell holon (line [12]). The CC/S also sends reject messages when its corresponding cell holon is not selected. TASK1 execution results are returned to the variable "result" of the shop from the cell through the CC/S (line [13]), the CC/C (line [14]), the CC/R (line [15]) and, finally, the shop holon (line [16]).

As shown by these source codes, the programs of the shop and cell holons are independent of the cooperation partners and protocols. The shop and cell holons can be programmed without knowing about their exact cooperation partners, because they communicate through CGs. They can be programmed also without knowing the contract-net protocol mechanisms, because the mechanisms are encapsulated in CCs. The shop and cell holons need only to have the methods "select_resource" and "bid", respectively, to utilize the contract-net protocol. The shop holon uses the method "select_resource" to select the appropriate cell that accomplishes TASK1, and cell holons use the method "bid" to evaluate their performance in executing given tasks. The two methods are not specific to the contract-net protocol, i.e. they are application mechanisms to evaluate resource performance, and generic mechanisms that are necessary for many kinds of client server-type cooperation.

The cooperation mechanism encapsulation achieved by CDs that consist of CGs and CCs improves the dynamically reconfigurable features of the holarchy. Cooperation groups of holons can be defined dynamically by defining a CD in a suitable holon, and adding relevant holons to it. In the process of adding holons to the CD, it is only necessary to define corresponding CGs. The cooperation protocols used in CDs can also be defined by attaching appropriate CCs to relevant CDs and CGs (but the current version of HMS-Shell does not support dynamic definition of CCs). Because the individual methods of holons are independent of cooperation protocols, there is no need to modify any holons when defining cooperation protocols. CDs also make the cooperation mechanisms of holons independent of communication platforms.

Implementations of cooperation mechanisms based on CDs and without CDs are compared (Fig. 10.8a uses a CD, and Fig. 10.8b does not). Here, the shop and cell holons behave in the same way as in Figs. 10.5 and 10.6, which are based on the contract-net protocol, i.e. the shop holon sends the task features as a task announcement to cell holons 1, 2 and 3. Then the cell holons evaluate their performance while comparing their abilities with the received task features, and return their performance as bids to the shop. Finally, the "select_resource" method of the shop holon sends "award" to cell holon 1 and "reject" to cell holons 2 and 3, and waits for the task execution results from cell holon 1.

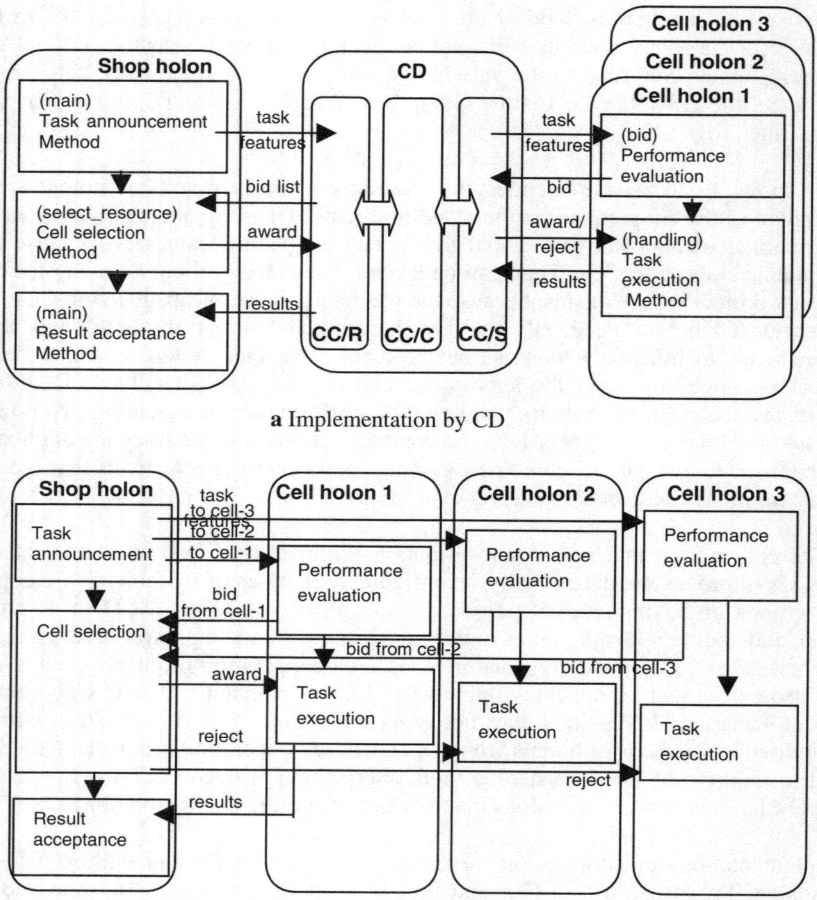

a Implementation by CD

b Implementation without CD

Fig. 10.8. Implementation of contract-net protocol

As shown in Fig. 10.8b, the methods of the shop holon must be aware of existing cell holons, and manipulate a lot of messages to and from many cell holons. Also, the methods of holons must include the exact contract-net protocol, e.g. the "select_resource" method of the shop has to keep track of message arrivals from cell holons. Therefore changes in cooperation protocols require modifications in many parts of individual holons. In the implementation using a CD, individual methods send messages only to the CD, i.e. the task announcement is distributed to every cell holon by the CD, and also the "award" message from the "select_resource" method is automatically reconstructed as "award" and "reject" messages by the CD, to be sent to cell holon 1 and cell holons 2 and 3, respectively. The important thing is that the methods of the holons are not aware of the exact contract-net protocol. For example, the "select_resource" method includes only the process to select one cell holon out of the list of bids received from the CC/R and the process to inform the CC/R about the selected cell. These are pure application processes and independent of communications with other holons.

10.4.2 HMS Editor

The HMS editor provides designers with graphical editing facilities for holons and holarchies. It has two primary windows, the holon and holarchy definition windows, as shown in Figs. 10.9 and 10.10, to design holons and holarchies, respectively. In the *holon definition window*, designers define cooperation parts of an individual holon through placing four kinds of entities (i.e. methods of the holon, CDs included in the holon, CGs and CCs implemented in CDs) and connecting them with lines. Figure 10.9 shows the design for the shop and cell holons illustrated in Figs. 10.5 and 10.6. Here, the cell holon executes tasks assigned by the shop holon. In the shop holon, a method "select_resource", a CD "CD-X", and a CG "CG-XA" are defined. A method "select_resource" is connected to CD-X that uses the CC/C as its cooperation protocol, through the CC/R and CG-XA. In the cell holon, two methods, "bid" and "handler", and a CG "CG-XB" are defined. These methods are connected to CD-X through the CC/S and CG-XB. The cooperation of holons described in Figs. 10.5 and 10.6 can be defined graphically in this way.

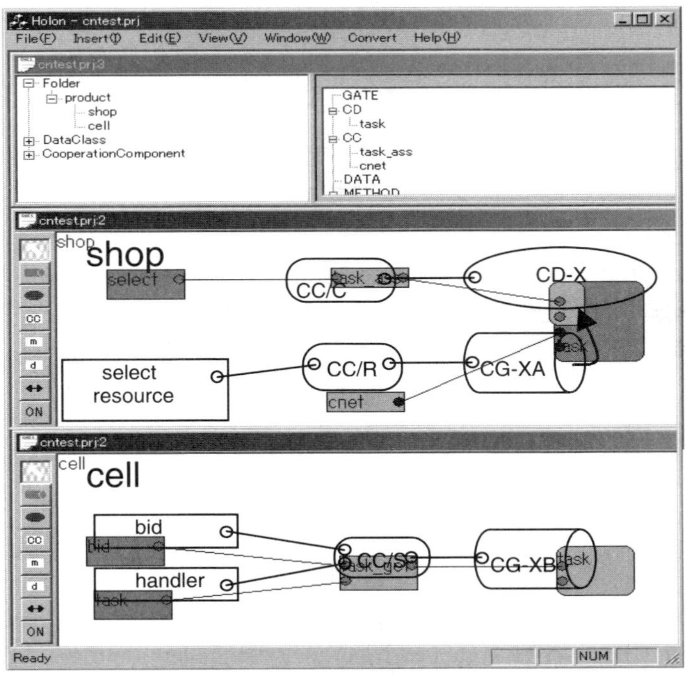

Fig. 10.9. Holon definition window

The *holarchy definition window* defines the initial holarchies configured by the holons. Designers can define a holarchy easily by allocating holons to suitable layers and connecting CDs and CGs. Figures 10.10 shows examples. In Fig. 10.10a, the cell holons A, B and C are allocated to the child holon layer of shop A, and the robot and sensor holons are allocated to child holon layers of individual cell holons. When a robot is a child holon of both shop A and cell A, the level where the robot is located occupies two layers, i.e. the child holon layers of shop A and cell A, as shown in Fig. 10.10b.

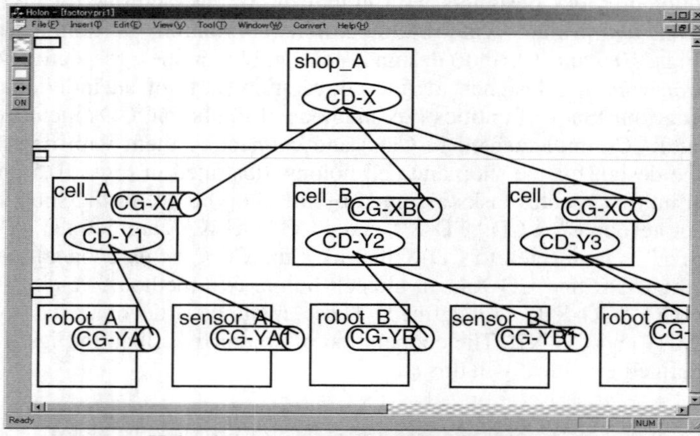

a Holarchy consists of single-layer holons

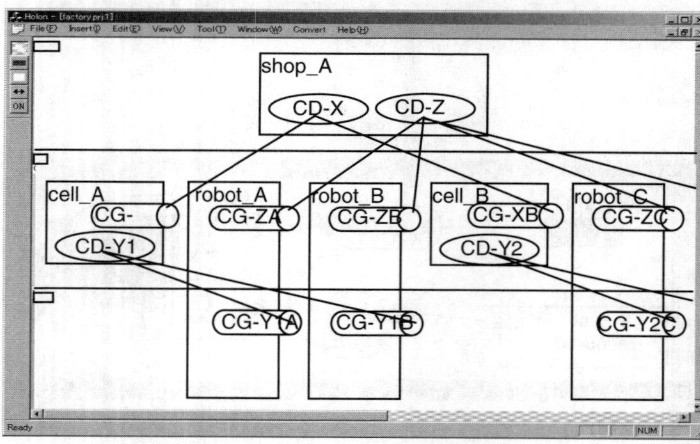

b Holarchy consists of multi-layer holons

Fig. 10.10. Holarchy definition windows

The HMS translator translates the outcomes of the HMS editor to skeletons of holon programs that can be compiled by ordinary compilers. Intrinsic functions of holons, which are independent of cooperation with other holons, are added to them by ordinary program editors before the compilation. Initial holarchies can be changed dynamically while the HMS is operating, by imbedding codes, which generate and handle CDs and CGs, into the programs of holons. The following are the functions supported by JDPS to handle holarchies in program codes:

- define CD/delete CD,
- define CG/delete CG,
- connect/disconnect CG to/from CD,
- connect/disconnect CG to/from method.

10.4.3 HMS Simulator

The HMS simulator is a distributed-clock discrete event simulator for supporting the development, operation and maintenance of HMSs. More efficient and safer behavior of manufacturing systems is required, along with an increase in the social responsibility of manufacturers. But the size and complexity of manufacturing systems have grown too large to allow one to capture their behavior correctly and quickly from design results. In many cases it is also difficult to evaluate the behavior of systems by running actual systems. The effects of experiments and the influence of interruptions to operations are becoming more and more serious, as manufacturing systems expand their activities. For these reasons, the importance of simulators is increasing rapidly. The behavioral correctness of manufacturing systems, their performance under various conditions, the effects of faults, the effects of improvements, etc. can be evaluated only through simulations. The role of simulators is especially critical in HMSs, because their behavior is essentially nondeterministic. However, simulator development requires tremendous effort, and in many cases only simple simulations are executed. Even when a simulator is developed, it is very rare to continuously improve it along with design changes.

The HMS simulator copes with these situations by means of a *self-simulation* capability. Self-simulators can be used for both actual operations and simulations by simply switching their running modes. Consequently, the development processes of simulators are completely included in HMS development processes themselves, i.e. designers can construct simulators and modify them continuously along with system improvements as pure design processes, not as additional processes. A HMS has a significant advantage in implementing self-simulation capabilities. Because every behavior of a holon affects the outside world and/or other holons through messages that the holon sends out, a self-simulation capability of a HMS can be implemented by a simple mechanism that switches the paths of messages according to the running mode. Also, the message-switching mechanism can be easily implemented in each holon by extending the holon class to a *sim-holon class*. Holons in a sim-holon class have two kinds of message send/receive functions: one is *simple send/receive*, and the other is *sim-send/receive*, which has abilities to handle the time at which an event occurs, etc., as described below. When a holon is invoked, it calls send/receive functions when it is in the normal mode and sim-send/receive functions in the simulation mode.

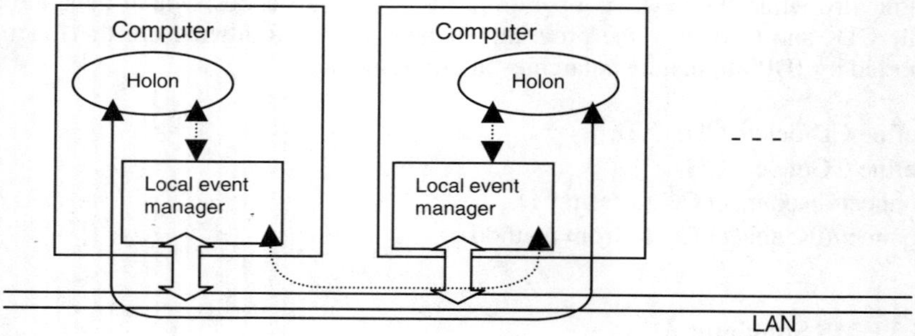

Fig. 10.11. HMS simulator

Figure 10.11 explains the message-switching mechanism. In the normal mode, holons exchange messages directly with other holons, as shown by solid lines in the figure, but in the simulation mode, messages are exchanged through local event managers, as shown by dashed lines. In the simulation mode, a message sent by a holon is considered as a simulation event, and a local event manager on the computer where the message-sending holon exists attaches the occurrence time of the event, based on the local simulation clock of the computer, to the message. Then, a local event manager at the computer to which the message receiver holons are allocated checks the occurrence time of the received event to put it into its event queue in chronological order, and it picks the first message from its event queue to deliver it to receiver holons, while updating the simulation clock of the computer to the time attached to the message. Because a local event manager is responsible for keeping the event occurrence order consistent only within a computer, a mechanism to keep the event occurrence order globally consistent is necessary. In the current implementation of HMS-Shell, a global clock manager takes care of this responsibility, but in this implementation, the effect of parallel processing among distributed computers is limited by the performance of the global clock manager. To simulate the behavior of a HMS efficiently, when the HMS may be a manufacturing system that expands continuously by integrating remote factories, pure distributed-clock simulation mechanisms are necessary [10.7].

10.5 JDPS

JDPS provides holons with complete location and multiplication transparency and simple communication mechanisms. To implement these features, JDPS adopts *asynchronous reliable broadcast* messaging as the primary communication method among holons. Basically, broadcast messaging has substantial advantages for achieving location and multiplication transparency of holons. Because broadcast messages arrive at every computer connected to a network, holons can send and/or receive messages independently of their location and of their multiplication degree (the existence of copies). However, extensive use of broadcast messaging has a serious problem, especially in applications that require reliable behavior. In broadcast communication, exchanging acknowledgments between senders and receivers to confirm the correct receiving of messages is inefficient and sometimes impossible, because the sender must wait for acknowledgments from many receivers and, in some cases, even from unknown receivers. To overcome this problem, JDPS provides a reliable broadcast messaging mechanism on bus-type LANs that does not require acknowledgment exchanges among holons. Thus, holons can use broadcast messages reliably and efficiently. Moreover, it simplifies communications among holons, e.g. holons can send messages to all relevant holons by sending a single message. Asynchronous messaging also simplifies the communication. A holon can send messages independently of the states of the receiver holons, and can execute the succeeding procedures continuously without waiting for responses from those holons. Sender holons receive responses eventually, if necessary.

Figure 10.12 shows the behavior of holons in JDPS, where holons are allocated over distributed computers connected by a bus-type LAN. Because holons are independent of the locations of other holons, the whole HMS can be extended by simply adding computers and loading new holons on to them. Response time can be improved by migrating holons on overloaded computers to less heavily loaded computers, and the reliability of individual holons can be enhanced by allocating their redundant copies any number of times according to the requirements.

In the figure, three copies of holon A are allocated to the three computers, holon B is replicated on computer-1 and computer-3, but holon C does not have a copy, because its roles are not so important. To reduce the computational load on computer-1, holon C is migrated from computer-1 to computer-2. These holons are platform-independent, because JDPS is implemented on the basis of JACK, a Java-based environment, and it is easy to integrate manufacturing systems developed on different platforms.

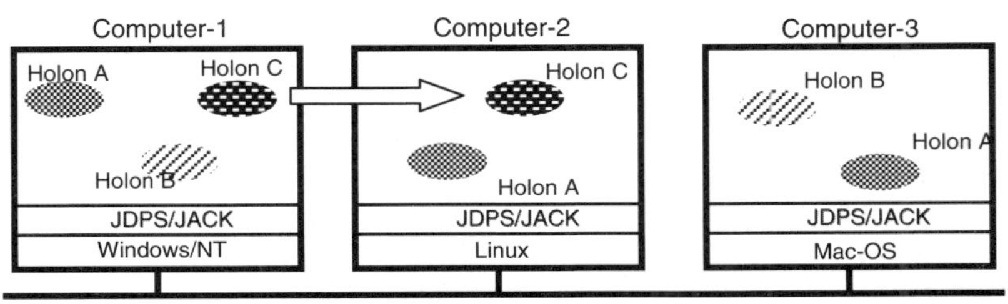

Fig. 10.12. Holon allocation

JDPS supports two types of communication, *method invocation* type and *data substitution* type, and both are asynchronous. The "InvokeMethod" function, and the "SendData" and "ReceiveData" function pair in Fig. 10.13 correspond to the method invocation and data substitution types, respectively (the exact arguments of the messages are omitted). In the figure, method-A1 of holon 1 is invoking method-B1 of holon 2 by calling the "InvokeMethod" function. Method invocation in JDPS is asynchronous, differently from the usual remote procedure call, and therefore method-A1 continues its execution after invoking method-B1 without waiting for a response from it.

Method-B1 of holon 2 sends data to holon 1 by calling the "SendData" function, so that method-A1 of holon 1 receives this data and substitutes it into the variable x by calling the "ReceiveData" function. Message-receiving methods identify their waiting data by comparing TAG names. This data substitution is also asynchronous, i.e. method-B1 continues the succeeding procedures without waiting for method-A1 to receive the data. If the data arrives at holon 1 before method-A1 calls "ReceiveData", the data is stored in the queue of "TAG1" to be read, and if method-A1 calls "ReceiveData" before method-B1 sends the data, method-A1 waits for the arrival of the data. In this way asynchronous messaging simplifies messaging procedures among holons.

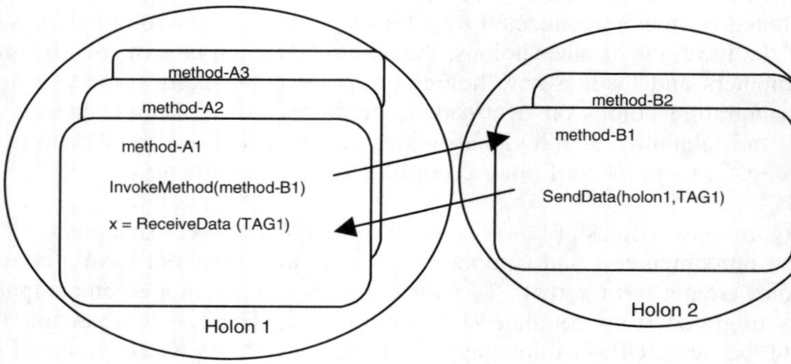

Fig. 10.13. Types of messages among holons

10.5.1 Reliable Broadcast

In JDPS, holons exchange messages without specifying their physical destination addresses, i.e. holons send messages specifying only the names of the destination holons. Figure 10.14 shows the mechanism.

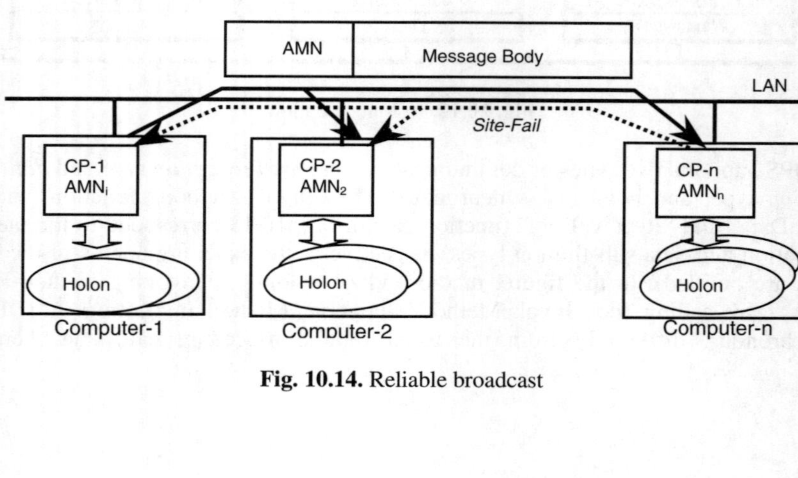

Fig. 10.14. Reliable broadcast

Each computer has its *communication process* (CP), a mechanism which is used to send and receive messages to and from other holons. A message sent from a holon in computer-*j* first arrives at CP-*j*, and CP-*j* broadcasts it to all computers through the LAN. Then CP-*k*, when it receives the message, delivers it to the receiver holons within computer-*k*. If there is no holon in the computer that has a name that coincides with the destination, CP-*k* ignores the message. Here, JDPS provides efficient and reliable broadcast communication that does not require acknowledgment between message senders and receivers, by attaching *Accumulated Message Numbers* (AMNs) to individual messages. AMNs work as follows. Each CP-*j* counts the number of messages that have traveled in the LAN by increasing the counter at every message-sending and message-receiving time, stores the result in its AMN and attaches its AMN to the messages that it sends. Then the message-receiving CP-*k* compares the AMN attached to the message with its own AMN. If either CP-*j* or CP-*k* is malfunctioning, the AMNs do not coincide. When CP-*k* finds differences between two AMNs, it sends a *Site-Fail* message to CP-*j*, and CP-*j* determines that it is malfunctioning when it receives Site-Fail messages from more than a predefined number (R) of CPs. CP-*j* can recover from the malfunction by requesting resending of the message, provided CP-*j* is still connected to the other CPs and it has not yet delivered any succeeding messages to the holons receiving the resent message. If the holons receiving the resent message receive succeeding messages before the resent one, CP-*j* abandons the recovery of these holons. If a holon has copies on other computers it can recover its state, even in this case, by acquiring its latest state from the copy. Even when CP-*j* cannot receive Site-Fail messages and continues wrong operations, the effect does not propagate to other CPs, because the other CPs assume that CP-*j* is malfunctioning and ignore messages from it.

Because the probability that a number of computers will fail simultaneously is very small, the mechanism above achieves highly reliable broadcast messaging without exchanging acknowledgment among computers, if the predefined number R is set appropriately. Namely, the order of magnitude of the probability that the system behaves incorrectly while some holons are failing to send and/or receive messages is p^R, where p is the probability that an individual computer fails to send and/or receive messages. However, if CPs are implemented outside of the LAN processors, they cannot count the total number of the messages that they have sent correctly. Let us assume that CP-*j* and CP-*k* are sending messages and both of their AMNs are n. If there is no message collision, CP-*j* adds 1 to its AMN before sending its message, and also CP-*k* adds 1 to its AMN after confirming that the AMN has been incremented by CP-*j* (it is assumed that CP-*j* sends its message before CP-*k*), and thus the contents of the AMN are correctly increased to $n + 2$. However, when messages collide on the LAN, both CP-*j* and CP-*k* send their messages unsuccessfully while simultaneously adding 1 to their own AMNs, and finally the LAN processors corresponding to CP-*j* and CP-*k* resend the messages successfully with the same AMN contents $n + 1$. Therefore CPs cannot maintain the AMN correctly, i.e. different messages have the same AMN. The reason for the incorrect AMNs is that CPs located outside of LAN processors cannot detect message collision so as to maintain the correct AMN contents, because LAN processors resend messages automatically after a collision. In order to handle these cases, in the current implementation of JDPS, each CP maintains the number of messages sent from individual computers separately, and the AMN is constructed as a set of these numbers.

It is straightforward to achieve location transparency of holons by the above broadcast mechanism. Multiplication transparency is also simple if the behaviors of CPs, when they receive and send messages normally, are ensured to be correct by mechanisms such as parity checking. Namely, a single broadcast message to a holon automatically arrives at all copies that have the same name. Message-receiving holons can accept one message out of the redundant messages sent by copies of holons, by checking the IDs attached to the messages and ignoring redundant ones. Thus holons can be developed without taking account of the existence of copies of themselves and of other holons, and their multiplication degrees can be changed dynamically with the reliability requirements. Individual copies of a holon can attach the same ID to redundant messages, as shown below, provided that copies of holons send exactly the same messages when they receive the same messages in the same order. In the following it is further assumed that copies of holons send the same sequences of messages when they receive messages in the same order from individual holons, i.e. it is not required that copies of holons receive all messages in the same order.

By constituting the message ID as a set {holon name, method name, accumulated time of method invocations, serial message number}, copies of holons can attach the same ID to messages that are sent in the same situation, even under the following conditions, when

- multiple methods act concurrently in a holon,
- redundant holons do not finish multiple method executions in the same order, because of differences in the states of the computers on which they are implemented (copies of holons start executions of multiple methods in the same order, i.e. they start method executions in the message reception order), and
- methods of holons are invoked reentrantly.

The holon name and the method name represent the name of the holon and the method that sends the message, respectively, and the message receivers can identify the message-sending holon and the method from these items. The accumulated time of method invocations (= r) indicates that the method is sending the message during its r-th invocation after system start-up, and message receivers can identify the context of the message-sending method by examining r even when the message-sending holon is invoked reentrantly, because copies of holons receive messages and consequently, are invoked in the same order. The serial message number (= q) is the identifier of individual messages in the same context, and represents that the method has sent (q - 1) messages before sending the corresponding message after the current invocation. The important thing is that individual copies of the holon can determine message IDs by themselves, i.e. without exchanging any information with other copies. If the assumption that the behavior of CPs which receive and send messages normally are maintained to be correct is not satisfied, some kinds of mechanisms to compare the behaviors of redundant holons are necessary, as discussed in Sect. 10.5.3.

In HMS applications, holons frequently broadcast messages without designating their receiving holons in order to deliver messages to unknown holons. In these cases, each broadcast message is physically sent once to the LAN, but when a CP receives it, it makes copies of the message to deliver them to all holons in the computer. Therefore many copies of a message are sent within individual computers. However, in many cases the actual message receivers are only a few of the holons, i.e. many other holons are interrupted in their execution because of the broadcast message, only to ignore it. To reduce overheads caused by broadcast messaging within individual computers, JDPS provides holons with *alias names*, which designate kinds of holons that receive copies of the message. Holons can have multiple alias names, and a message sent to an alias name arrives at only holons that have that same alias name. Therefore the number of copies of the message is reduced to the number of holons that have the corresponding alias name. JDPS further provides mechanisms to manage hierarchies of alias names to restrict message-receiving holons in a systematic way. Individual CPs maintain *connector trees* that represent hierarchical relations among the alias names of holons, to efficiently control message-arriving areas.

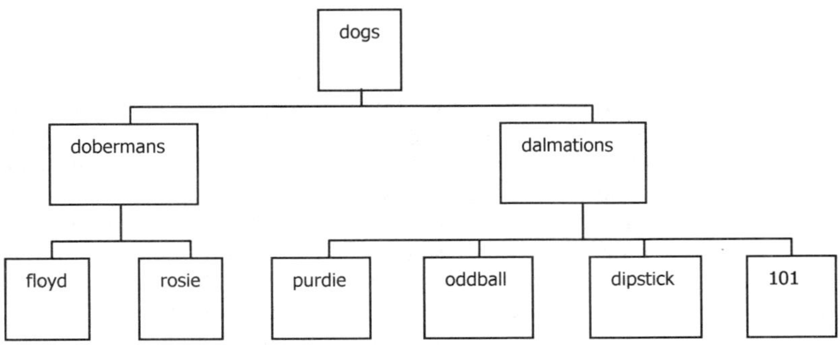

Fig. 10.15. Name hierarchy of holons

Figure 10.15 shows an example. Although messages addressed to the name "floyd" or "rosie" are delivered only to holons that have the name "floyd" or "rosie", respectively, messages addressed to "dobermans" or "dalmations" are delivered to "dobermans", "floyd" and "rosie" or "dalmations", "purdie", "oddball" "dipstick" and "101" respectively. In the same way, messages to "dogs" arrive at all holons in the figure. In this way, the hierarchy of names restricts the area at which messages arrive, and reduces the number of messages within computers and unnecessary interruption of holons' behaviors. A holon can be involved in multiple connector trees simultaneously, and which connector tree can be used in a particular communication is specified by designating connector names as a part of the message address.

Here hierarchies of connector trees are defined only for massage-addressing purposes, and are completely independent of those in holarchies. Connector trees are not centralized mechanisms, i.e. it is not necessary that individual computers have the same connector trees. Individual computers maintain only connector trees that consist of holon names allocated to them. Therefore changes of connector trees in a computer do not imply changes of connector trees in other computers directly. Connector trees and alias names can be defined dynamically in the program codes of holons.

Table 10.1. Performance of JDPS communication

Communication mode	Message length (bytes)	Time required for communication (ms)
Within a computer	32	5.6
	128	4.8
	512	5.4
	1024	5.6
Between computers	32	7.5
	128	7.3
	512	8.3
	1024	9.5

Table 10.1 shows the performance of JDPS broadcast messaging. The time required for message exchanges between holons was measured while varying the message length. The values in the table are averages of 200 communications. Communication mode represents whether the messages are exchanged among holons within the same computer or among holons on different computers. The table shows that communications among holons within a computer are more efficient than those among holons on different computers, but the difference becomes small when the message length is short, i.e. the performance of communication among holons within a computer is less sensitive to message length than for holons on different computers. The time required for message exchange among holons on different computers is reasonably fast compared with the reported performance of several Java-based systems.

10.5.2 Message Arrival Order

A communication system is called *source order consistent* when all holons in a system receive messages from the same holon in the same order, and it is called *total order consistent* when all holons receive all messages in the same order. The communication system shown in Fig. 10.16 is not total order consistent, because in the figure holons A and B receive messages in a different order, i.e. holon A receives messages in the order messages 1, 2, 3 and 4, and holon B receives messages in the order messages 1, 3, 2 and 4. But holons A and B receive messages in the same order from individual holons C and D, i.e. they receive messages from holon C in the order message 1 and 3, and receive messages from holon D in the order message 2 and 4. Therefore it is source order consistent.

Because CPs on individual computers send and receive messages in a first-come first-out order, and the order of messages on the LAN is preserved, it is easy to achieve source order consistency in LAN environments. However, total order consistency is not automatically ensured even in LAN environment. The reason is that a broadcast massage has two different paths to receiver holons; one is to holons on other computers through the LAN, and the other is to holons within the same computer directly without using the LAN. Figure 10.17 shows an example. A message m-1 sent by holon A on computer-X arrives at holon C on computer-Z earlier than a message m-2 sent by holon B on computer-Y, provided that m-1 is put onto the LAN earlier than m-2, but it is possible that holon D on computer–Y receives m-2 earlier than m-1 directly. In order to avoid such cases, CPs must deliver m-2 from holon B to holon D, which are both on computer-Y, after confirming the successful dispatching of m-2 to the LAN. However, it is difficult to implement an efficient confirmation mechanism for message dispatching to the LAN unless functions of CPs are implemented on LAN processors. The only way to detect successful dispatching of messages to the LAN from outside of the LAN processor is to receive acknowledgment from receiver holons, and this sacrifices the efficiency of JDPS broadcast messaging completely.

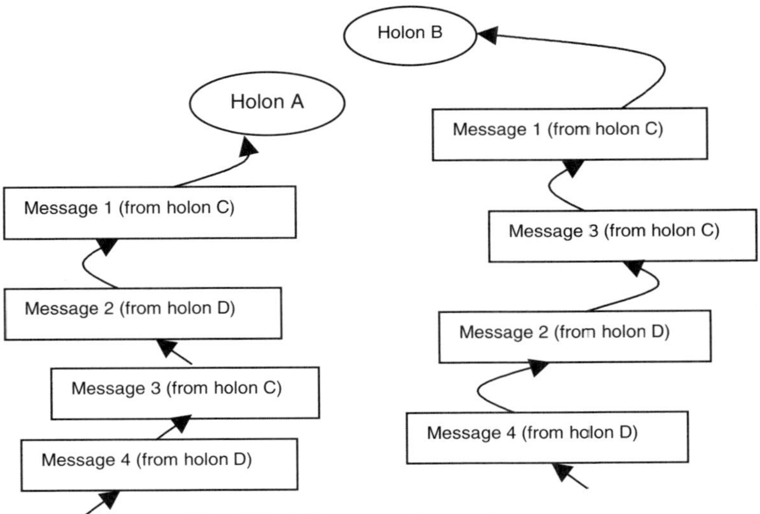

Fig. 10.16. Message order consistency

Sequencers are frequently used to achieve total order consistency of broadcast messages, i.e. individual computers send messages to a sequencer, and the sequencer broadcasts them to receiver computers while attaching sequence numbers to the messages. In this mechanism holons can not only detect missing messages but also receive messages in the same order as other holons, by checking the sequence numbers attached to received messages. However, such mechanisms are centralized, and are causes of various kinds of bottlenecks, e.g. in performance and reliability. To achieve its objectives, a HMS must avoid centralized mechanisms as much as possible, although the LAN itself is already a centralized mechanism. Sequencers also decrease the efficiency of communications, because individual messages must be sent twice, from sender holons to sequencers and from sequencers to receiver holons. Therefore the current implementation of JDPS does not adopt sequencers, and ensures only source order consistency of messages. In future JDPS development, the most simple and preferable implementation of total order consistency is expected to consist of embedding the current mechanisms in LAN processors.

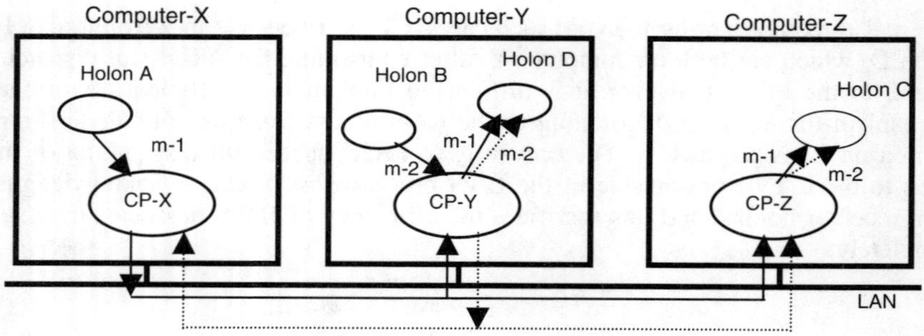

Fig. 10.17. Broadcast message paths to holons

10.5.3 Holon Multiplication

JDPS provides two types of mechanisms for holon multiplication, *active* and *passive* (or *parallel* and *standby*) multiplication, and holons are independent of not only their multiplication degrees but also their multiplication mechanisms, i.e. holons can be developed without one knowing their multiplication degrees and mechanisms. Multiplication degrees and mechanisms can be determined at the loading time of the holons, and they can be changed during system operations. Figure 10.18 explains the behavior of redundant holons. In the figure, three copies of holon A (A1, A2 and A3), two copies of B (B1 and B2), three copies of C (C1, C2 and C3) and two copies of D (D1 and D2) are defined; A1, A2, B1, B2, C1 and D1 are active, and the others are passive copies. Active copies of a holon execute their *processings* and send their output messages to the outside world redundantly in parallel. Therefore a holon can maintain normal operation if at least one copy is working. On the other hand, passive copies are standby holons and do not execute their actual processings. They only store messages that have arrived at them and receive the current states of the corresponding active holons so that they can take over the role of an active holon if the latter fails. Namely, active copies send their current states to the passive copies at specified time, and passive copies store the latest current states as their checkpoint states. When all active holons fail, one of the passive copies starts its execution from the checkpoint state by using its stored messages to take over the roll of the active copies.

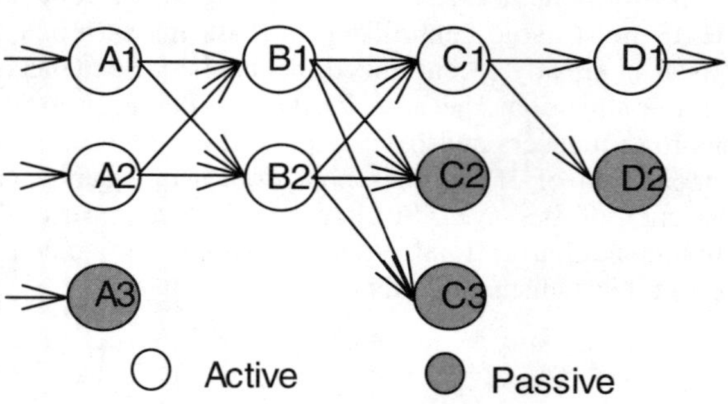

Fig. 10.18. Object replication

To hide the mechanisms required for passive multiplication from individual holons, a passive copy requests active copies to send their current states when its message storage buffer becomes full. Therefore designers can develop holons as non-redundant ones, i.e. without specifying checkpoint places within the programs. Here multiple passive copies request the current state independently and active copies receive multiple requests, but because these request messages have the same ID, active copies can accept only one of them, which reduces the frequency of state information transfers. Namely, each active copy broadcasts its current state only once, and ignores a request if the ID of the request is identical to the ID attached to a request that the copy already received. It is also possible to avoid multiple sending of current states from different active copies; for example, an active copy can skip sending its current state when it receives the current state from another copy.

Under conditions of source-order-consistent broadcast messaging, all passive copies can recover to the correct and the same state from the state that they were in when they received a message by using messages they have saved. It is possible that the stored messages which the individual copies of a holon have saved during a particular interval between checkpoints may differ between copies, because only source order consistency is ensured. However, because a message which is received by some copy at time *t*, is received also by other copies around time *t* (the exact time depends on the computational load of the computer on which the copy is located), all copies save the same messages provided that they have saved several messages that were sent before the latest interval between checkpoints.

Failures of active copies are detected when passive copies cannot receive their checkpoint states within a predefined duration. In the current JDPS implementation, priorities are attached to all copies, and when a failure is detected in an active copy, the passive copy with the highest priority is selected as a new active copy. At this time, a passive copy may become an active one even if active copies are working normally, because of the message delay. But the new and old active copies behave consistently, because all of them are receiving the same messages.

In the above, it is assumed that a holon which sends and receives messages normally behaves correctly. If this assumption is not satisfied, some additional mechanisms are necessary, and usually systems adopt a majority-based redundant architecture, namely, the outcome of a holon is accepted when the majority of its active copies produce the same outcome. However, majority-based methods require mechanisms to keep track of the multiplication degrees of individual holons, and in some case designers must know the exact multiplication degrees of individual holons, i.e. the design is not independent of the multiplication degrees. JDPS adopts the *first CN-come* method to achieve multiplication transparency of active holons. In this mechanism, a message-receiving holon accepts one message from a set of redundant messages when it receives a predefined number of messages with the same ID and the same content. Here the predefined number is called the *confirmation Number (CN)*. It is not necessary that individual holons have the same CN value, and it can be defined dynamically in either message-sender holons or receiver holons.

The advantage of the first CN-come method is that message-receiver holons do not need to know the exact multiplication degrees of other holons, and that they can decide what are the correct messages without waiting for messages from the majority of copies, unlike the case for majority-based methods. When a holon has N active copies, the set of copies of the holon fails when more than $(N - CN + 1)$ copies fail out of N copies. Therefore the order of magnitude of the probability that the holon survives is $\{1 - p^{N-(CN-1)} - q^{CN}\}$, where p is the probability that a single holon fails. The last term q^{CN} is the probability that CN holons generate completely the same wrong outcome. Because p and q are small, the reliability of the HMS can be improved by increasing N and CN.

The first CN-come method has the possibility that different copies of the same holon accept different messages in an environment where only source order consistency of broadcast messages is ensured. For example, let us consider the case where two active copies of holon A (A1 and A2) receive message-X from four copies of holon B (B1, B2, B3 and B4). If total order consistency is ensured, copies A1 and A2 receive copies of message-X from copies of holon B in the same order, but when only source order consistency is ensured, A1 and A2 may receive them in a different order. If $CN = 2$, A1 receives messages from B1, B2, B3 and B4, A2 receives messages from B4, B3, B2 and B1, in those orders, and if the content of message-X from B1 and B2 is "a", and that of the message from B3 and B4 is "b" ("a" and "b" are different), then A1 and A2 receive different contents of message-X, i.e. "a" and "b." However, inconsistent behavior can be neglected for practical purposes when CN is large. This inconsistency can be removed easily and completely when total order consistency of broadcast messages is ensured or when majority-based methods are adopted.

10.5.4 Connection with Legacy Systems

To support communication among holons in JDPS (JDPS holons) and those in legacy systems (legacy holons), JDPS prepares *JDPS-Gateway Managers* (GW Managers) and *JDPS-Gateway holons* (GW holons). Using GW managers and GW holons, JDPS holons and legacy holons can communicate with each other without knowing the platforms on which the message-receiving holons are implemented. Holons on different platforms exchange messages as shown in Fig.10.19, i.e. GW holons are defined corresponding to individual legacy holons, and these GW holons translate JDPS form messages (JDPS messages) to non-JDPS form messages (legacy messages), and vice versa. In the figure, holons X and Y are JDPS holons, and holons A, B, C and D are legacy ones. Therefore GW-A, B, C and D are defined. Legacy holon A sends a legacy message to JDPS holons or legacy holons through the GW holon corresponding to holon A, then the GW holon translates the legacy message to a JDPS one and delivers it to the holons designated by the message address. When the message-receiving holons are actual JDPS holons (not GW holons), they simply receive the delivered message, but when they are GW holons, these GW holons translate the JDPS messages to legacy messages, and transfer them to the final message-receiving holons. GW holons and legacy holons exchange messages through appropriate facilities, e.g. TCP/IP sockets.

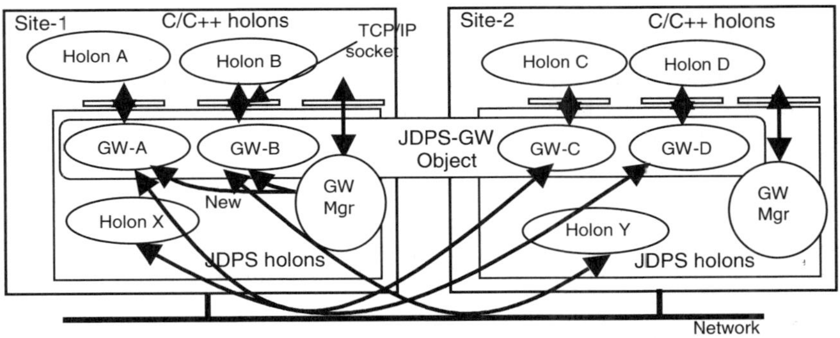

Fig. 10.19. Connection with legacy systems

GW managers on individual computers are responsible for generating GW holons and relevant communication measures between GW holons and legacy holons, at the request of applications. Thus, dynamically generated legacy holons and JDPS holons can exchange messages regardless of the platforms on which the message sender and receiver holons are implemented. In this way, JDPS can also be used to integrate heterogeneous legacy holons. When multiple platforms coexist in a computer, multiple kinds of GW managers are necessary on the computer, corresponding to multiple kinds of platforms.

10.6 Conclusion

In this chapter the implementation structure of the Cooperation Domain (CD) is proposed to achieve the objectives of HMSs; its development and implementation support HMS-Shell and JDPS. A CD is a communication channel through which holons exchange messages, and encapsulate cooperation partners and cooperation protocols, to enable easy reconfiguration and incremental development of manufacturing systems. Using the proposed CD structure, it becomes easy to constitute a holarchy dynamically in order to manufacture given products, and to add and migrate holons with the growing requirements for functionality, reliability and computational performance in manufacturing systems, without stopping the operations of the system. The encapsulation of the cooperation protocol also enables easy reuse of cooperation protocols.

HMS-Shell is a set of tools for developing manufacturing systems based on CDs. Its graphical user interfaces for editing holons and holarchies enable quick and reliable HMS design, together with the reuse capabilities of existing holons and cooperation protocols. Its self-simulation capability (discrete event simulation) enables programs developed for actual manufacturing systems to be used also for simulation by simply switching operation mode. Therefore simulators can be continuously modified to adapt to various changes in design and operation of manufacturing systems, and consequently, more efficient and reliable design and operation can be achieved.

The CD implementation platform is JDPS. The efficient and reliable broadcast messaging mechanism of JDPS enables holons to use broadcast messaging as their primary communication paths, and the complete location- and multiplication-transparencies of holons based on broadcast messaging make CD implementation simple, easy and reliable. IDPS, the predecessor of JDPS, were successfully used for a part of a HMS testbed system that had been developed in Japan for manufacturing induction motors. But, at the same time, it was found that IDPS required relatively heavy resources, i.e., only small numbers of holons and/or methods of holons could be implemented. JDPS overcomes this defect by implementing holon methods as Java threads. It also makes integration of HMSs developed on different platforms easy.

Acknowledgements

We are thankful to James Cole and Rashid Jayousi of the DAKE Group of the Computer Science Department at Keele University for assisting us with corrections and formatting, and to Shingo Ando of Yasukawa Electric Corporation for providing performance data of JDPS.

References

[10.1] Christensen J (1994) Holonic Manufacturing Systems – initial architecture and standards directions, 1st Euro. Conf. on HMS, Hanover, Germany

[10.2] Hasegawa T, Seki T, Tamura S (2000) Environment for developing Holonic Manufacturing System applications – HMS Shell – Proc. 6th Intl. Conf. on Intelligent Autonomous Systems, Venice, Italy

[10.3] Seki T, Hasegawa T, Okataku Y,Tamura S (1991) An operating system for Intellectual Distributed Processing System – an object oriented approach based on broadcast communication -, J. Information Processing, Vol.14, No.4

[10.4] Leeuwen E,.Norrie D (1997) Intelligent manufacturing: holons and holarchies, Manufacturing Engineer, Vol.76, No.2

[10.5] Howden N, Ronquist R, Hodgson A, Lucas A (2001) JACK Intelligent Agents – summary of an agent infrastructure, 5th International Conference on Autonomous Agents, Montreal, Canada

[10.6] Tamura S, Hasegawa T,. Seki T (2001) Cooperation mechanisms of Holonic Manufacturing Systems, Trans. on IEE Japan, Vol.121-C, No.6 (in Japanese)

[10.7] Imasaki N, Tharumarajah A, Tamura S (2002) Distributed self-simulation of Holonic Manufacturing Systems, Knowledge and Technology Integration in Product and Service, Kluwer Academic Publishers

11. Transport Agents – Specification and Development

A. Ritter, A. Braatz, M. Höpf, C. Schaeffer and E. Westkämper

Fraunhofer Institute of Manufacturing Engineering and Automation (IPA)
Stuttgart, Germany
Email: ritter@ipa.fhg.de

Abstract. Transport agents, or holonic automated guided vehicles (H-AGVs), are important sub-systems of agent-based manufacturing systems such as holonic manufacturing systems (HMS). In this chapter concepts for the development of transport agents are presented, including taxonomy, agent models, architectures, realisations and the concept of a prototyping suite for the development of transport agents. Related work began at the start of the HMS project and has continued within several other projects.

11.1 Introduction

Modern technological systems, such as manufacturing systems, consist of many different complex structures. According to Lüth, the elementary units of complex technological systems can be classified as fractals, robot cells, cells, software objects, actors, agents and holons [11.10]. Traditional, centrally controlled manufacturing systems have reached their limits. In terms of flexibility and stability to perturbations, or even programmability and testability, improvements are demanded. In this context, intelligent manufacturing systems (IMS), which are based on intelligent material flow systems, are gaining much more importance because they provide the improvements demanded. For about ten years, paradigms in manufacturing engineering have been changing towards agent-based manufacturing systems, which can also consist of transport agents. Some representatives of those concepts are holonic manufacturing systems (HMS), the fractal factory, and random, bionic, genetic or agile manufacturing systems [11.5, 11.6, 11.7, 11.9, 11.14, 11.24, 11.33, 11.34]. The main advantages of agent technology for manufacturing are simple engineering, robustness and flexibility. The first agent-based industrial applications are already successfully in use [11.13, 11.28, 11.29, 11.32].

The term "holon" is very often referred to in agent theory, and sometimes is equated with agent [11.3]. Several holonic architectures have been presented in the past [11.1, 11.2, 11.11, 11.13, 11.25, 11.30, 11.31]. The term "holon" was originally created by Koestler. In his book The Ghost in the Machine [11.8] Koestler suggested a concept of open hierarchical systems (OHS). "Holon" is a combination of the Greek word *holos*, meaning "whole", and the Greek suffix *on*, meaning "particle" or "part", as in "electron" and "neutron". A holonic system is characterised by autonomy, cooperation, intelligence and self-similarity. Holons form open, dynamic hierarchies, which are called holarchies. Even human beings with holonic behaviour can be part of holarchies. At the end of the 1980s, scientists transferred the idea of holons to manufacturing systems. They expect that agent-based manufacturing systems such as HMS will have a number of convincing advantages in comparison with hierarchical and centrally controlled manufacturing systems, and especially in comparison with CIM technology. Expected benefits of HMS are [11.14]:

- more flexible, more capable, more reliable and less expensive automation systems;
- more modular systems and easier integration into multiple operating environments.

Further benefits resulting from the above properties are flexible extendibility and stability to perturbations.

Holonic manufacturing systems are an integral agent-based approach for manufacturing systems consisting of humans, hardware and software systems, e.g. mechatronic systems used as robots or components for MES or SCADA systems (MES = manufacturing execution systems, SCADA = supervisory control and data acquisition) [11.2, 11.13]. In this context, mobile transport systems play a special role. The HMS consortium [11.14] has identified holonic automated guided vehicles (H-AGVs) as an important component of HMS. This was proved by an internal study performed by the HMS research partners.

In the HMS work-package 7 (Holomobiles), with industrial and scientific partners such as Simon Fraser University (Canada), DaimlerChrysler AG, Fraunhofer IPA, Softing AG (Germany) and VTT (Finland), models, concepts and simulations for H-AGVs, as well as components for wireless communication, have been developed [11.14]. The aim of the Holomobiles work-package was research on H-AGVs and their further development to optimise material flow and productivity in production units.

Up to now, the distribution of agent technology has been prevented not only by a rather conservative point of view in the field of manufacturing, but also by a plenitude of different aspects, such as lack of standardisation for:

- proprietary agent languages or content languages, e.g. Knowledge Query and Manipulation Language;
- interaction protocols, e.g. the contract net protocol (CNP);
- ontologies;
- agent platform; and
- agent models.

Owing to the specification work of FIPA (Foundation for Intelligent Physical Agents), there are now standards, at least for some parts [11.20]. These include FIPA-ACL as an agent (conversation) language, semantic language (SL) or knowledge interchange format (KIF) as a content language, the FIPA agent platform, the FIPA interaction protocols including CNP, interactive CNP, and Dutch and English auctions. However, at present there are no manufacturing-specific interaction protocols. Further approaches do exist, such as modelling languages, e.g. the Unified Modelling Language (UML) and Agent UML (AUML) [11.16, 11.22, 11.23], and even tools, especially open source software tools, such as FIPA-OS, JADE and ZEUS [11.17, 11.19, 11.21].

Although this has been a great success, there are still some questions left open:

- interaction protocols for manufacturing;
- interoperability between multi-agent-systems (MAS) and external systems (conventional manufacturing systems, MES, production planning systems, etc.);
- integration within overall manufacturing systems of humans, i.e. workers, engineers, software developers, operators, etc.

As a consequence, the EFAPS (Engineering Framework for Agent-Controlled Production Systems) project was initiated by the European HMS partners. Furthermore, agent technologies, especially mobile transport agents, have been developed within the AgenTec project [11.15, 11.28]. Research on modern software specifications, e.g. for agent-based systems or holonic transport systems, has been done by the DFG priority programme, especially in the project SPP IOSIP [11.18, 11.35].

11.2 Taxonomy of General Agent Models

In general, an agent is somebody who deals on behalf of someone else. Transferred to technical systems, agents can be understood as autonomous cooperating entities in partially decentralised systems [11.10, 11.26]. In principle, there is a difference between technical agents, such as transport agents, and software agents; see Fig. 11.1. In contrast to technical agents, software agents are pure software entities. Software agents can be split up into desktop agents, Internet agents and intranet agents, according to their tasks [11.4]. On the basis of a definition by Lüth [11.10], a technical agent can be regarded as an autonomous, interactive entity which is directed at process optimisation and stabilisation. A technical agent is provided with intelligence, for instance the ability to learn and the ability to cooperate and coordinate.

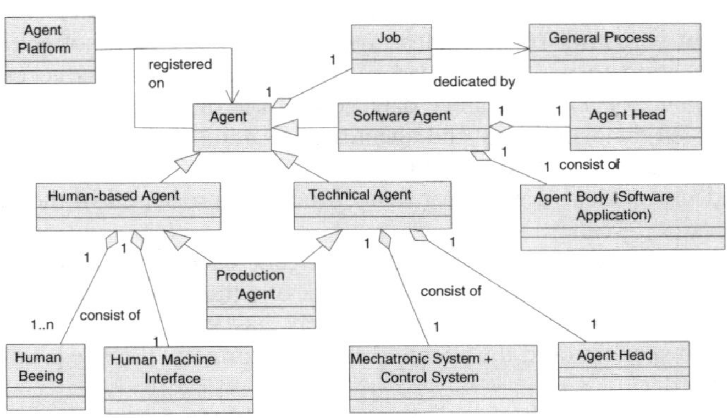

Fig. 11.1. Taxonomy of agents within manufacturing systems

Unlike pure software agents, a technical agent has mechatronic components, including control systems. Examples of technical agents are interactive industrial robots working mutually on one workpiece, and mobile, autonomously cooperating transport robots which negotiate transport orders among themselves. Such mobile robot agents or AGV agents can be called transport agents. A holonic AGV can be regarded as a special case of a transport agent. If a human being is an active part of an agent system, for instance an active negotiating party or contractor (e.g. an assembly contractor), that means the human being plays the role of an agent in the agent platform; in this case the human being, and his/her tools and human–machine interface can be called a human-based agent (see Fig. 11.1.).

In the case of interaction and cooperation of more than one agent, we talk about a multi-agent-system (MAS). Unlike a homogenous MAS, an inhomogeneous MAS contains different types of agents with different functionalities.

An agent can be regarded as a capsule; the agent can be implemented internally and independently from the developer. Concerning the internal realisation and architecture of an agent FIPA specifications do not exist. An agent has to follow the appropriate management, e.g. for defined login and logout procedures, to be applicable on an agent platform, e.g. for defined login and logout. It also has to follow interaction protocols depending on the respective application, e.g. FIPA-CNP, to be able to cooperate with others. The same agent language (FIPA-ACL is proposed), the same content language and the same ontologies are further requirements for interaction.

As the agents communicate among themselves and have to integrate themselves into the agent platform, each agent needs its own communication interface. To fulfil those requirements, the communication interface has to enable communication and cooperation on the basis of FIPA-ACL and the corresponding FIPA interaction protocols.

Software agents and technical agents consist of an agent head and an agent body. The agent head and agent body have to be connected via an interface. For the technical agent, the agent head is the control system, whereas the agent body represents the mechatronic system (the whole technical agent consists of a mechatronic system). The mechatronic system, e.g. an AGV or an industrial robot, has its own control system, for example an industrial robot control system. The body of a software agent is a software application to which the head of the software agent has access. Software agents and technical agents exchange data in the form of events and messages via an interface between the agent body and agent head. Additionally, a technical agent also needs an interface to the technical process for events and messages, as well as for analogue and binary data, e.g. sensor data, which are processed in the agent body's control system.

Each agent has its concrete task within the agent system, which is determined by the corresponding process. For example, the task of a transport agent is determined by material flow. As an autonomous unit, an agent has various alternative courses of action. For each specific task, the agent takes on one or more roles, e.g. as an executor of a transport order.

In contrast to the currently established agent models, a precise model of an agent designed to control a production system will be necessary in order to implement an agent-based control system. In the following, such an agent will be called a production agent (PA) (see Fig. 11.1) [11.28]. A PA is a specification of a human-based or technical agent. A network of production agents is called a PA-MAS. The following features characterise a PA:

- A PA is autonomous, which means internal data and states are encapsulated, and it uses only message-oriented communication (asynchronous events and messages, realisation according to FIPA). Furthermore, a PA assumes a manufacturing-specific task, for which it is provided with a number of alternatives in action.
- A PA is supposed to be persistent, which means it is an integral part of the production system. A PA always consists of hardware and software components. A PA can also be a human being acting as an operator, worker, planner and decision maker.

11.3 Transport Agents – Asimov-Architecture for H-AGVs

During the HMS project an H-AGV was modelled as an Asimov holon [11.25]. The overall concept of a technical MAS with Asimov-holonic characteristics is illustrated in Fig. 11.2. It is an open architecture. The Asimov-holonic control systems consist of the following elements: communicator (COM), negotiator (NEG), cooperator (COP), coordinator (COR), executor (EXE), observer (OBS), organiser (ORG) and interpreter (INT).

As every Asimov H-AGV (AH-AGV) has the same control architecture and the ability to execute the same transport orders, they form a holarchy, even if the AH-AGVs have different characteristics in terms of size, mechanical and electrical set-up, and power supply. In terms of their behaviour and functionality, they are also self-similar.

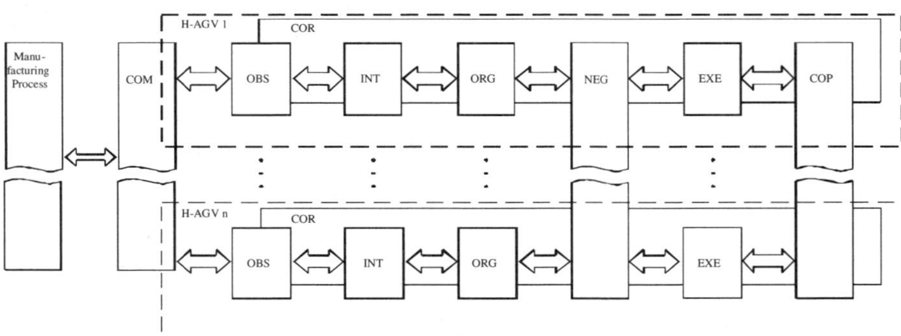

Fig. 11.2. Asimov-holonic control system for AH-AGVs

AH-AGVs are mobile robots. Therefore they have to communicate wirelessly. Distributed communication will avoid the possibility of total breakdown caused by failure of the central communication system. Two classes of interfaces exist between the holarchies of AH-AGVs and the total manufacturing process:

- material flow interfaces (e.g. docking stations between AH-AGVs and conveyor belts) and
- communication interfaces.

Cooperation is made possible by a negotiator (negotiation of orders). Process interactions such as collision detection and avoidance are supported by an observer and executor.

The interpreter and organiser are responsible for developing optimal plans and the selection of the right strategies for executable orders. These are the systems with most autonomous and intelligent characteristics.

The communicator is a communication interface between the manufacturing process and several AH-AGVs. General information, process states, internal states of the AH-AGVs, tasks of the manufacturing process (i.e. transport orders) and task states can be communicated. Technically, the communicator is realised as a wireless network, e.g. wireless Ethernet.

The observer observes all relevant information and states for AH-AGVs, such as general information, process states, internal states (e.g. battery charge and positions of AH-AGVs) and task states.

In contrast to traditional task scheduling, the AH-AGVs are able to derive a demand for orders by observing information and states as mentioned above.

The interpreter is like a kind of filter. The interpreter ignores new transport orders executing old orders, interrupts action for special orders (e.g. interruptions for battery charging) and creates an order list (for all new orders).

The organiser produces executable plans for all executable orders. This is the result of different strategies and an optimisation process. To each executable order and plan, a degree of priority (measure of quality) is given, which can be compared by the negotiator.

Negotiation is the most important part of cooperation. Each AH-AGV receives the same information about process states and orders. The orders are negotiated by each AH-AGV. The ability to execute an order is checked individually for each AH-AGV. Only the optimal order (the order that is optimally executable by a certain AH-AGV) is allowed to be executed. The result of negotiation is one executable order for certain AH-AGVs (this order will be executed), a list of orders that are not executable (e.g. if there are more orders than available AH-AGVs), and a list of AH-AGVs which are not executing an order (e.g. damaged AH-AGVs).

The executor will execute a certain order if the AH-AGV has won the negotiation for this order (a comparison of measures of quality). The executor puts those plans into action which have been produced in the organiser. The execution can be interrupted by the interpreter if special conditions occur (e.g. battery charge too low) and so special orders have to be executed (e.g. moving to a battery-charging station).

Some tasks have to be executed at the same time. For this purpose, a coordinator coordinates tasks (and even orders) and processes. The coordinator is part of the system of communicator, observer, interpreter, organiser, negotiator and executor. It coordinates internal processes of AH-AGVs.

The coordination of external processes is managed by a cooperator. This system prevents and resolves deadlock conflicts, for example in the case of a traffic jam. This process is similar to negotiation, but negotiation is the process of transport order creation and distribution before execution of transport orders and cooperation takes place.

11.4 The Transport Agent as a Special Realisation of a Production Agent

An example of the realisation of different production agents is given in Fig. 11.3. The PA-MAS realised here consists of two different MAS, namely a transport agent MAS consisting of two transport agents, an access agent, a visualisation agent and a database agent supplied by Fraunhofer IPA, and a machine MAS, consisting of two industrial robots with two machine agents and a workpiece agent (Factory Broker™) supplied by Schneider Automation GmbH. The two MAS interoperate by using the same FIPA-ACL-based interaction protocol (see also [11.13, 11.27, 11.29]). In this context, a machine agent is the agent head, and an industrial robot, in form of a mechatronic system together with the control system, is the agent body of a PA. Furthermore, the transport agent is a PA. The other agents can be regarded as software agents, apart from the visualisation agent, which can be regarded as a human-based agent [11.15].

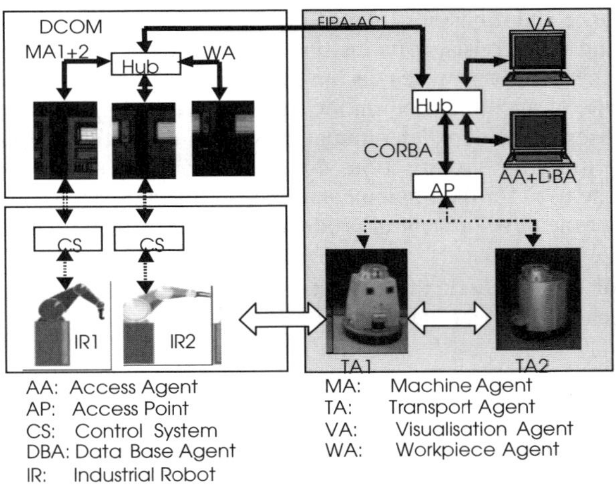

Fig. 11.3. Example of production agents

11.5 Prototyping Suite for Production Agents

The totality of all tools for developing agent control systems for specific production applications may be called a prototyping suite for production agents; see Fig. 11.4. [11.28].

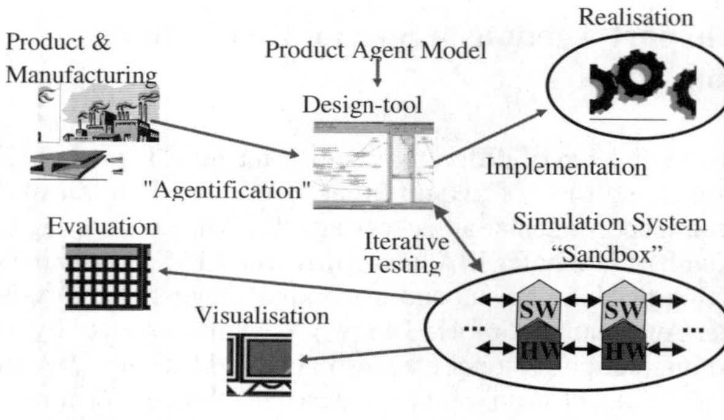

Fig. 11.4. General concept of a prototyping suite

One of the most important and decisive tasks in the development of PA systems, e.g. transport agents, is the design of the overall structure. The basic characteristics, such as performance and robustness, depend mainly on this design.

The prototyping suite consists of a method to agentify the systems and processes of a manufacturing system. It provides design tools for agents and MAS that are able to validate the results of agentification on the basis of detailed agent and MAS models. Furthermore, it uses a "sandbox", a simulation system consisting of agents and agent platforms, using tools for visualisation, documentation and evaluation. Finally, the design tools can be used for the implementation of production agents as control units.

The process for the development of production agents is as follows:

- Application of one method: throughout, namely an object-oriented development method for technical MAS (ODEMA) [11.35].

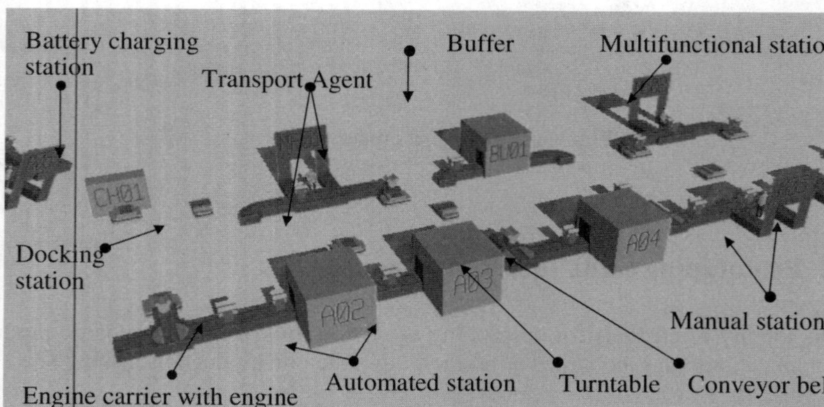

Fig. 11.5. Process and system model for agentification

- Agentification: identification and modelling of technical processes and systems, e.g. industrial robot cells, buffers, control systems and databases, including human beings as agents, see Fig. 11.5. For example, an AGV is identified as a technical agent forming a transport MAS. Furthermore, a multifunctional station or a manual station can be agentified as a (human-based) PA, and also automated stations and buffers can be agentified as PAs. The conventional methods for structuring in manufacturing follow two philosophies:
  - Material flow: line control is mainly determined by the transport system.
  - Work flow: production line control is mainly determined by the manufacturing processes.
- In conventional systems, the control hierarchy does not allow optimisation and adaptation of the ratio of material flow to work flow. Agent control fills this gap.
- Design of communication: the MAS model incorporates a comprehensive communication model, consisting of:
  - an agent platform,
  - FIPA interaction protocols or application-specific extended interaction protocols (see Fig. 11.6),
  - agent language, content languages and ontologies (according to FIPA),
  - the design of or decision about an agent platform (according to FIPA),
  - agent types (software and production agents, e.g. transport agents).

  In Fig. 11.7, an example of an extended agent platform is given; this platform consists of FIPA systems and services such as an agent management system, directory facilitator and message transport system, and additional production-specific agents, e.g. a supervisor agent for control of production agents, access agents for connections which are not compatible with FIPA systems, and database agents for job data and resource data. Further steps of the development process are as follows:
- Design of agent model: a capsule, consisting of the agent head and body and interfaces, taking account of the communication model, and the models and architectures for hardware and software. (For simulation, hardware is modelled and simulated. For implementation, the simulation model is replaced by a mechatronic system, e.g. an industrial robot or AGV).
- Verification with the "sandbox", consisting of a model of the final agent platform and a communication model. The interaction protocols are used in development, in simulation and in the final, implemented production system, which consists of agent heads (unchanged later), and simulated hardware, i.e. agent bodies of PAs, later replaced by real systems. Simulated systems can be substituted step by step by real systems, i.e. a continuous transition through development, simulation and testing of the real system, and finally operation of the system.

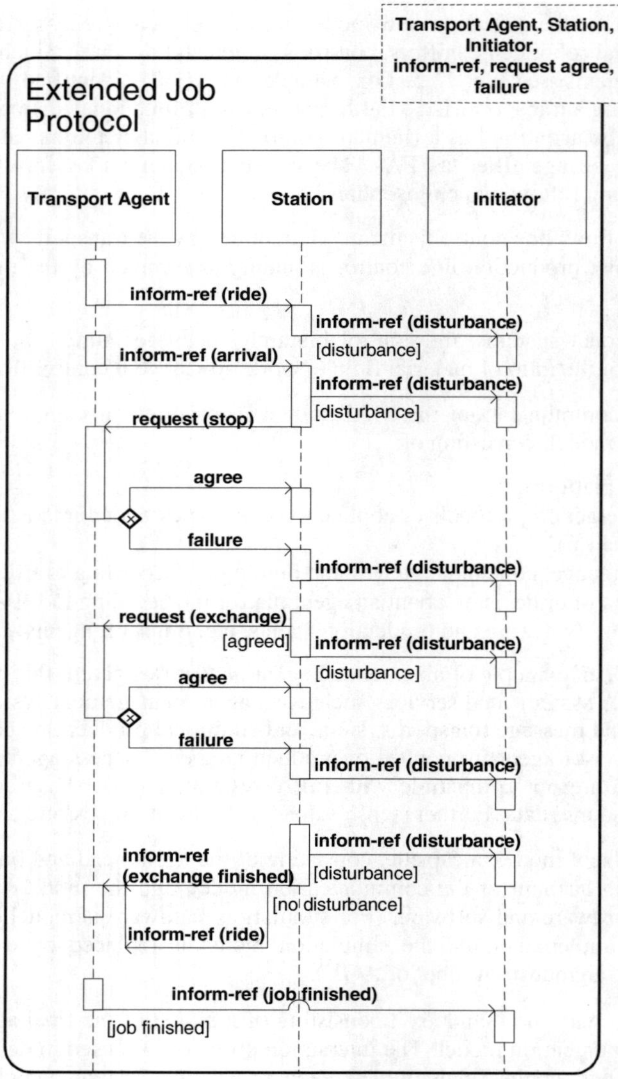

Fig. 11.6. Example of an application-specific interaction protocol

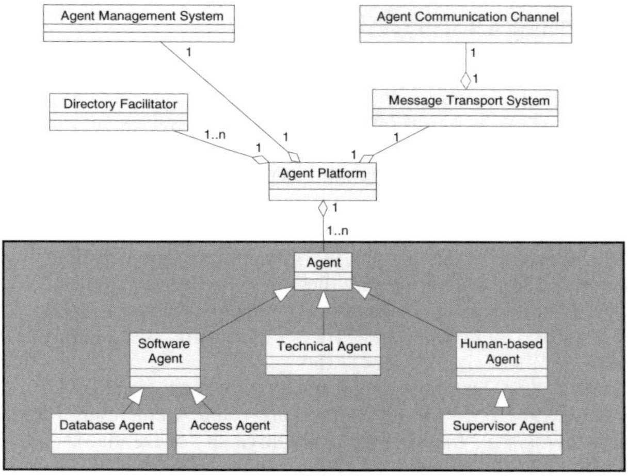

Fig. 11.7 Extended FIPA agent platform [11.27]

11.6 Conclusion and Outlook

In this chapter, a specification for agents within a manufacturing MAS was presented. A realisation of a PA-MAS for transportation was described. The general concept of a prototyping suite for production agents was described. This is a solution for transparent, consistent and effective development of MAS-based control systems for manufacturing systems and their sub-systems, e.g. transport agents, which includes a UML-based development and analysis method (ODEMA), methods for agentification, agent models, and modules for the development, simulation and testing of production agents.

Acknowledgements

We would like to express our thanks especially to Schneider Automation GmbH and Fraunhofer AiS, for supporting our academic and industrial work. The AgenTec project is funded by BMBF, Germany (Grant No. 01AK905A/B) and the HMS project was funded by the European Commission under the BRITE-EURAM Programme, HMS-BRPR-CT97-9000.

References

[11.1] Bongaerts L, Valckenaers P, Brussel H Van, Wyns J (1995) Schedule Execution for a Holonic Shop Floor Control System. Pre-prints of the Advanced Summer Institute 95 of the N.O.E. on Intelligent Control of Integrated Manufacturing Systems, Lisbon, Alfamiicro, June 25–28 1995

[11.2] Brussel H van et al (1998) Reference architecture for holonic manufacturing systems: PROSA. Computers in Industry 37, pp. 255–274

[11.3] Bussmann S (1998) An Agent-Oriented Architecture for Holonic Manufacturing Control. First Open Workshop, IMS Europe, Lausanne, Switzerland, ESPRIT Working Group on IMS

[11.4] Caglayan A, Harrison C (1998) Intelligent Software-Agents. Carl Hanser, Munich, Vienna

[11.5] Iwata K, Onosato M (1994) Random Manufacturing Systems: a New Concept of Manufacturing Systems for Production to Order. Annals of the CIRP, vol. 43, 1: 379–384

[11.6] Kadar B, Monostori L, Szelke E (1997) An Object Oriented Framework for Developing Distributed Manufacturing Architectures. Proceedings of 2nd World Congress on Intelligent Manufacturing, Processes and Systems, Budapest, 1997: 548–554

[11.7] Kimura F (1993) A Product and Process Model for Virtual Manufacturing Systems. Annals of the CIRP, vol. 42, 1:147–150

[11.8] Koestler A (1989) The Ghost in the Machine. Arkana, London

[11.9] Langer G (1999) HoMuCS – A Methodology and Architecture for Holonic Multi-cell Control Systems. PhD thesis, Technical University of Denmark

[11.10] Lüth T (1998) Technical Multi-Agent-Systems – Distributed Autonomous Robotic or Manufacturing Systems. Carl Hanser, Munich, Vienna

[11.11] McFarlane D, Marett B, Elsley G, Jarvis D (1995) Application of Holonic Methodologies to problem Diagnosis in a Steel Rod Mill. IEEE: 940–945

[11.12] Miyamoto T, Kimagai S (2000) Multi Agent Net Modelling and Design of Autonomous Distributed Systems. Intelligent Autonomous Systems 6:164–171

[11.13] Neubert R, Colombo W, Ax U, Schoop R (2001) An Approach to Integrate a Multi-agent-based Production Controller into a Holonic Enterprise Platform. Proceedings of the 1st International Conference on Information Technology in Mechatronics (ITM'01), Special Session, Istanbul, 2001: pp 65–70

[11.14] HMS Consortium: Frequently Asked Questions (FAQ) – Holonic concept. http://hms.ifw.uni-hannover.de (July 2000)

[11.15] AgenTec: Agent Technology for Robots in Production and Services. http://www.agentec.de (July 2001)

[11.16] FIPA: Agent UML Web site, http://www.fipa.org, (July 2001)

[11.17] British Telecommunications plc: BT Intelligent Agent Research – ZEUS. http://www.labs.bt.com/projects/agents/zeus (July 2001)

[11.18] DFG: Priority Programme – Software Specification – Integration of Software Specification Techniques for Applications in Engineering. http://tfs.cs.tu-berlin.de/projekte/indspec/SSP/softspez-eng.html and http://tfs.cs.tu-berlin.de/ projekte/indspec/SPP/iosip.html (2001)

[11.19] Emorphia – About FIPA-OS. http://fipa-os.sourceforge.net (July 2001)

[11.20] Foundation for Intelligent Physical Agents. http://www.fipa.org/ (July 2001)

[11.21] JADE (Java Agent DEvelopment Framework). http://sharon.cselt.it/projects/jade/ (July 2001)

[11.22] Object Management Group. http://www.omg.org (July 2001)

[11.23] Odell J, Parunak H, Bauer B (2000) Extended UML for Agents. AOIS Workshop at AAAI

[11.24] Okino N (1993) Bionic Manufacturing Systems. Flexible Manufacturing Systems – Past, Present, Future. Ljubljana, Slovenia, pp. 73–95

[11.25] Ritter A, Neugebauer J, Schraft RD, Westkämper E (1999) Holonic Task Generation for Mobile Robots. Proceedings of the 30th International Symposium on Robotics, Tokyo, October 27–29, 1999, pp. 553–560

[11.26] Ritter A, Braatz A, Winz G (2000) Agent systems in manufacturing. Tagungsband/SPS IPC DRIVES Nürnberg 2000, 11. Fachmesse und Kongress. Hüthig, Heidelberg

[11.27] Ritter A, Braatz A, Baum W (2001) Transport Agents in Manufacturing. Tagungsband/SPS IPC DRIVES Nürnberg 2001, 12. Fachmesse und Kongress, Hüthig, Heidelberg

[11.28] Ritter A, Braatz A, Höpf M (2001) Models and Modules for Production Agents. Proceedings of the 1st International Conference on Information Technology in Mechatronics (ITM'01), Special Session, Istanbul, 2001: pp. 47–52

[11.29] Schoop R, Neubert N, Süssmann B (2000) Flexible Fertigungssteuerungen durch SPS, CNC und Software-Agenten. Tagungsband/SPS IPC DRIVES Nürnberg 2000, 11. Fachmesse und Kongress. Hüthig, Heidelberg

[11.30] Sousa P, Ramos C (1997) Proposal of a Scheduling Holon for Manufacturing. http://www.dei.isep.ipp.pt/~psousa/papers/paam97/html (1997)

[11.31] Sugi M, Maeda Y, Aiyama Y, Arai T (2001) A Flexible Robot System for Assembly with a Concept of Holon. Proceedings of the 32nd ISR (International Symposium on Robotics), vol. 2, April 19–23 2001, pp. 844–849

[11.32] Sundermeyer K, Bussmann S (2001) Einführung der Agententechnologie in einem produzierenden Unternehmen – ein Erfahrungsbericht. Wirtschaftsinformatik 43, 2: 135–142

[11.33] Ueda K (1993) A Genetic Approach Toward Future Manufacturing Systems. Flexible Manufacturing Systems – Past, Present, Future, Ljubljana, Slovenia, pp. 221–228

[11.34] Warnecke H-J (1995) Aufbruch zum Fraktalen Unternehmen – Praxisbeispiele für neues Denken und Handeln. Springer, Berlin, Heidelberg

[11.35] Westkämper E, Braatz A (2001) Eine Methode zur objektorientierten Spezifikation von dezentralen Automatisierungssystemen mit der Unified Modelling Language (UML)". at Automatisierungstechnik, 49. Jahrgang, Heft 5: pp. 225–233

12. A Holonic Shot-blasting System

Tapio Heikkilä[1], Leila Rannanjärvi[2], Mikko Sallinen[2] and Mika Rintala[3]

[1]University of Oulu
Systems Engineering Laboratory
Finland
Email: Tapio.Heikkila@nethawk.fi

[2]VTT Electronics
Advanced Interactive Systems
Finland
Email: {Leila.Rannanjarvi| Mikko.Sallinen }@vtt.fi

[3]Blastman Robotics Ltd
Finland
Email: Mika.Rintala@blastman.fi

Abstract. Holonic features such as cooperative and autonomous behaviour can ensure the flexibility and robustness of a manufacturing system. When a holonic system is to be implemented, several important issues have to be resolved: How to describe the task to be completed? How to decompose a task for different production devices? How to optimise the workflow and co-ordinate the production?

Our holonic shot-blasting system is a demonstration, where two separate robots work together to complete surface treatment tasks. A 3D design model of a product, with selected surface treatment operations, is the starting point of holonic shot-blasting. The workpiece (the product) must be localised, with an optical sensor in the robot cell, before any treatment can proceed. The treatment plan is transformed into TCP paths. The final decision of the active robot depends on the kinematics of the robot, on the tool in hand, on the workload and on the free space in the cell. In our system, the decision mechanism is a blackboard and robots are active and willing to complete tasks.

12.1 Introduction

The requirements for modern manufacturing control systems are rapidly changing. Manufacturing systems become intelligent and inherently more complicated as the decision making processes are distributed closer to the factory floor. A variety of interfaces to product models, users and sensors, all in varying levels of abstraction will, be needed also, the "Grand Challenge of Intelligent Robots" [12.8] identifies these same features. The critical issues are not related to intelligence in the traditional sense of Artificial Intelligence, but to human-robot interfaces, modularity in general, and system integration. Self-similarity within the system modules is expected to be advantageous for introducing autonomy and intelligence.

Intelligent manufacturing as an application domain, has some common features. "Intelligence" is implemented in one way or another, throughout a set of multiple manufacturing system components (machine tools, transportation systems, robots etc.). In Holonic Manufacturing Systems, the system is constructed from distributed intelligent entities, with two main characteristics: autonomy and cooperation. The definitions for holon, autonomy and cooperation by the HMS consortium [12.25] are:

- A *Holon* is an autonomous and cooperative building block of a manufacturing system for transforming, transporting, storing and/or validating information and physical objects. The holon consists of an information processing part and, often, a physical processing part. A holon can be a part of another holon.
- *Autonomy* is the capability of an entity to create and control the execution of its own plans and/or strategies.
- *Cooperation* is a process whereby a set of entities develops mutually acceptable plans and executes them.

The holonic paradigm has been created for managing and controlling of very flexible manufacturing systems. It has its background in the analysis of social organisations and biology [12.5, 12.7, 12.10]. Holons in a holonic system are organised in dynamic hierarchies, but at the same time they also cooperate to achieve global goals, and represent autonomy and distributedness to obtain local features, such as computing power, reactivity and adaptivity. Negotiation is considered to be a key feature in holonic systems, and a holonic approach may even yield an ever-changing structural organisation in the manufacturing system based on cooperative activities. In this sense there is a clear correspondence between holonic systems and agent societies: holons can be seen as an agent society specialising in manufacturing, i.e., having constraints for structures and behaviour specific to manufacturing [12.11]. The holonic paradigm has been applied in a variety of manufacturing applications, and here we consider shot-blasting.

Shot-blasting is very dangerous and dusty work for human beings and robot automation has recently been used more and more. The shot-blasting process is of a very dynamic nature. The process parameters are time variant: the cleaning properties of the blasting material have effects on the final results of the blasting/cleaning. The blasting material is recycled, which causes the particle size to change with time. Also, nozzle wear affects the shape and size of the blasting cone.

The Holonic Shot-blasting System described here [12.11, 12.12] specifies a flexible multi-robot application for surface treatment applications. It is based on autonomous, cooperative units, i.e. shot-blasting robot holons, which are integrated into product models. Product models support the planning activities of the robots. Task sharing for the robots is carried out through a blackboard and contract-net-based negotiation, including task allocation, detailed motion planning (in time and space) to prevent collision, and sensing planning for accurate location of the workpieces. The robots support each other by providing sensor information to locate workpieces, and by sharing the workspace by gradually constructing the motion plans for the shot-blasting paths. Human integration is considered of key importance; that is, a human prepares the "missing elements" for product models and corrects or fine tunes the product models or even the instantiated task descriptions, i.e. the paths of the robots' motions. The system also supports instantiation and maintenance by flexible management of the system resources (creating and maintaining cooperation domains, holons, etc.). Here we describe an experimental holonic shot-blasting system, and report some experiences on using the holonic principles in a real robot environment. The test system consists of one real Blastman shot-blasting robot and one virtual robot. Blastman robots [12.4] are hydraulic robots and are designed for difficult blasting conditions. Blastman robots are used, for example, in casting bays, in the railway industry and in other metalworking industries, which need effective blasting equipment. The tasks of the test system include automatic, cooperative creation and execution of integration of humans for system guidance by teleoperation interfaces.

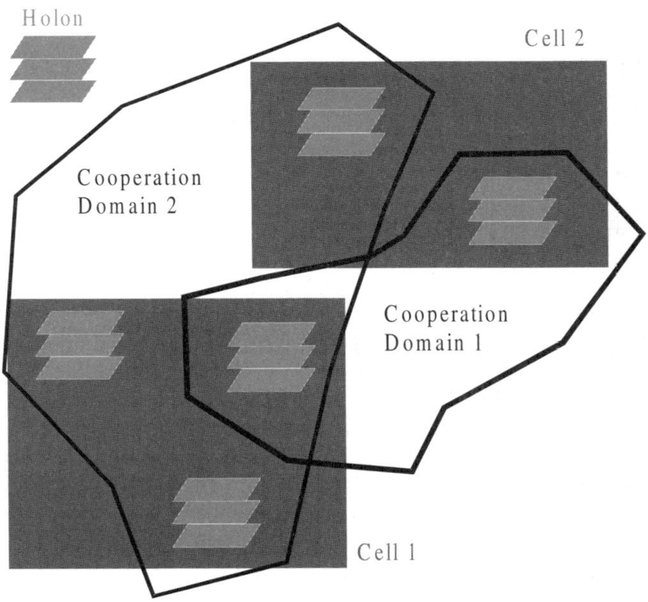

Fig. 12.1. Holons in cooperation domains

12.2 Holonic System Structure

Conceptually the system is constructed from dynamic cooperation domains (Fig. 12.1). A cooperation domain is composed of a set of holons that are cooperating to perform tasks efficiently. A cooperation domain specifies the set of holons, which negotiate to establish the common goal [12.11]. A robot is a holon, which exists when it

- belongs to a cell (is physically located in a cell) and
- is a member of one or more cooperation domains, i.e. is able to complete a task or a sub-task of a cooperation domain.

Cooperation domains can be created and deleted on-line, depending on what kind of tasks are to be performed, and what kind of resources are available.

12.2.1 Holon Structure

A robot holon is organised in three layers, each of which handles information and knowledge specific to the layer – the lower the layer is, the more specific and detailed the information handled is and it is handled with higher frequency or a shorter time span of actions. The layers are the part layer, the feature layer, and the motion primitive layer (Fig. 12.2). Each layer consists of one or more control classes. Each class shares a similar basic structure, i.e., the PEM structure [12.13]. Negotiation based communication is possible for the two highest layers (the part and the feature layers), but only for external communication; internally, the modules in different layers communicate by passing command and response messages.

The lowest control layer, the motion primitive layer, includes functionalities, related to a time span of hundreds of milliseconds or less. Typically this relates to the activities of dynamic control, e.g. servo control and trajectory control. In the next layer, the feature treatment layer, blasting of a specific spatial feature of the part is considered. Typically, this is a region of a surface. The time span is in seconds, and covers motion-planning and execution activities. In the highest layer, the part (or object) treatment layer blasting of the whole target part is considered. The time span is within tens of seconds or minutes. Typical activities include symbolic task planning. The holon structure is illustrated in detail in the class diagram shown in Fig. 12.3.

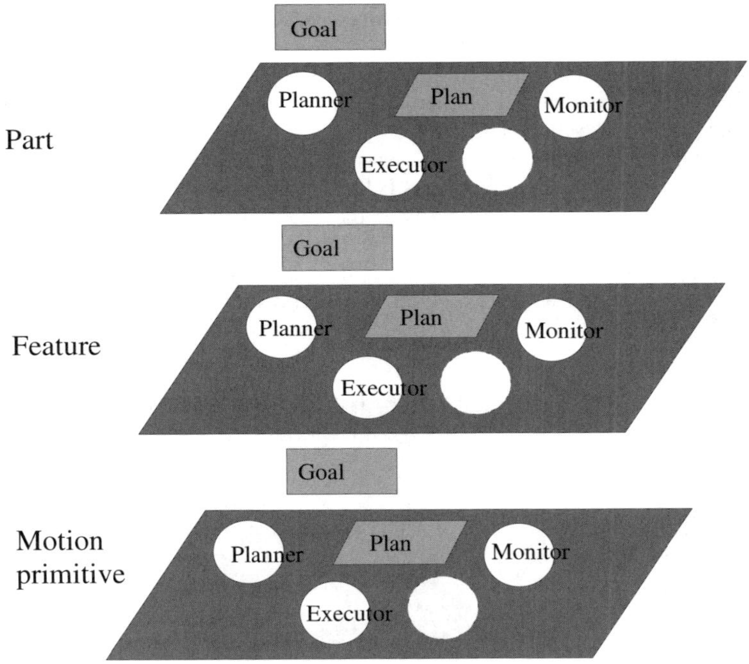

Fig. 12.2. PEM structure on three levels. Similar functionalities and data stores

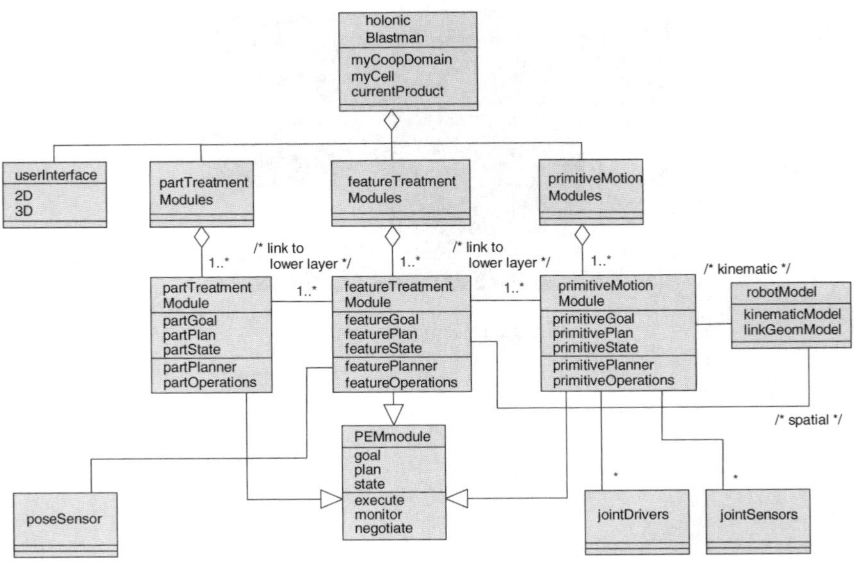

Fig. 12.3. Detailed structure of a holonic shot-blasting system

12.2.2 Architecture of the Holonic Surface Treatment Cell

The holonic surface treatment cell is capable of controlling two robots simultaneously. In our test system one of them is a real shot-blasting robot, model BWS (Blastman for Wheel Sets) [12.4] (
Fig. **12.4**) and the other one is a simulated robot in a workstation (Fig. 12.5). The control system runs in a PC (Windows NT), which is connected by TCP/IP sockets to the simulation and by I/O-cards to the robot. The real robot requires hard real time control and that is provided by INtime ™ developed by RadiSys ® [12.14]. An optical range scanner, connected by a serial bus, is used to check if a workpiece to be treated where expected. A video camera presents the situation in the cell. (Fig. 12.6)

Fig. 12.4. Demonstration robot

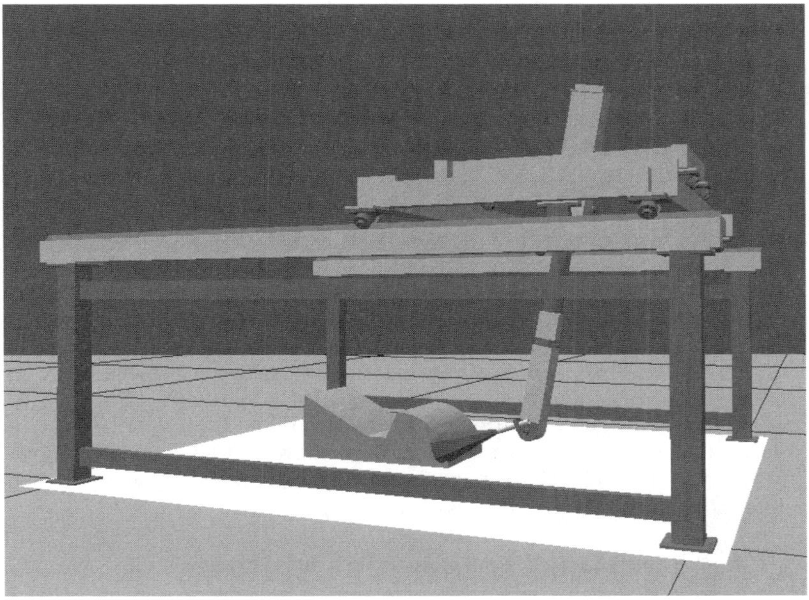

Fig. 12.5. Simulated robot and workpiece

Fig. 12.6. Demonstration robot and workpiece

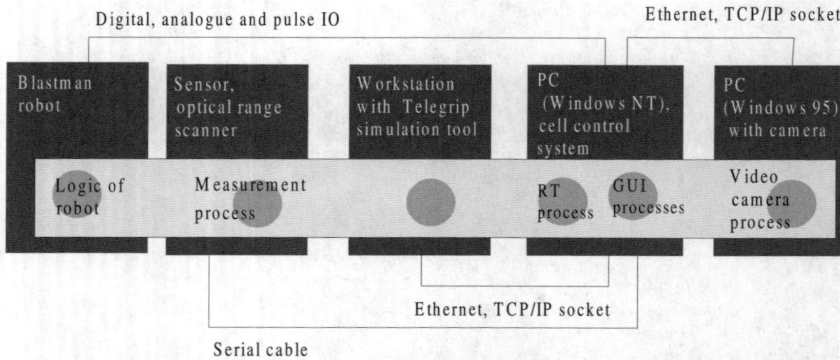

Fig. 12.7. Nodes and processes in the control system of a holonic cell

The control of the cell is distributed among several nodes (Fig. 12.7): the robot, the optical range sensor, the workstation, a Windows NT PC and a Windows 95 PC. In the Windows NT PC there are several separate processes, several graphical user interfaces and a real-time process (Table 12.1).

Table 12.1. Nodes and processes in a surface treatment cell

Process	**Node**	**Tool**
Interactive Co-ordinator (i.e. Graphical User Interface with Task List);	Windows NT	Java
Planner for real robot	Windows NT	C++
Executor for the real robot, a real-time process with Windows NT and INtime	Windows NT and INtime	C
Planner for the simulated robot	Windows NT	C++
Executor for the simulated robot	Workstation	Telegrip
Collision check for transfer paths	Windows NT	C++
Positioning	Windows NT	C++
Video camera process for monitoring or for teleoperation	Windows 95/98	
Part planner generating paths for a path library	Windows NT	C++

12.2.3 Holonic Control in the Surface Treatment Cell

The three-layer structure of a holon (Fig. 12.2) also indicates the layered structure of the main functions in a holonic surface treatment cell. These are Part planning, Part execution, Feature planning, Feature execution, Path or motion primitive planning and Path or motion primitive execution. Figure 12.8 illustrates these functions. The user can trigger any of these functions and select the required input from libraries.

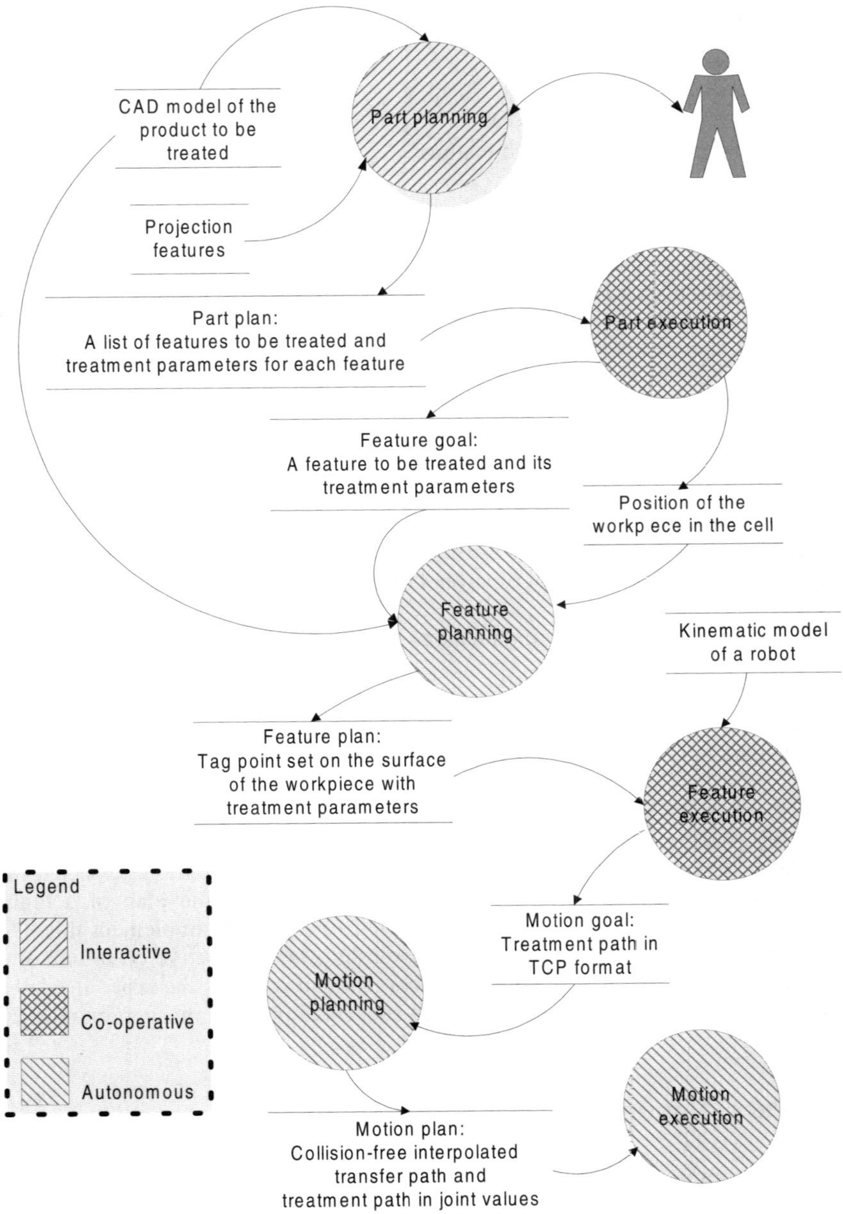

Fig. 12.8: Planning and execution functionalities in three levels of a holon: part, feature and motion primitive

We consider part planning to be interactive or semi-autonomous, because the user interacts with a CAD model of the product to be treated in order to create a part plan. The part plan includes

- a list of features to be treated and
- treatment parameters for each feature.

Part execution is cooperative, because it starts a sequence of autonomous, cooperative functions. The user always triggers part execution. So a robot or a cell cannot decide which workpieces are to be treated. The user has to trigger an atomic function "position the workpiece" before entering on the feature level.

Feature planning is autonomous, because now we have

- a feature to be treated
- treatment parameters for this feature, and
- the position of the workpiece in the cell.

This information is required before it is possible to generate a path in TCP (tool control point) format. The TCP format is attached to the workpiece but not attached to any kinematic model of a robot (holon).

Autonomous motion planning includes sub-functions like

- Generate a collision-free transfer path from the current point to the first point of the path.
- Calculate joint values for this robot and interpolate the path.

If these sub-functions succeed, motion execution can be triggered automatically. Motion execution is a function (servo control loop) which controls hydraulic valves so that the planned feature will be treated and the expected motion path followed. Intra-holon communication transports the plan of a higher level of the holon to the goal of a lower level. File pointers implement this. One holon in a cooperation domain also has the role of an interactive co-ordinator. It manages the task list of the cooperation domain and reserves the space needed to complete a treatment task in a cell. Stream sockets implement this communication.

12.3 Part Planning

Product planning in a holonic surface treatment cell answers the question of how to create a blast program from a CAD model of a workpiece? Planning affects all three levels of the holonic control system: the part level, the feature level and motion primitive level (Fig. 12.9.).

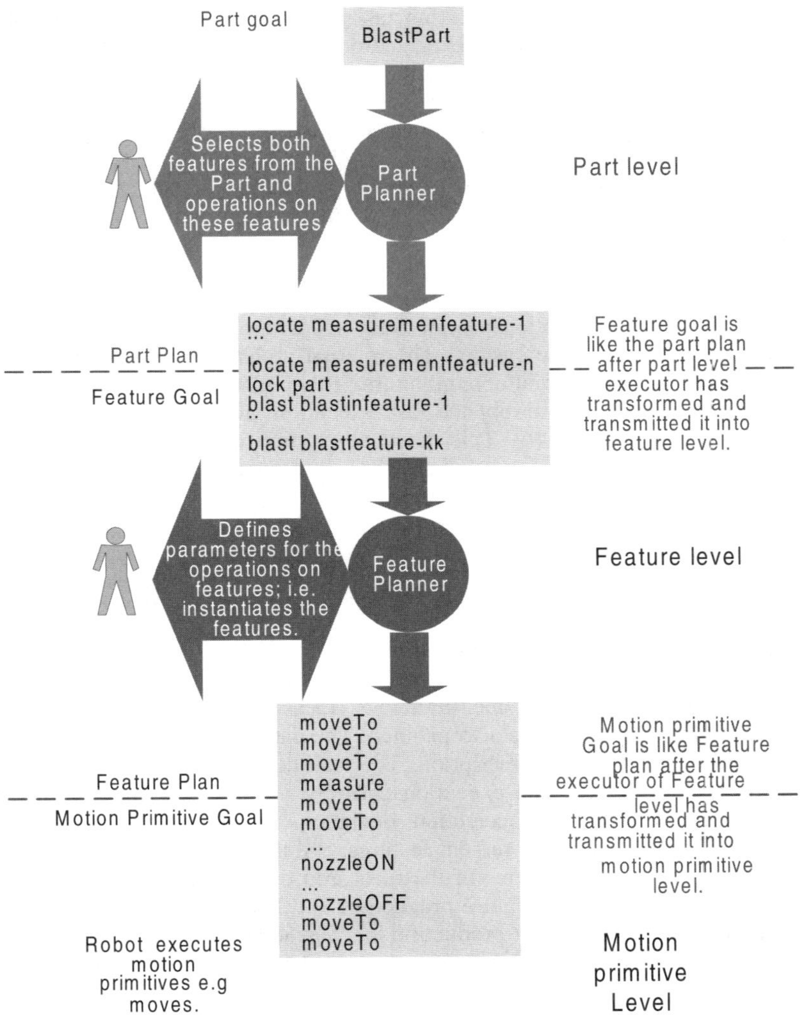

Fig. 12.9. Plans on three levels of a holonic control system

12.3.1 Task Descriptions

A complete task consists of sub-tasks, which mainly consist of motions that transfer the robot to starting positions of blasting paths, blasting along given paths, and transferring away from the end point of the blasting path. Complexity arises from the shapes of the paths: often the surfaces are curved in two directions, and even irregular, free-form surfaces are possible.

During execution of the shot-blasting tasks, the environment is rather stable. Changes that take place are mostly expected production changes, and typically no unexpected events occur, except in possible device failure situations. In such failure situations the most important thing is to run the system down safely and then consider some recovery procedures, either automatically or assisted by a human operator.

The task descriptions are related to the representation of part goals and part plans in the part layer, and feature goals and feature plans in the feature layer of the holon structure shown in Fig. 12.2. In both layers, we specify task descriptions, as described above, as a linear set of operations typical of that level of abstraction. These operations and related control structures make up plan languages, which together with the motion primitive layer operations make up the task decomposition structure of the system. The internal architecture takes care of the flow of control in failure situations: monitors, if they observe the deviations from expected behaviour, will interrupt the task execution, and report an error to the main controller of the activity. This drastically simplifies the necessary control structures of the plan languages. Figure 12.9 illustrates the task decomposition structure.

12.3.2 Task Sequence

The complete task sequence is simple: nominal shot-blasting paths are planned on the basis of a CAD model of the product. To adapt to spatial uncertainties (location of the product) targets for measurements (measurement features) and reference data for measurements are also planned. The data specifying this in the two highest layers form the task description. The task description has to be planned before the task execution, and execution can rely on task descriptions planned earlier. During the actual task execution, deviations in the product locations are measured, using planned measurement features, and then paths are updated in the plan by means of transformations of the measured deviations. If the path deviations are deterministic, e.g. within a production lot, then updated path points can be stored, to be used within that production lot. This means that the measurements can be made only for the first product, and to shorten the treatment time, the rest of the batch is done without measurements.

The task description language and the principles of planning and controlling apply to execution with or without negotiation. The tasks can be shared by negotiation, but in practice this means instantiating the plans for the individual holons in a top-down manner. The plans still have the same structure and content if they are negotiated, and so it does not make any difference to the principles of the task description. However, if synchronisation of execution between several holons is necessary, then handling of external events in the plans is necessary.

12.3.3 Part Plan as Operations on Features

The task descriptions in the part level are simply lists of feature treatment operations. In the part layer, there is no need to explicitly represent the flow of events, because they are restricted to confirming the correct completion of an operation or notification of failure of an operation (replies from the feature layer).

An operation keyword and parameters identifying the product type and the method of handling the product give the part goal. The part plan consists of a list of feature-level operations i.e. operations on features. The part plan is a task template maintained at the part level. The template includes plan-dependent variables, which must be instantiated before the plan can be executed. This instantiation is carried out by a human operator, and is supported by the control system as explained later on. The operations in the feature layers cover shot-blasting operations on different types of features (edge, seam and surface) and locating the products with different measurement features (plane, cylinder and sphere).

12.3.4 Feature Plan as an Action on the Motion Primitive Level

The task descriptions in the feature level are also simple. A task description consists of a list of primitive operations. These include motion primitive operations, (part or pose) sensing operations, (shot-blasting) nozzle control operations, and flow of control with internal and external events. Also, in the feature-level layer there is, in principle, no need to explicitly represent the flow of events, because they are restricted to notifying the successful or unsuccessful completion of operations, and relaying the values of acquired sensor data.

The feature goal is taken directly as the operation of an instantiated part-level plan. This is transformed and transmitted by the part-level executor to the feature-level module. A feature-level goal describes an operation for either locating the product or treating the surface.

12.3.5 Blast Features

There are four types of blast features: seam, surface, edge and corner.

- The seam (Fig. 12.10) is the simplest blast feature. It is described points following a constraining curve along which the path is located. The simplest geometric form of a seam is a straight line in space.
- Points following a constraining surface on which the path points are located describe a surface (Fig. 12.11). The simplest form of a surface is a plane in space.
- Points following the constraining surfaces on which the path points are located describe an edge (Fig. 12.12). The angle between the surfaces can be acute or obtuse. The simplest form of an edge is the composition of two planes in space.

- Points following three constraining surfaces on which the path points are located describe a corner (Fig. 12.13). The angles between the surfaces can be acute or obtuse. The simplest form of a corner is the composite of three planes in space.

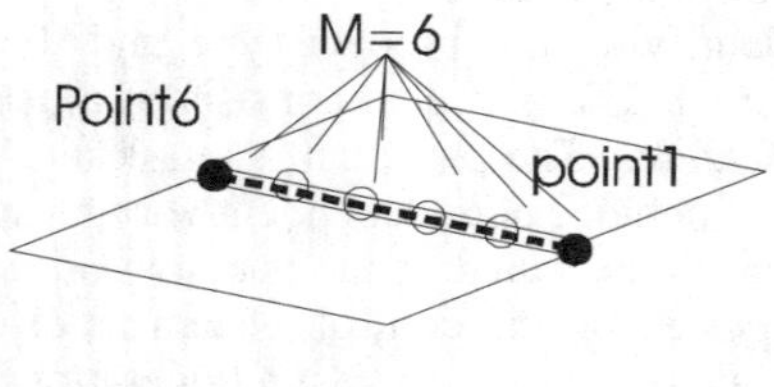

Fig. 12.10. Seam

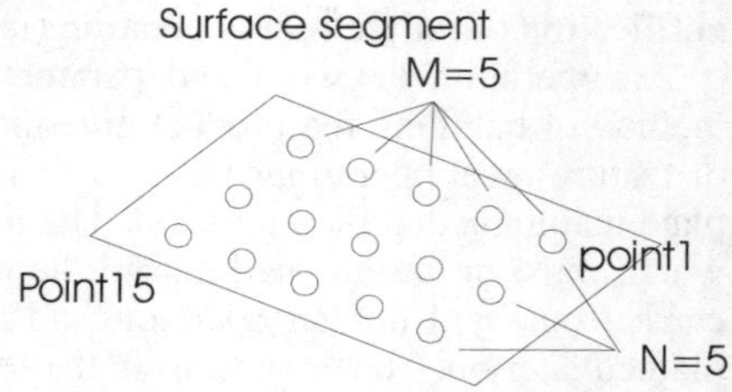

Fig. 12.11. Surface

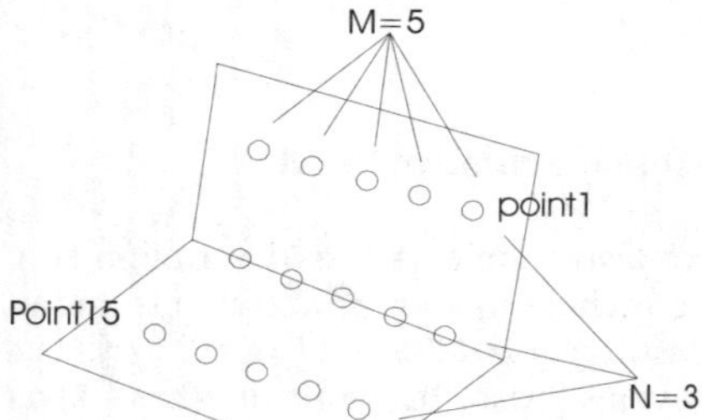

Fig. 12.12. Edge

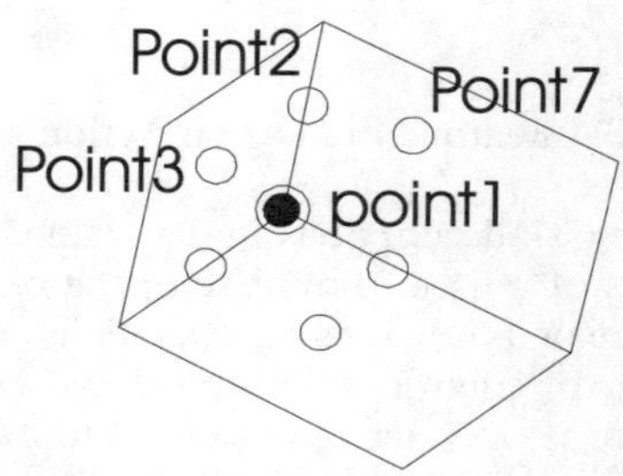

Fig. 12.13. Corner

Our current implementation supports interactive usage of 3D product models only in the feature level. Part planning is done, instead, by simply specifying the part plans textually in task lists, and the surface feature tasks can be planned interactively on the basis of product models. This is described in the next section.

12.4 Feature Planning

Feature planning is based on creating projected point patterns on the surface of the CAD model of the part (here the STL format is used) and gathering tag points from the projected points. These path points make up a blasting path specific to the type of feature (edge, a sweeping path; seam, a simple, straight path; surface, a scanning path). This means that the operator defines the paths for a robot and the features to be treated within those parts. The user has several selections to make:

- Select the CAD model (STL) of the product to be treated.
- Select the feature, which may be an edge, corner or surface, or some more complicated composite. Different properties of patterns can be used for different features.
- Move the projecion pattern into a position where it intersects properly with the part. Projection points are visualised, and the user may select and save the set of tag points for instantiating the feature path.
- Define treatment parameters.

The user activates the planning procedure with all this input about the feature. The result is a task expressed in workpiece co-ordinates, which can also be visualised for the user. Co-ordinate transformations from the workpiece to the robot are done before execution.

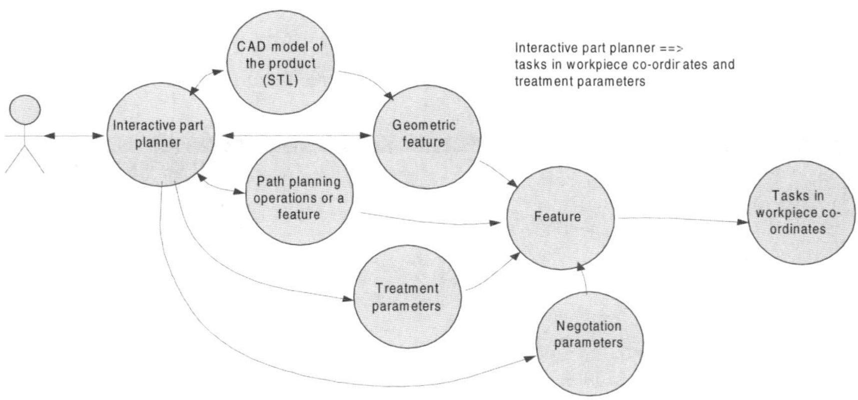

Fig. 12.14. Simple feature planning

In simple planning (Figs. 12.14 and 12.15) the user selects tag points from the surface of a workpiece. This selection takes place in a feature planner, where the user can graphically see possible tag points when the workpiece intersects with one or more projection planes (Fig. 12.16). Intersection creates a set of tag points for a feature path. After the definition of tag points, the user defines treatment parameters such as painting, blasting or vacuuming. When the method of treatment has been selected, the user has an option to define more parameters for the treatment process. Also, the parameters for stirring are asked for. These parameters are the angle and the frequency of the stirring. The angle can be typed in degrees. The frequency means how many times the nozzle stirs per second. If blasting has been selected as the treatment process, treatment parameters such as grade of roughness and cleanness can be defined.

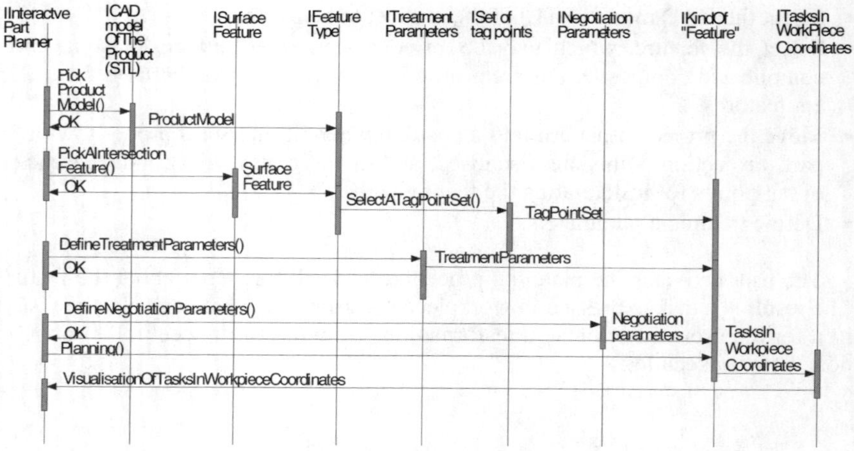

Fig. 12.15. Interaction diagram for simple feature planning

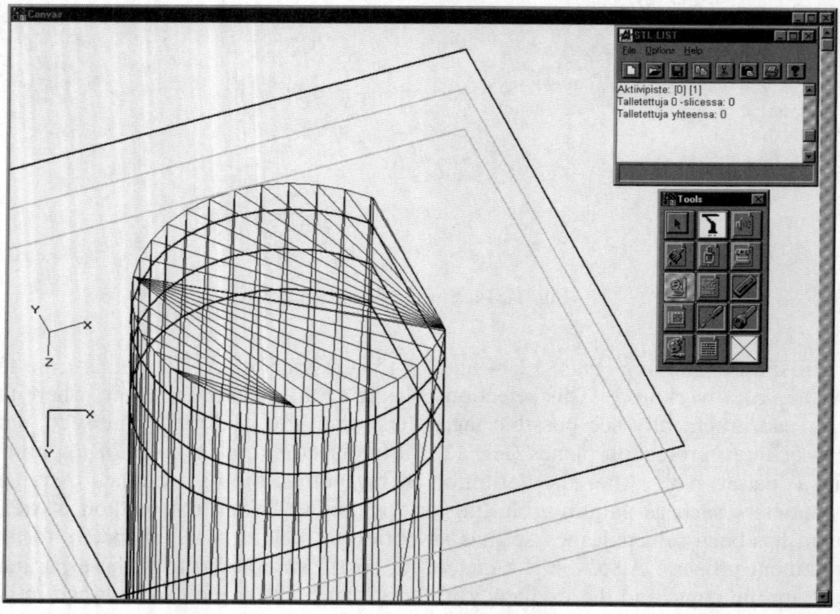

Fig. 12.16. User interface for selecting tag points for the feature instance

The treatment parameters for a task are:

- Char[6] TypeOfFeature; {Casing|Corner}
- int FeatureID; increasing number

- char[20] NameOfMasterProduct
- bool SweepingOnOff
- bool VacuumOnOff
- bool SandOnOff
- bool PaintOnOff
- int Amplitude [mm]
- int Frequency [mm]
- int DistanceFromSurface [mm]
- float Velocity[m/s]
- int Pressure[mBar]
- char[20] Material
- float GrainDiameter [mm]
- int Status {Created|Reserved|Working|Ready}
- int Priority, max 254.

Simple feature planning is not so straightforward, because

- It is not sophisticated. It does not use any kind of real features effectively, as all the information has to be added manually. The user has to define the paths for a workpiece interactively; they cannot be generated automatically.
- There might be some problems with complicated features. If a planner adds the distance of a tool directly from the surface normal of a facet the system might not be able to blast corners or other complicated surfaces.
- Collision avoidance might be/is difficult if sweeping is only a parameter defined by amplitude and frequency.

Before a path can be planned further, the path has to be addressed to a suitable robot and a kinematic solution found. In addition, the workpiece to be treated has to be positioned in the physical work cell (Sect. 12.6). After positioning the workpiece the collision check (Sect. 12.7) finds a path to the starting point and also evaluates whether it is possible to run the treatment path without hitting any static obstacles. These three phases are controlled by the Feature Executor described in the next section.

12.5 Feature Executor

Feature execution selects the most suitable robot for the feature to be treated. Feature execution has several phases. First of all, there has to exist a list of tasks (features) to be treated. We have developed a user interface, which enables a user to add tasks to be treated (see Sect. 12.5.1). The motion planner for either a simulated robot or a real robot has to ensure the robot is capable of completing the task and has many variables to consider (see Sect. 12.5.2).

12.5.1 Interactive Co-ordinator

The interactive co-ordinator is a self-standing process that enables the user to add both treatment tasks and robots. The robots pick a suitable task one at a time to be completed. The user interface looks as in Fig. 12.17. It was created with Java, and robots can interact with it via a network interface (sockets).

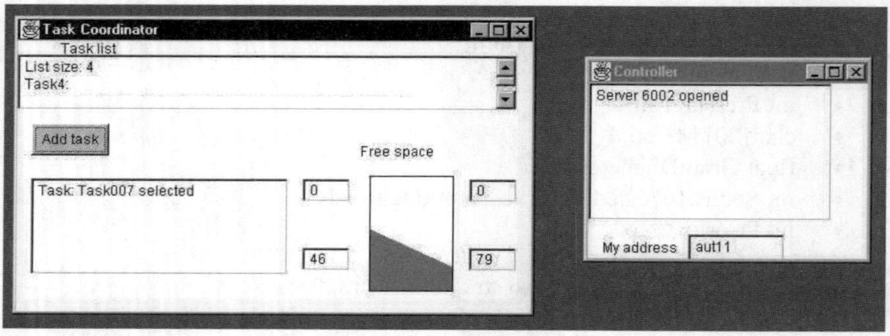

Fig. 12.17. Graphical user interface for the interactive co-ordinator

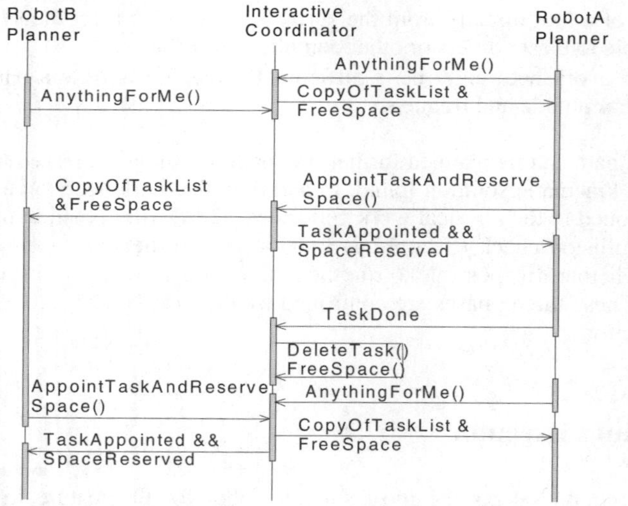

Fig. 12.18. Interactive co-ordinator manages the task list and gives only one copy of it to a robot planner considering a task

After the robot has been connected to the interactive co-ordinator, it starts to request tasks with a message "AnythingForMe". The interactive co-ordinator replies with the current task list and current free space in the cell. The robot considers which task to complete and requests one task, with space reservation. If the space is free, the interactive co-ordinator makes the reservation and the robot starts to work. After the task is completed, the robot acknowledges this fact and the task will be removed from the task list and the space reservation cancelled (Fig. 12.18).

The interactive co-ordinator needs to have several parameters for a task before it is able to make a proposal. The information required is:

- The task name.
- A link to a file carrying the program for a feature.
- The status of the task (waiting/working/ready/created).
- The space in the cell requested to complete this task (line/2D polygon/tube or 3D polygon). Time slice may be either all the time requested to complete the task or a constant time.
- The robot (intending to complete the task).
- The time for execution for the willing robot.
- The status of negotiation if using contract net (request/proposal/contract).
- The priority of the task.
- Previous tasks (which task has to be completed before this one?).

12.5.2 Robot Considering a Task

When a robot considers which task to select, it can have several criteria (with different weights):

- Can the robot reach (blast) the task (kinematic solution, position of workpiece)?
- Is the robot able to do the task (correct tool in hand or reachable)?
- Is the space required to do the task available?
- Is there already too much to do (load level)?
- Is the movement after the previous task acceptable (transfer time)?
- How much time does this robot need to complete the task?
- Preferences for this task; is a higher-priority task waiting.?

Positioning of the workpiece and collision checking for paths are explained further in the following sections. Here we continue with a description of our test system, i.e. the states of a simulated robot when interacting with the interactive co-ordinator (Fig. 12.19). The simulated robot opens a connection (when the user requests it). If the connection is OK, the simulated robot asks for a task list. After the task list and free space are received, the simulated robot selects a task to be completed and offers to complete it (Fig. 12.20). If OK is received, the task will be completed and new task list requested.

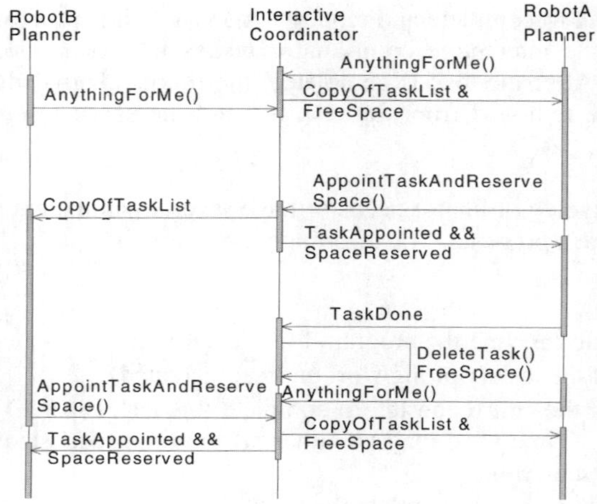

Fig. 12.19. Messages between the Interactive Co-ordinator and the Robot planner considering and executing a suitable task

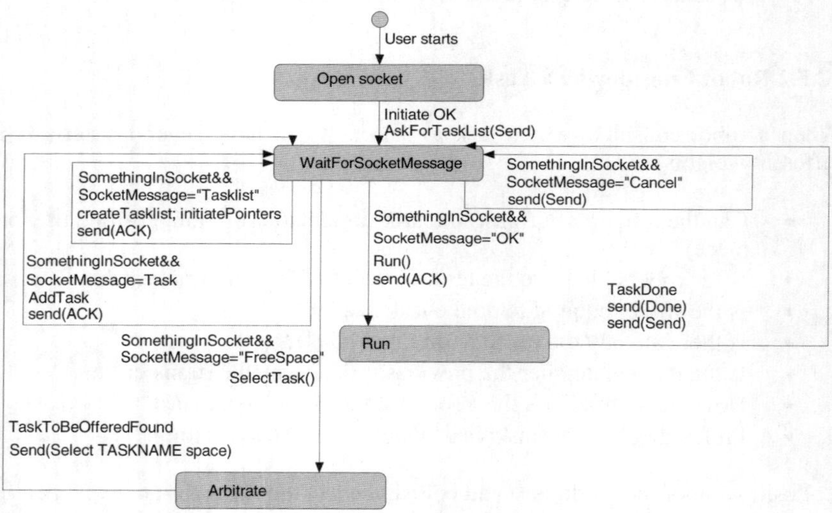

Fig. 12.20. Simulated robot interacting with the task list

The states of the real robot (Fig. 12.21) start a little earlier and describe all activities to be done before the robot can connect to a cooperation domain/interactive co-ordinator. First, the real-time operating system [12.14] has to be started, and the installation function (measurement cards and joints) initialised. Then the operator can drive the robot manually with three joysticks, select a single path from the library or connect to the task list. The interaction between a robot and the task list is similar to that between the simulated robot and the task list. The user interface of the robot looks like as in Figure 12.22 when the robot is running an appointed task.

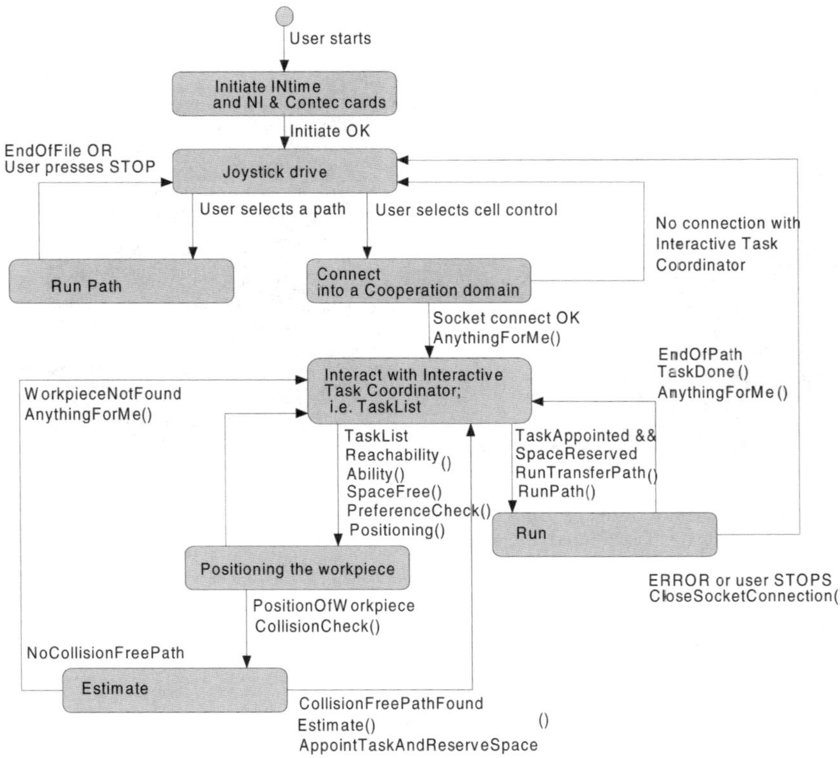

Fig. 12.21. Robot interacting with the task list

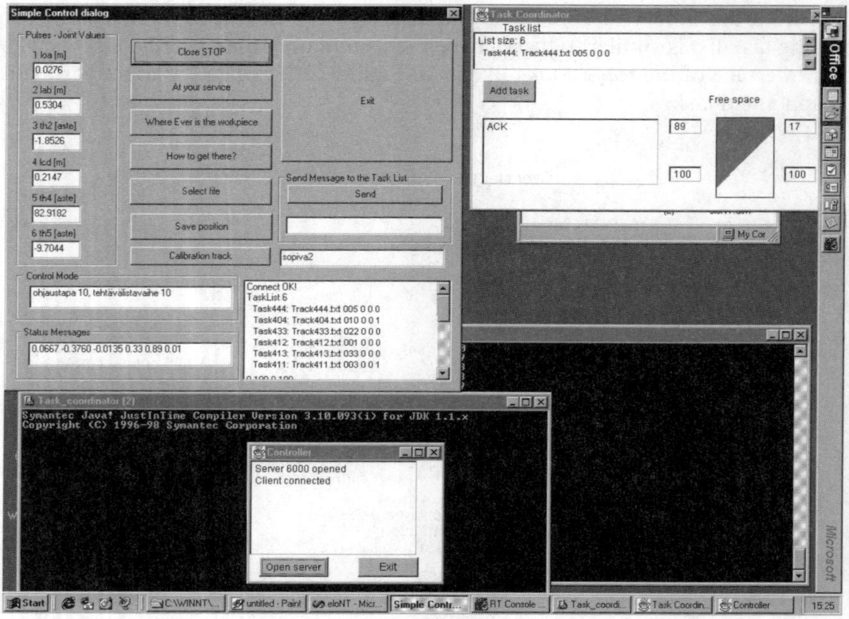

Fig. 12.22. Snapshot of the user interface after the robot has been connected to its task list, a task has been appointed, the space has been reserved and the robot is running

12.6 Positioning

12.6.1 Localising the Workpiece

In the work cell there is the requirement to locate the work object to enable accurate surface treatment. Measuring the known surface forms of the work object and estimating the workpiece location in the work cell constitutes the localization task. In a multi-robot system, localisation can be shared between different robots, and both the information and the uncertainties it includes can be combined. To facilitate this, we have introduced an interactive measurement planner into the system that enables the observation of spatial uncertainties during the location process.

A distributed multi-robot system for cooperative sensing has been studied by Sakane et. al. [12.26]. Their system has 3D model based agents which utilise 3D model information about the task environment for distributed sensor planning. The holonic work cell uses the same kind of task sharing [12.12, 12.31]. To estimate spatial uncertainties, we use linearised models with posture covariances, i.e. Bayesian -form estimation (posture will be further abbreviated as a pose). The approach has been used in [12.9, 12.15, 12.16]. Localisation of multiple forms of a work object has been studied in [12.17, 12.29], and different surface forms in CAD-based robotics have been studied in [12.3].

The accurate position of the workpiece has to be obtained before shot-blasting is possible. The location of a workpiece is determined by measuring it from different directions. In our surface treatment test cell there are two robots, so the tasks are divided. If the workpiece is very large, it may even be impossible to measure it using a single robot. When there are two robots, sensing tasks can be divided so that there is the minimum number of measurements and they can even be performed simultaneously, which may speed up the whole blasting process.

12.6.1.1 Measurement Process

The planned process can be simulated and uncertainties in blasting points can be estimated, to reduce the number of actual measurements. When the accepted level of uncertainties has been reached, the plan is fixed and actual measurements can be performed. The principle of the plan simulation and of the actual measurements is illustrated in Fig. 12.23.

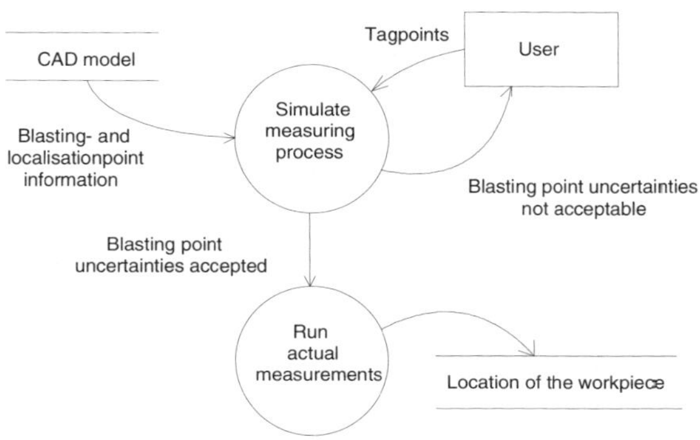

Fig. 12.23. Phases of simulation and actual measurements

The whole localisation process in described in detail in Fig. 12.24. First, the workpiece location uncertainty covariance matrix is calculated. Second, using the workpiece location uncertainty and given the blasting point information, uncertainties for each blasting point are calculated. If the level of uncertainty obtained is not low enough, more points from the CAD model surface have to be selected until as the required level is reached.

In the measuring process, only the simulation is run at first. When the uncertainties of the blasting points are small enough, actual measurements can be made. The user defines the limit of the point uncertainties.

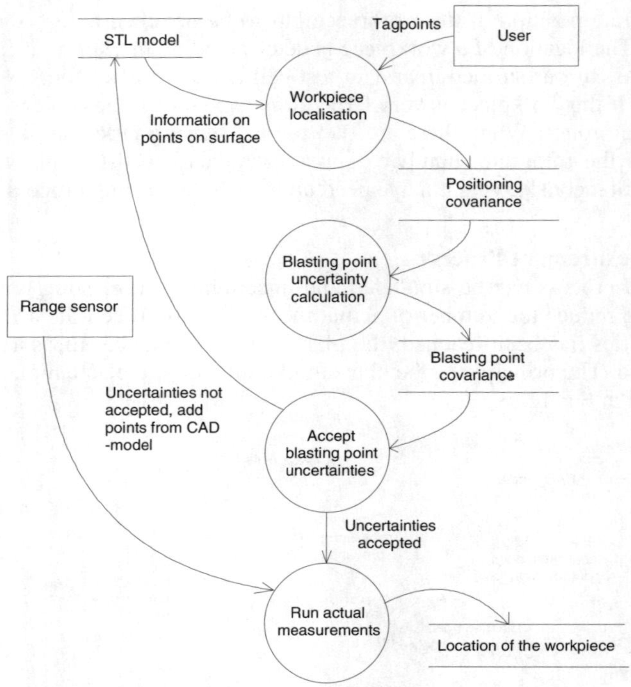

Fig. 12.24. Workpiece positioning flow chart

12.6.1.2 Workpiece Localisation

To measure the target object we use an optical sensor, developed by VTT [12.19]. The sensor gives 20 measurement values, in the form of x and y co-ordinates. The co-ordinate system is set so that the x axis is the distance and the y axis the horizontal axis. The sensor manufacturer gives nominal information about the range values of the sensor. In a typical industrial application, when the tool correction is known, manufacturer information can be used as source data to link the sensor to the whole measuring system. If the spatial position and orientation of the sensor are not known, the sensor calibration has to be done separately. Also, the spatial uncertainty of the sensor calibration must be added to the uncertainty analysis of the whole measurement system.

12.6.2 Object Parameterisation

In the STL CAD format, the surface forms of the workpiece are defined by triangular facets with three corner points and a surface normal vector. For our measuring and estimating requirements, we define the workpiece with planar, cylindrical or spherical surfaces. The surface model parameters are estimated from a set of points selected from the workpiece surface. The user of the Planner software needs only to select the form of the surface. The parameters will be calculated automatically using the algorithms given in Appendix A.

12.6.2.1 Planar Surface

We define a planar surface using a surface normal vector and the shortest distance from the origin of the workpiece. The equations for estimations of a planar surface are presented in Appendix A. When estimating the surface parameters, we need an error function to be minimised. Using an error vector we try to fit the set of points to the surface model. The error vector for the calculation of a planar surface is the dot product of the vector of one point and the surface normal vector minus the shortest distance from the co-ordinate origin. The equations for the method are presented in the Appendix A, as well as the equation for localisation of the planar surface.

12.6.2.2 Cylindrical Surface

A cylindrical surface is defined by the vector of the cylinder axis $\bar{o}$, the distance d of the cylinder origin from the work object origin, the cylinder direction vector $\bar{v}$ and the radius r of the cylinder. The estimation of these surface parameters is carried out using the iterative method described in Sect. 12.6.4.2. The error vector is calculated from the cross product between the cylinder direction vector and two points of the cylinder surface. The partial derivatives of the error vector with respect to the estimated parameters make up the Jacobian matrix for parameter estimation.

12.6.2.3 Spherical Surface

A sphere in the object frame is defined by the vector of the sphere origin $\bar{o}$, the direct distance d of the sphere origin from the work object origin, and the radius r of the sphere. The error vector is a shortest distance between point on the surface of a sphere and middle point of the sphere decreased by the radius of the sphere. In sphere estimation, we use the same principle to calculate the Jacobian matrix as in the case of plane and cylinder estimation. For more details, see Appendix A.

12.6.3 Uncertainties in the Measuring System

The spatial uncertainties of the measuring system which includes sensors, the workpiece and two robots located in a world frame, is illustrated in Fig. 12.25.

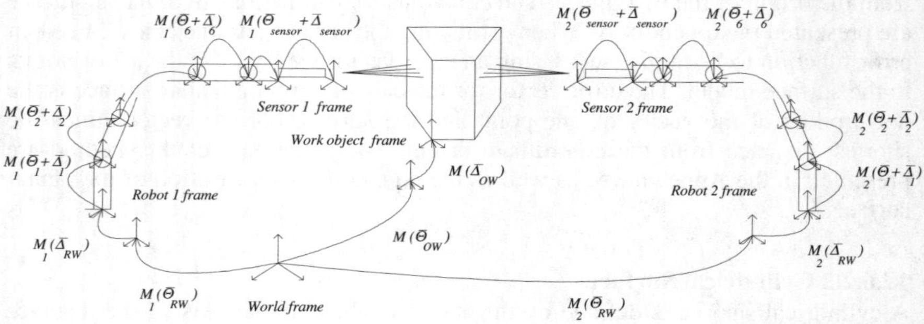

Fig. 12.25. Spatial uncertainties represented by error covariances with two robots and a workpiece

The uncertainties that every co-ordinate frame has can be transformed into another frame. When robot 1 has measured the target object, the location of the object still has some uncertainty. This uncertainty can be transformed into the frame of robot 2, etc.

The uncertainty in the robot positioning (the uncertainty of the robot TCP) has been calculated on the basis of the uncertainties in every joint. Each robot includes three revolute joints and three prismatic joints, and it has been assumed that each joint has some positioning accuracy. Using these initial assumptions, the combined uncertainty of all robot joints has been calculated, and in further calculations, this combined uncertainty is used.

12.6.4 Object Localisation

The principle of object localisation is to measure the object from several directions and on the basis of each surface form, fit the measured data to the corresponding surface form. The surface forms that we use are the plane, cylinder and sphere. Cylindrical and spherical surfaces do not have to be entire cylinders or spheres, they may be only a part of such a surface form or a or close approximation to it. A workpiece may include several of these surface forms and they may be located anywhere in the workpiece. The location of each sub-object is calculated from the CAD model.

12.6.4.1 Bayesian Estimation

To estimate the location of the target object, we use Bayesian-type estimation. This is an iteration method, in which we create the weight matrices from partial derivatives of all measurement sources. In Bayesian estimation, the reliable measurements are weighted more than the measurements with more noise. When the number of iteration steps is compared with least-squares estimation, there are usually more steps in the Bayesian method because of the smaller increments, especially with noisy data [12.18].

To use standard linear Bayesian estimation, we have to make the following assumptions:

- The correction parameter vector $\overline{\Delta}_{correct}$ and the initial covariance matrix of the object pose uncertainty $\overline{P}_0$ are Gaussian random variables with sero mean.
- The measurements are of the form $\overline{m}_i = \overline{m}_i^* + \delta\overline{m}_i$, where the $\overline{m}_i^*$ are true values, and the $\delta\overline{m}_i$ are Gaussian random errors with sero mean.

Spatial uncertainties in the measurement process increase the measurement uncertainty and the sensor calibration uncertainty.

12.6.4.2 Estimation Algorithm

The estimation of the surface model parameters and of the location of the workpiece is divided into the following eight steps. The estimation algorithms are based on linearised models and the iterative algorithm is very similar to conventional Newton iteration:

1. Read measured plane, cylinder or sphere information ($\overline{v}$, $\overline{c}$ or $\overline{s}$), measurement data $\overline{m}_i$, uncertainty data for the measurement, and sensor tool correction.
2. Transform the measured points into the object frame.
3. Calculate the distance vector (error vector) $\overline{e}$ from the points to the plane in the object frame.
4. Calculate the partial derivatives (Jacobian matrix) for the point/plane pairs $\frac{\partial e_i}{\partial \overline{\Delta}}$ and the covariance matrix $\overline{Q}$.
5. Estimate the increments of the parameters by using a weighted least-squares estimation ($\overline{\Delta} = -(\overline{J}^T \overline{Q}^{-1} \overline{J} + \overline{P}_0)^{-1} \overline{J}^T \overline{Q}^{-1} \overline{e}$).
6. If the increments are close enough to sero, invert the transformation and stop.
7. Perform the corresponding incremental transformation to the *nominal* object. Update the covariance matrix $\overline{Q}$.
8. Repeat from step 3.

The required number of iteration steps is dependent on the noise of the measurement. Typical values are between 3 and 5 with low-noise data [12.19].

12.7 Motion Planning

Planning of collision free-motions is an essential property of holonic robot systems. Although collision-free motion planning has been a central problem of robotics research since its early stages, no complete solution has yet been found [12.20]. There is strong evidence that a complete planner, a planner which finds a path whenever one exists and otherwise indicates that none exists, will take a time which increases exponentially as the number of degrees of freedom (DOFs) of the robot increases [12.21]. Most motion-planning algorithms developed in recent years are based on the use of *configuration space* [12.22]. The basic idea is to represent the robot as a point in the robot's configuration space, with the obstacle entirely mapped in this space; the obstacle, when mapped in this space, is called a *C-obstacle.* This mapping transforms the problem of planning the motion of a spatial robot into a problem of planning the motion of a point from an initial configuration to a goal configuration without entering the C-obstacle. Our solution is based on the use of the configuration space method.

12.7.1 Planning of Collision-free Motions

Collision-free motion planning has to be completed from the approach point of the blasting path to the start point of the blasting path, from the start point of the blasting path to the end point of the blasting path, or from the end point of the blasting path to the departure point of the blasting path.

Many methods have been proposed to solve the problem of collision-free motions. Although there are many external differences, most methods are based on a few different general approaches: the *roadmap, cell decomposition* and *potential field* methods [12.23]. These approaches are briefly described below.

The *roadmap method* consists of capturing the connectivity of the robot's free space in a network of one-dimensional curves. These curves are called the *roadmap.* Thus, the path planning is reduced to connecting the initial and goal configurations to points in the roadmap, and scanning the roadmap for a path between these points. The paths are calculated by using a simple but very fast local path planner [12.24]. The roadmap can be constructed using several different techniques, for example

1. uniform sampling followed by enhancement in difficult regions [12.24],
2. using random reflections at C-space obstacles [12.22], and
3. sampling on contact surfaces in the configuration space. [12.1]

The algorithm developed in this study can also be a part of roadmap generation.

Some of the most extensively studied motion-planning methods are *cell decomposition* methods. These methods consist of decomposing the robot's free space into small, simple regions, called *cells*. Because of this, a path between any two configurations in a cell can be easily generated. Therefore, a non-directed graph, the connectivity graph representing the connecting relation between the cells, is constructed and searched. The nodes of this graph are the cells, which are extracted from the free space. Two nodes are connected if the two corresponding cells are neighbours. The output of the search is a sequence of cells. A continuous path can be generated from this sequence of cells [12.23].

In the *potential field* approach the robot is represented as a point in configuration space that corresponds to a particle moving under the influence of an artificial potential. For example, in the bitmap method, the principle is to construct *collision-avoiding attractive potential fields* over the workspace of the robot. Each of these potentials applies to a selected point in the robot, called a *control point*, and pulls this point toward its goal position among the obstacles. The work space potentials are then combined into a new potential function. This function is defined over the configuration space of the robot. This new potential pulls the whole robot toward its goal configuration [12.26].

In order to develop an effective path planner for robots with many degrees of freedom, it seems that it is necessary to drop the completeness requirement. Starting from the initial configuration of the robot, the *Randomised Path Planner* (RPP) applies first a best-first algorithm. In other words, it descends the potential until it reaches a local minimum. Such a motion is called a *gradient motion*. If the global minimum is found the problem is solved and the planner returns the constructed path. If not, it attempts to escape from the local minimum by executing a series of *random motions* starting from the local minimum. At the terminal configuration of every random motion, the algorithm executes a gradient motion until it reaches a new local minimum. From each local minimum, if none of them is the goal, the algorithm performs another series of random motions. Thus, a graph of the local minima is built up incrementally. A randomised search of this graph is performed until the goal configuration is reached or the planner gives up. If the path is successfully generated, the path is transformed into a smoother path [12.26].

Path planning by *Robot Decomposition* is based on a special decomposition of the robot manipulator using an off-line preprocessing stage and a two-phase on-line path planning scheme. The first step, the decomposition of the robot, consists of separating the robot into several chains where a chain consists of a number of consecutive links and joints. In the first on-line phase, local planning is performed independently and simultaneously in the joint subspaces associated with robot chains. In the second on-line phase the paths which have been found for the chains are associated with obtaining a collision-free path for the robot. The main purpose of this decomposition of the robot is to reduce the dimensionality of the search problem significantly. As a result, it is possible to attain short path planning times for robots with many DOFs [12.27].

12.7.2 Modelling of the Robot's Links and the Obstacles

Here the obstacles and the links of the robot are modeled as polyhedral objects. Joints connect the robot's links into a chain of rigid bodies in order to enable a flexible motion. Hence, co-ordinate systems are established for all robot links. As shown in Fig. 12.26, the location of the base link (the zeroth link) of the robot is referred to as (x_0, y_0, z_0). The fixed world co-ordinate system is referred to as (x_w, y_w, z_w). The other link locations are also referred to their own co-ordinate frames.

When each robot link is modeled by a polyhedron, the equations of the facets and vertices of the polyhedron are first determined with respect to the robot's local co-ordinate system. Then co-ordinate values of all vertices of every polyhedron in the world co-ordinate frame are determined. Fig. 12.26 shows the reference configuration when the joint variable θ_k for the kth frame is zero for k = 0, 1 or 2. The value for the prismatic joint is also zero.

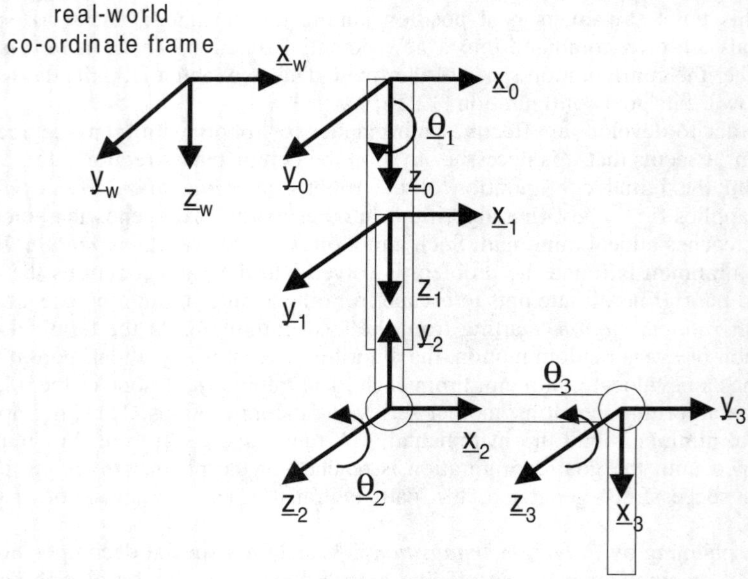

Fig. 12.26. The co-ordinate systems for a four-link robot ($\theta_k = 0$, k = 0, 1 or 2 in the configuration above)

When polyhedral modelling is applied to the obstacle, facet geometric models are also defined. By using the suitable form of the result given in [12.17], the facet equations and vertex co-ordinates expressed in the world co-ordinate frame, can be found. For more information, see Appendix B.

The value of the prismatic joint is always sero if the robot can reach the target point (the start point or the end point). If the robot is not able to reach the target point, the value of 10 at each trial will raise the telescope. The trials must be repeated until the robot reaches the target point or until the value of the telescope is the same as the maximum value of the telescope. If the value of the telescope is the maximum value and the robot is still not able to reach the target point, the inverse kinematic process will return an error, which means that the target point is not in the workspace of the robot.

12.7.3 Transfer Paths

In the approach that we are following, the transfer path planning is carried out in the configuration space. By a transfer path, we mean here the path from the approach position to the start point of the path, or from the end point of the blasting path to the departure position. The concept of a C-obstacle is retained, but only the relevant part of its surface requires evaluation, not the entire C-obstacle. This arrangement reduces the computation time enormously. Instead of depending on a potential function to guide the motion, this method uses a chain code technique to arrive at collision checking functions in order to find the C-obstacle boundaries on relevant planes. This proposed motion planning technique can then be conveniently applied to effective path searching. There is no question of reaching a "local minimum". An application of the proposed method will be described for a three-joint robot. The fourth joint of the robot that is used in this work is a telescope in the base link of the robot, and it will be utilised only in reaching problems.

12.7.3.1 C-obstacle Contour Determination

The spatial robot is reduced to a point without dimensions and the obstacle in Cartesian space is transformed to a C-obstacle in the configuration space. The configuration space used here is a three-dimensional joint space. These three joints control the translation of the robot end effector. The control of three joints is enough for gross motion movements of a Blastman robot. Because the time taken for determining a whole solid C-obstacle or set of C-obstacles is very long, this planning method is unsuitable for real-time on-line motion planning applications. The approach proposed here restricts the exploration of possible collisions of the robot with the C-obstacle surface to particular C-space planes only. In the tests we have performed, there were 18 searching planes in the joint space, and thus the angle between two planes was 10°. This plane system, proposed in [12.28], reduces the computation time enormously.

The only points which have to be given are the start point and the end point in Cartesian space, the inverse kinematic process will find the start point (p_s q_s r_s) and the end point (p_e q_e r_e) in the configuration space (or in the joint space). Now there can be more than one solution. When this is the case, the path planning process is applied for every solution. The exploration of the C-obstacle is initiated at the straight line in the configuration space (the joint space) between the start point (p_s q_s r_s) and the end point (p_e q_e r_e). Then 18 planes containing this straight line are determined, and searching for the C-obstacle contour is restricted to those planes.

12.7.3.2 Searching Plane

The first procedure is to detect a plane, which contains the straight line which passes through the start point and the end point. The plane is perpendicular to the p-q plane in the configuration space [12.28]. For more information, see Appendix B.

12.7.3.3 Contour in the Searching Plane

Because of the 2-D contour searching the technique of direction numbers is used [12.28]. This technique originates from image processing for labeling purposes, where the technique of directional chain codes is used. All possible direction numbers are shown in Fig. 12.26. In the searching plane, the x co-ordinate is p' and the y co-ordinate is q'. Thus the point in the searching plane is (p' q'). Direction number 3 is the same as the direction from the start point to the end point.

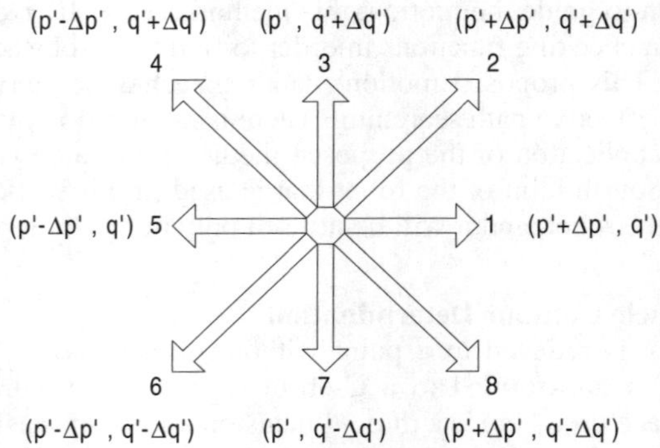

Fig. 12.27. Direction numbers in the searching plane. Direction number 3 is from the start point to the end point. (p' q') is one point of the path between the start and the end points [12.28]

Search movements on the searching plane are governed by certain fundamental rules:

1. The selection of a new direction number always starts at a collision-free point next to the point with a collision.
2. The very first direction number is always 3, which is the original direction from the start point to the end point. If everything is right, there *has to be an intersection* in direction 3. Tests show better results when direction number 3 is used first. After each trial, the direction number is increased by one, and the trial is repeated until there is no intersection between the robot and the obstacle.
3. When the first collision-free direction is found, a one-step change in this direction gives a contour segment for the C-obstacle in the configuration space.
4. The new point is now marked as a path point and the searching trial is repeated at this point. The previous point is marked as a "denied" point because it is not useful to go back to it.
5. The rotation of the obstacle is continued until the original line between the start point and the endpoint is reached.
6. If all eight directions have been tried and there are no free directions, the algorithm returns an error, and changes the searching plane and starts again on the new plane.
7. The searching is terminated if the robot is not able to access the point under consideration in the search step or if the point is outside the working space.
8. This loop is repeated until every searching plane has been checked.

In [12.28] the very first direction number was selected by using a mathematical equation. This caused many errors and in our case the first direction number is always 3. The direction of rotation of the obstacle was selected in [12.28] by using the first direction number. If there was a collision in the first direction the direction number was increased by one, and if there was no collision, the direction number was decreased by one. Thus, the direction of rotation of the obstacle depends on the first direction number in [12.28].

The rules governing the selection of direction numbers for contour searching are illustrated in Fig. 12.28.

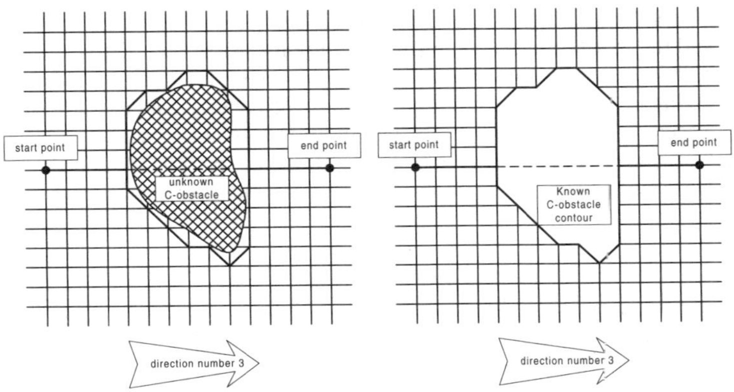

Fig. 12.28. C-obstacle contour searching. The left part illustrates the direction number-searching-process. The right part shows the C-obstacle when the contour has been found

The searching strategy is summarised below:

(a) The first direction number for the first trial when a collision has been detected, is found by rule 2.
(b) Rules 3–5 are applied to find the correct direction for the next movement of the point that will not lead to direct collision.
(c) Rules 6–9 are applied to determine the general searching rules of the process.
(d) The list of points in the path corresponds to the contour co-ordinates of the C-obstacle on the plane under consideration.
(e) Rule 1 is applied to start the path planning process.

12.7.3.4 The Shortest Path Detection

Generally the detection of the shortest path is an easily solved problem after the C-obstacle contours have been found. One of the simplest and fastest methods is proposed below, again using the direction number, the assignment rules and the searching rules discussed above. In [12.28] the detection of the shortest path is only mentioned and no solution is given for this problem. Our proposed method is as follows:

1. First, take the point list that is made in the first successful searching plane. The first point is taken as a base point for calculation.
2. The original straight line between the start point and the end point is taken as a base line. The next step is to determine all the lines between the start point and all the other points of the collision-free path.
3. The angles between the base line and all the other lines have to be calculated. The line that gives the greatest value of the angle is marked as the next shortcut line.
4. The end point of the shortcut line is now taken as a new base point for the calculation. The new base line is the line between this new base point and the original end point. Other lines have to be determined between the new base point and the rest of the points until the end point is reached.
5. The angle between the new base line and every other new line is calculated, and the line that gives the greatest value of the angle is taken as a new shortcut line.
6. This procedure is repeated until the end point is reached.
7. In the searching plane the value of the angle is always positive on the left side of the base line. An example of shortcut detection is illustrated in Fig. 12.29.

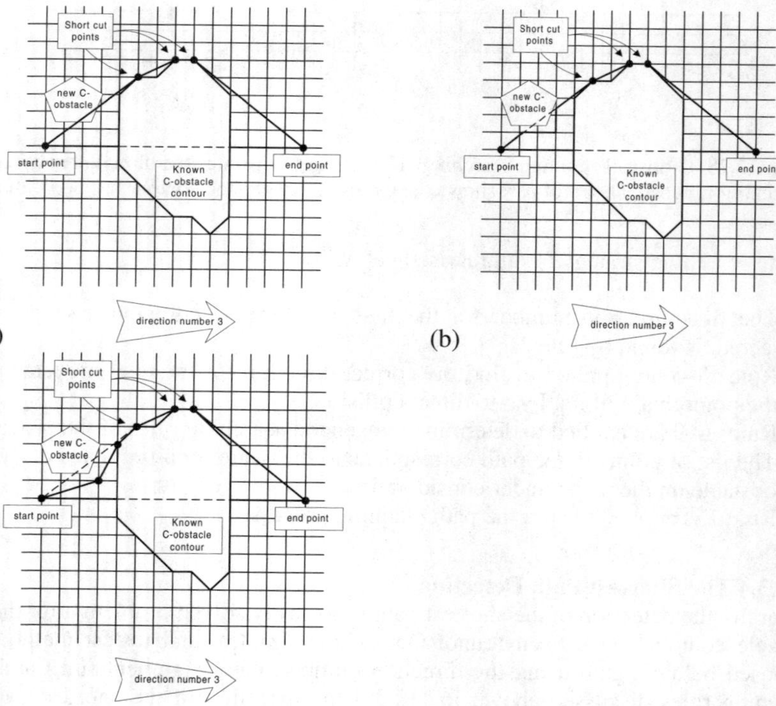

(a) (b) (c)

Fig. 12.29. (**a**) In the final interpolation a new unknown C-obstacle is found (**b**) Contour searching for a new C-obstacle. (**c**) The final shortcut path is detected

The next procedure is to interpolate between all the points in the shortcut path. If this interpolation detects new collisions, the whole path planning process has to be repeated between the shortcut points. This kind of situation is illustrated in Fig. 12.29. Here, a new unknown C-obstacle is found in the space between the start point and the first shortcut point. In [12.28], a new C-obstacle is not considered at all.

In this way, the interpolation distances become shorter and shorter until the whole path is explicitly detected. When all the possible paths have been generated, the length of every path can be calculated, and the shortest path in the joint space will be the final collision-free path.

12.7.4 Surface Following Path

In generating paths for continuous surface treatment, i.e. from the start point of the blasting path to the end point of the blasting path, the basic principles are that the distance of the TCP from the surface of the object and the angle between the tool and the object remain almost unchanged throughout the path. In practice this means that there are limits, defined by the user, for this distance and this angle and the object and they must stay within these limits. The result of this is that the number of different joint configurations for robots with six DOFs for each Cartesian point is limited.

In order to enable generation of the blasting path, the non-exact path that has been saved in the model of the object must be represented in the form of tag points, and the centre of the surface treatment spray must be aimed at the tag point in question. A path is formed between the tag points by interpolating between the Cartesian tag points so that the distance between the newly formed tag points is what is wanted. For example, if the wanted TCP speed is 100 mm/s and the time between the set values is 200 ms, then the distance between the tag points must be 0.2 s x 100 mm/s = 20 mm. This is the way to form a path which consists of successive tag points on the surface of the object. This path illustrates the progress of the centre of the surface treatment spray over the surface of the object during blasting.

The next task is to define the robot's joint configurations for each Cartesian tag point so that the blasting path becomes as flexible as possible and use the robot's motion space in the most logical way. After this, a collision check must be done for each point's joint configuration in the path. This is done in order to find out whether any of the robot's links will hit the object or any other obstacle in the environment. If there are no collisions, the point in question can be saved as a path point and then it is possible to move on to study the next point.

If there is a collision a new joint configuration must be found for the Cartesian point. This point is one where there are no collisions. The search for new joint values is worth doing near a point where there is a collision, because this helps to avoid moving the robot unnecessarily, and the movement to a collision-free point is carried out with the minimum amount of joint movements.

The controlled search for a new collision-free joint configuration is carried out by setting limits on inverse kinematics. This means that since the robot has redundant DOFs, the joints that are not needed are locked in order to find a new collision-free point as close to the original point as possible. The locked joints are chosen on the basis of inverse kinematics, but still bearing in mind that the joint that is located before the link with a collision is one of the locked joints.

The search for a new point is carried out by moving one or more of the locked joints as much as is wanted. The search begins by moving the joint before the link with a collision. This movement is a very small step. It is then possible to try to find values for the free joints. The values should provide a solution for the tag point under consideration and none of the links should collide with the object. The next task is to check the next tag point. If the new point that was found the search can be finished, but if the point has a collision, the search is continued by varying the same locked joint until an appropriate point is found. If it is impossible to find a point in spite of small changes of the joint, the changes of the joint can be made bigger in order to enable searching a larger area of the robot's working space. If even this does not help in obtaining a solution, the joint that is varied can be changed for another joint until a solution is found. If a solution still cannot be found by varying one joint, it is possible to expand the search by varying more than one joint at the same time, either randomly or step by step until a solution is found. If it is still difficult to find a solution, the only thing that can be concluded from this is that there is no solution in the robot's working space for this tag point.

12.8 Holonic Aspects of the Application

In our application, we have only one type of (robot) holon, and it has three layers. The product to be treated is not a holon in our application. A robot holon is organised into three layers, each of which handles information and knowledge specific to the layer – the lower the layer is, the more specific and detailed the information is handled, and with higher frequency. The layers are the part layer, feature layer and motion primitive layer.

A holon or a cooperation domain itself is not *autonomous*, because the user has to give an order; a workpiece to be treated. However, on different levels, there are autonomous functions which need no human interaction, for example feature planning, motion planning and motion execution (Fig. 12.8).

Holons (i.e. robots) *cooperate* to complete the treatment of a workpiece. One holon in a cooperation domain manages the list of treatment tasks (user defined). Holons are active and ask for something to do. Cooperation arises at the feature level when a holon completes feature execution. Also, part execution is considered cooperative or semi-autonomous, because it starts a sequence of autonomous or cooperative functions.

In reality, it is not possible to *reconfigure* robots on-line; it has to be done manually. So, in our application, the cell is fixed (the kinematics and position of the robot are fixed) before any treatment. But if a robot is able to switch the tool in hand, it is able to take part in another cooperation domain.

In one cooperation domain there is one active three-layered holon (part, feature, motion primitive), and the others have the third level pending, i.e. waiting for an opportunity to take over. The part-planning process is not dependent on any issues in the cell or cooperation domain and it is possible to execute it anywhere outside a holon.

A robot (i.e. a holon) is active and asks for tasks to complete or paths to run. There are several optimising criteria which influence the sequence of treatment tasks to be completed by a robot, which we do not consider here. Motion planning guarantees that a robot will not hit the workpiece or environment or blast another robot. In addition to this, the monitoring process on each level allows execution to continue if there are no disturbances.

12.9 Conclusions

Cooperative and autonomous behaviour of a shot-blasting system consisting of two robots has been described here. The test system contained one real robot and one virtual robot. The demonstrations completed within this holonic shot-blasting system included automatic and cooperative creation and execution of surface treatment tasks, and human integration for system guidance by teleoperation.

Acknowledgements

This work was financed by the commission of the European Union (Contract No. BRPR-CT97-9000), VTT and Blastman Robotics Ltd; this finance is gratefully acknowledged by the authors. The authors would also like to thank the persons in the HMS consortium, especially those of the HANDS workpackage, for fruitful cooperation.

References

[12.1] N. Amato and Y. Wu: "A Randomised Roadmap Method for Path and Manipulation Planning", *Proceedings of the IEEE International Conference on Robotics and Automation*, 1996, pp. 113–120

[12.2] P. Besl and N. McKay: "A Method for Registration of 3-D Shapes". IEEE Trans. on Pattern Analysis and Machine Intelligence, **14**(2). 239–256 (1992).

[12.3] Blastman Robotics Ltd: "*Blastman Robots*", http://www.Blastman.fi (2001).

[12.4] H. Van Brussel: "The Vision Matching the Problem". *Proceedings of the 1st European Conf. on Holonic Manufacturing Systems*. Hannover, Germany, 1994.

[12.5] J.H. Christensen: "Holonic Manufacturing Systems: Initial Architecture and Standards Directions", *Proceedings of the 1st European Conf. on Holonic Manufacturing Systems*. Hannover, Germany, 1994.

[12.6] J.D. Crisman and G. Beckey: "Grand Challenge for Robotics and Automation: The 1996 ICRA Panel Discussion", IEEE Robotics and Automation Magazine, **3**(4), 10–16 (1996).

[12.7] H.Durrant-Whyte: "Uncertain geometry in robotics", *Proceedings of the IEEE International Conference on Robotics and Automation*, Scottsdale, Arizona, (1989), pp. 646–652.

[12.8] A. Koestler: *The Ghost in the Machine*, Arkana, London (1967).

[12.9] L. Rannanjärvi and T. Heikkilä: "Software development for holonic manufacturing systems", Computers in Industry, **37,** 233–253 (1998).

[12.10] T. Heikkilä, N. D'Agostino, L. Rannanjärvi and P. Salonen: "Feature-based product modelling for holonic surface treatment robots", in *Progress in Simulation, Modelling, Analysis and Synthesis of Modern Electrical and Electronic Devices and Systems,*ed. N. Mastorakis, World Scientific and Engineering Society, Athens, Greece (1999).

[12.11] T.Heikkilä and J. Röning: "PEM-modelling: A Framework for Designing Intelligent Robot Control", Journal of Robotics and Mechatronics **4**(5), 437– 444 (1992).

[12.12] TenAsys Corporation: "*INtime ™ software overview guide*", http://www.tenasys.com/intime.htm (2000)

[12.13] R. Smith.and P. Cheeseman: "On the representation and estimation of spatial uncertainty", Int. Journal of Robotics Research **5**(4), 56– 68 (1987).

[12.14] A. Sanderson: "Assemblability Based on Maximum Likelihood Configuration of Tolerances", IEEE Trans. on Robotics and Automation, **15**(3), (1999).

[12.15] Z. Li, J. Gou and Y.Chu: "Geometric Algorithms for workpiece Localization", IEEE Transactions on Robotics and Automation, **14**(6), (1998).

[12.16] M. Järviluoma: *Object Model Correction with Iterative Estimation.* Report C6, University of Oulu, Systems Engineering Laboratory (1991).

[12.17] J. Marszalec: "*Photoelectric Range Scanner*", VTT Optoelectronics Laboratory, Project Report, Form PO 6-1/1.2 (1993).

[12.18] A. Lazanas: "*Reasoning about Uncertainty in Robot Motion Planning,*" Standford University, Standford, CA. (1994).

[12.19] J.H. Reif: "Complexity of the mover's problem and generalizations", *Proceedings of the 20th IEEE Symposium on Foundations of Computer Science (FOCS)*, pp. 421–427 (1979).

[12.20] T. Horsch, F. Schwarz and H. Tolle: "Motion Planning for Many Degrees of Freedom – Random Reflections at C-Space obstacles", *Proceedings of the IEEE International Conference on Robotics and Automation,* pp. 3318–3323 (1994).

[12.21] J.-C. Latombe: *Robot Motion Planning*, Kluwer Academic, Boston (1991).

[12.22] L.E. Kavraki, P. Švetska, J.-C. Latombe and M. H. Overmars: "Probabilistic Roadmaps for Path Planning in High-Dimensional Configuration Spaces", IEEE Transactions on Robotics and Automation, **12**(4), 566–580 (1996).

[12.23] E.H. van Leeuwen and D. Norrie: "*Holonic Manufacturing Systems*", internal report of the IMS/HMS project, HMS consortium (1996).

[12.24] S. Sakane, T. Sato, H. Okoshi and M. Kakikura: "Distributed Sensing System with 3D Model-Based Agents", *Proceedings of the IEEE/RSJ International Conference on Intelligent Robots and Systems*, Yokohama, Japan (1993).

[12.25] M. Tarokh and A. Hourtash: "Path Planning by Robot Decomposition and Parallel Search", *Proceedings of the 1997 IEEE, International Conference of Robotics and Automation,* Albuquerque, New Mexico, pp. 562–568 (1997).

[12.26] S.K. Tso and K.P. Liu: "A Fast Motion Planner Based on Configuration Space", *Proceedings of the 1993 IEEE/RSJ International Conference on Intelligent Robots and Systems*, Yokohama, Japan (1993).

[12.27] M. Sallinen, T. Heikkilä and M. Rintala,: "Sensor Fusion with Spatial Uncertainties in a Holonic Multirobot Workcell", *Proceedings of the IEEE International Conference on Systems, Man, and Cybernetics*, Nashville, Tennessee, (2000).

[12.28] T. Heikkilä, L. Rannanjärvi, M. Rintala and P. Salonen: "A Holonic Shot-blasting System", *Proceedings of the Scandinavian Symposium of Robotics 1999*, Oulu, Finland, pp. 193–200 (1999).

APPENDIX A

12.A.1 Planar Surface

To define the parameters of a planar surface, we use the normal vector of the plane and the shortest distance from the origin to express the measurement plane, as follows:

$$\bar{v} = f(\bar{n}_i, d) \quad , \tag{12.1}$$

where $\bar{n}_i$ is the normal vector of the plane and
d is the direct distance from the origin of the object frame.

To calculate the difference between the measured points and those of the reference plane taken from the CAD model, we have to determine the error vector. The error vector is non-linear, and when a least square-based estimation is used to determine the object location, it has to be linearised.

Each digitised point belongs to a plane selected from the CAD model. The plane parameter information includes the plane normal vector $\bar{n}$ and the distance from the origin d. For calculation of the plane parameters, the error e between a digitised point and the reference plane can be written as:

$$e = \bar{n} \cdot \bar{p}_i - d \quad , \tag{12.2}$$

where $\bar{p}_i$ are the co-ordinates of the measured point.

To fit the set of measured points to the respective planes, we use a Bayesian model. In the situation where $\bar{\Delta} = \bar{0}$, i.e. the measurements fit the target object model, the derivative of the error $\bar{e}_i$ for one point in a known frame can be calculated by using the equation

$$\frac{\partial e_i}{\partial \bar{\Delta}} = \frac{\partial e_i}{\partial \bar{p}_i} \frac{\partial \bar{p}_i}{\partial \bar{\Delta}} \tag{12.3}$$

The combined derivative, known as the Jacobian matrix $\bar{J}$, is obtained from (12.3):

$$\frac{\partial e_i}{\partial \overline{\Delta}} = \begin{bmatrix} n_x & n_y & n_z \end{bmatrix} \overline{R}_O^T \overline{R}_R \begin{bmatrix} 1 & 0 & 0 & 0 & z_i & -y_i \\ 0 & 1 & 0 & -z_i & 0 & x_i \\ 0 & 0 & 1 & y_i & -x_i & 0 \end{bmatrix}, \tag{12.4}$$

where x, y and z are points in the object frame,
$\overline{R}_O$ is the rotation matrix of the target object frame and
$\overline{R}_R$ is the rotation matrix of the robot frame.

12.A.2 Cylindrical Surface

The parameters are defined by the vector of the central axis $\overline{o}$, the distance d of the cylinder origin from the work object origin, the direction vector of the cylinder $\overline{v}$ and the radius r of the cylinder. This definition is expressed as follows:

$$\overline{c} = f(\overline{o}, d, \overline{v}, r), \tag{12.5}$$

where $\overline{o}$ is the unit direction vector of the cylinder origin,
d is the direct distance from the origin of the object frame to the cylinder origin,
$\overline{v}$ is the unit direction vector of the cylinder and
r is the radius of the cylinder.

For calculating parameter estimates for a cylindrical surface, the error can be calculated from the direct distance of the measured point from the line (= cross product of the vectors of the line and the measured point). Here the line is cylinder direction vector. To express the position of the direction vector in space, we use the middle point of the bottom of the cylinder:

$$e = \left| \overline{v} \times \overline{p_0 p_1} \right| - r, \tag{12.6}$$

where $\overline{p}_0$ are the co-ordinates of the cylinder origin and
$\overline{p}_1$ are the co-ordinates of the measured point. Here

$$\overline{p_0 p_1} = \left[p_x^o - p_x^m \quad p_y^o - p_y^m \quad p_z^o - p_z^m \right], \tag{12.7}$$

where p_x^o, p_y^o, p_z^o are the x,y and z values of the point in the middle of the cylinder and

p_x^m, p_y^m, p_z^m are the x,y and z values of the point in the surface of the cylinder.

In cylinder estimation, we use the same principle to calculate the Jacobian matrix as in plane estimation. Here, in addition to the work object pose parameters, we can estimate the cylinder radius r (e.g. $\overline{\Delta} = \left[x \quad y \quad z \quad \phi_x \quad \phi_y \quad \phi_z \quad r \right]$). The error derivative for cylinder estimation can be written in the following form:

$$\frac{\partial e_i}{\partial \overline{\Delta}} = \left[D_x \quad D_y \quad D_z \right] \overline{R_O}^T \overline{R_R} \begin{bmatrix} 1 & 0 & 0 & 0 & z_i & -y_i & \frac{1}{D_x} \\ 0 & 1 & 0 & -z_i & 0 & x_i & 0 \\ 0 & 0 & 1 & y_i & -x_i & 0 & 0 \end{bmatrix}, \tag{12.8}$$

where D_x, D_y, and D_z are partial derivatives as follows:

$$D_x = \frac{\partial \left| \overline{v} \times \overline{p_0 p_1} \right| - r}{\partial x}, \tag{12.9}$$

$$D_y = \frac{\partial \left| \overline{v} \times \overline{p_0 p_1} \right| - r}{\partial y} \tag{12.10}$$

$$D_z = \frac{\partial \left| \overline{v} \times \overline{p_0 p_1} \right| - r}{\partial z} \tag{12.11}$$

where x, y and z are points in the object frame,

$\overline{R_O}$ is the rotation matrix of the target object frame and

$\overline{R_R}$ is the rotation matrix of the robot frame.

12.A.3 Spherical Surface

A sphere in the object frame is defined by the vector of the sphere origin $\overline{o}$, the shortest distance d of the center point of the sphere from the work object origin and radius r of the sphere. The definition is expressed as follows:

$$\bar{s} = f(\bar{o}, d, r) \tag{12.12}$$

where $\bar{o}$ is the vector of the center point of the sphere,
d is the direct distance from the origin of the object frame to the center point of the sphere and
r is the radius of the sphere.

The error in the localisation of the sphere can be calculated from the direct distance of the point from the origin of the sphere:

$$e = \sqrt{\left(p_x^o - p_x^m\right)^2 + \left(p_y^o - p_y^m\right)^2 + \left(p_z^o - p_z^m\right)^2} - r, \tag{12.13}$$

where $\bar{p}^o$ are the co-ordinates of the sphere origin and
$\bar{p}^m$ are the co-ordinates of the measured point.

In sphere estimation, we use the same principle to calculate the Jacobian matrix as in the case of cylinder estimation. The error derivative for sphere estimation can be written in the following form:

$$\frac{\partial e_i}{\partial \bar{\Delta}} = \begin{bmatrix} D_x & D_y & D_z \end{bmatrix} \bar{R}_O^T \bar{R}_R \begin{bmatrix} 1 & 0 & 0 & 0 & z_i & -y_i & \dfrac{1}{D_x} \\ 0 & 1 & 0 & -z_i & 0 & x_i & 0 \\ 0 & 0 & 1 & y_i & -x_i & 0 & 0 \end{bmatrix}, \tag{12.14}$$

where

$$D_x = \frac{\partial \sqrt{\left(p_x^o - p_x^m\right)^2 + \left(p_y^o - p_y^m\right)^2 + \left(p_z^o - p_z^m\right)^2} - r}{\partial x}, \tag{12.15}$$

$$D_y = \frac{\partial \sqrt{\left(p_x^o - p_x^m\right)^2 + \left(p_y^o - p_y^m\right)^2 + \left(p_z^o - p_z^m\right)^2} - r}{\partial y}, \tag{12.16}$$

$$D_z = \frac{\partial \sqrt{\left(p_x^o - p_x^m\right)^2 + \left(p_y^o - p_y^m\right)^2 + \left(p_z^o - p_z^m\right)^2} - r}{\partial z}, \tag{12.17}$$

where x, y and z are points in the object frame,
$\bar{R}_O$ is the rotation matrix of the target object frame and
$\bar{R}_R$ is the rotation matrix of the robot frame.

APPENDIX B

12.B.1 Modelling of the Robot''s Links and the Obstacles

Before carrying out collision checking, the variables x_k, y_k and z_k for k = 0, 1, 2 or 3 are expressed in terms of the world co-ordinates (x_w, y_w, z_w), with the aid of homogenous transformation matrices as below:

$$[x_w \quad y_w \quad z_w]^T = {}^{w}_{k}T \times [x_k \quad y_k \quad z_k]^T \quad , \tag{12.18}$$

$${}^{w}_{k}T = {}^{w}_{0}T \ldots {}^{k-2}_{k-1}T \times {}^{k-1}_{k}T \,, \tag{12.19}$$

where ${}^{k-1}_{k}T$ is the homogenous transformation matrix to transform a representation of co-ordinates of the kth frame $[x_k \quad y_k \quad z_k]^T$ to representation of co-ordinates of the (k-1)th frame $[x_{k-1} \quad y_{k-1} \quad z_{k-1}]^T$.

12.B.2 Searching Plane

This plane is described by:

$$p\cos A + q\cos B - C_1 = 0 \,, \tag{12.20}$$

in which

$$\cos A = \frac{q_e - q_s}{ModuPQ} \,,$$

$$\cos B = \frac{p_e - p_s}{ModuPQ} \,,$$

$$ModuPQ = \sqrt{(p_e - p_s)^2 + (q_e - q_s)^2} \,,$$

$$C_1 = p_s \cos A + q_s \cos B \,.$$

Now we have the equation of the first plane, and what remains to be done is to adjust the equation of the plane so that this plane is perpendicular to the r–p plane, containing the line between the start point and the end point [12.28]. This plane is described by

$$r\cos C + p\cos D - C_2 = 0 \quad , \tag{12.21}$$

where

$$\cos C = \frac{p_e - p_s}{ModuRP},$$

$$\cos D = \frac{r_e - r_s}{ModuRP},$$

$$ModuRP = \sqrt{(r_e - r_s)^2 + (p_e - p_s)^2},$$

$$C_2 = r_s \cos C + p_s \cos D \cdot$$

Equations (12.20) and (12.21) include the basic elements for forming the general equation of a searching plane which contains the line through the start point and the end point. The general equation is shown in (12.22) and it can be reduced as in (12.23):

$$(p \cos A + q \cos B - C_1) + K(r \cos C + p \cos D - C_2) = 0, \tag{12.22}$$

$$(\cos A + K \cos D)p + (\cos B)q + (K \cos C)r - (C_1 + KC_2) = 0, \tag{12.23}$$

$$\lambda = \tan^{-1}\left(-\frac{\cos B}{K \cos C}\right), \tag{12.24}$$

where K is an arbitrary factor which can change the inclination of the searching plane.

The inclination angle λ of the searching plane relative to the p–q plane can be calculated from (12.26). By changing the value of the factor K (K=0 → K=10000, implies 90° rotation), it is possible to obtain a set of planes, each of which contains a straight line between the start point and the end point and each of them satisfying (12.23) As a result of using the equation of the plane only two of the three variables in (12.23) are independent, and one is dependent. If we examine one plane at a time in the configuration space, the searching problem becomes two-dimensional. The next section shows a logical method for 2-D contour searching, which is used in the tests.

12.B.3 Intersection of Polyhedra

It is impossible for an object to be totally enclosed in the robot body without touching any surface. If there is a collision between two polyhedral objects there must also be an intersection between the facets of the two polyhedral models of the items [12.28]. According to [12.6], there can be three types of virtual intersections between convex polyhedral objects. These types are characterised as follows:

- A type 1 intersection is between a vertex of the obstacle model and a facet of the model of the kth link.
- A type 2 intersection is between a facet of the obstacle model and a vertex of the model of the kth link.
- A type 3 intersection is between an edge of the obstacle model and an edge of the model of the kth link.

By distinguishing between the types of virtual intersections, it is possible to determine the direction for the robot to move in order to avoid hitting a moving obstacle. More about path planning with moving obstacles is given in [12.28].

Checking the intersections of the facets can be done in three steps. First, the signed distances from all vertices of the polyhedron of one link or obstacle to all the facets of the polyhedron of another object. If there are V vertices and F facets for each of the two polyhedra, there will be 2 x V x F distance calculations. Next, the vertices of the two facets in the test are picked up from a look-up table, the first facet from a robot link and the second from the obstacle. The facet equation in the world co-ordinate frame takes the form

$$Ax + By + Cz + D = 0 \tag{12.25}$$

where (x y z) are expressed in the world co-ordinate frame.

The distance function is then defined as

$$F(x,y,z) = Ax + By + Cz + D \tag{12.26}$$

where A, B, C and D describe the plane of the facet of one object, and (x y z) are the co-ordinates of the vertex of the other object.

After calculating distances between all vertices of the other object and the plane of the facet being considered, collision detection can be performed by the following procedure [12.28]:

- if all the distance functions have the same sign, there is no intersection;
- if the distance functions do not have the same sign, the vertex co-ordinates of the other object and the roles of A, B, C and D of the plane are reversed for the facets under consideration (change of roles);
- if all the distance functions have the same sign, again there is no intersection;
- an intersection may occur only if both positive and negative signs are found in each set of distance functions;
- if an intersection may occur, the intersection between the border lines of one facet and the other facet is calculated; an intersection takes place only if the point just calculated is inside the first facet.

One way of performing the localisation of the intersection point is to take vectors from the intersection point to all vertices of the facet and then calculate the angles between all four of these vectors. When the absolute values of the angles are added up, it is possible to make a decision as to whether there is a collision. If the combined value of the angles is 360° or 2π radians a collision is detected. If the value is less than 360° or 2π radians there is no collision. Localisation of the intersection point is illustrated in Fig. 12.30.

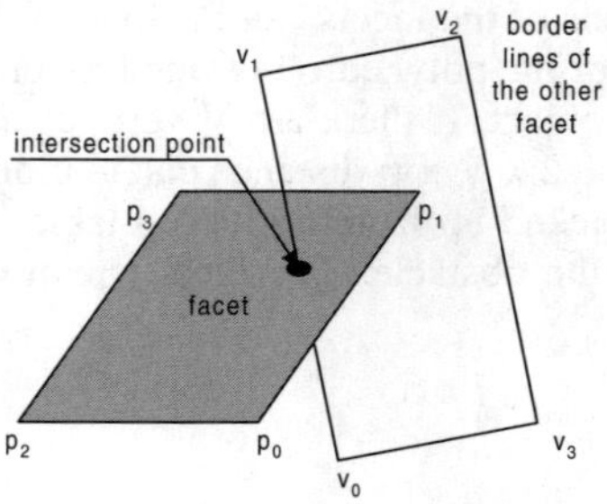

(a)

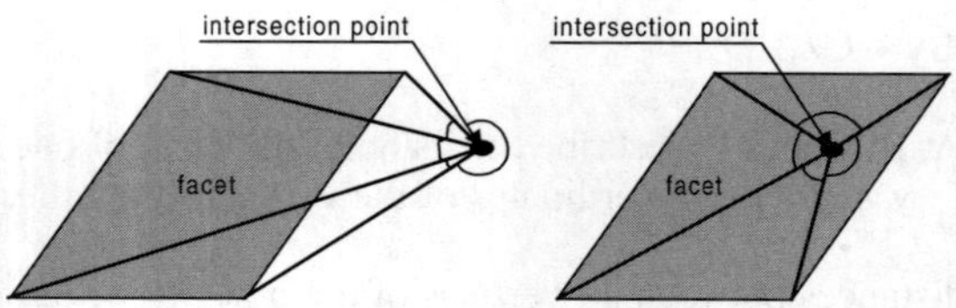

(b)

Fig. 12.30. Placement of the intersection point can be detected by angle rules because arccos(x) can only provide values between 0 and 180°, when $x \in [-1,1]$: (**a**) The intersection between two facets has been found and the intersection point can be calculated; (**b**) Two possible placements of the intersection point: collision, and no collision

13. Holonic Manufacturing Control: Rationales, Developments and Open Issues

Duncan C. McFarlane[1] and Stefan Bussmann[2]

[1] Institute for Manufacturing
University of Cambridge
United Kingdom
Email: dcm@eng.cam.ac.uk

[2] DaimlerChrysler AG
Research and Technology
Berlin, Germany
Email:Stefan.Bussmann@daimlerchrysler.com

Abstract. Holonic manufacturing systems support a more plug-and-play approach to configuring and operating manufacturing processes, and thereby address increasing efforts to meet the needs for market responsiveness and mass customised products. This chapter is primarily concerned with the control approaches associated with holonic manufacturing systems. It addresses three key issues. Firstly, a clear business rationale for a holonic approach to manufacturing control is outlined. Secondly, a number of the key developments in holonic control systems are summarised. Finally a number of outstanding issues for the design and implementation of holonic control systems are highlighted.

13.1 Introduction

The concept of holonic manufacturing was introduced by Suda in the early 1990s [13.58,13.59] to address emerging challenges for manufacturing in the 21st century. It is intended to enable a "plug and play" approach to designing and operating a manufacturing system. In the last ten years, an increasing amount of research has been devoted to holonic manufacturing over a diverse range of both theoretical issues and industrial applications. This chapter draws on and extends previous research by the authors examining the rationale for holonic control and its developments [13.13], [13.47]. It describes a vision for holonic manufacturing and assesses how that vision matches the current needs of manufacturing businesses. It then briefly reviews current research in holonic control systems development and outlines a number of open issues that must be addressed before holonic control systems can be deployed industrially.

13.1.1 Background to Holonic Systems

The holonic concept was proposed by the philosopher Arthur Koestler in order to explain the evolution of biological and social systems [13.37]. He made two key observations:

1. These systems evolve and grow to satisfy increasingly complex and changing needs by creating stable "intermediate" forms which are self-reliant and more capable than the initial systems.
2. In living and organisational systems it is generally difficult to distinguish between 'wholes' and 'parts': almost every distinguishable element is simultaneously a whole (an essentially autonomous body) and a part (an integrated section of a larger, more capable body).

These observations led Koestler to propose the word *"holon",* which is a combination of the Greek word *holos* meaning whole and the Greek suffix *on* meaning particle or part as in proton or neutron. Suda's observation [13.58,13.59] was that such properties would be highly desirable in a manufacturing operation which is subject to increasingly stringent demands and faster changes. He therefore proposed a building-block or "holon"-based model for designing and operating elements comprising manufacturing processes similar in concept to the one defined in [13.12] which is outlined in Fig. 13.1.

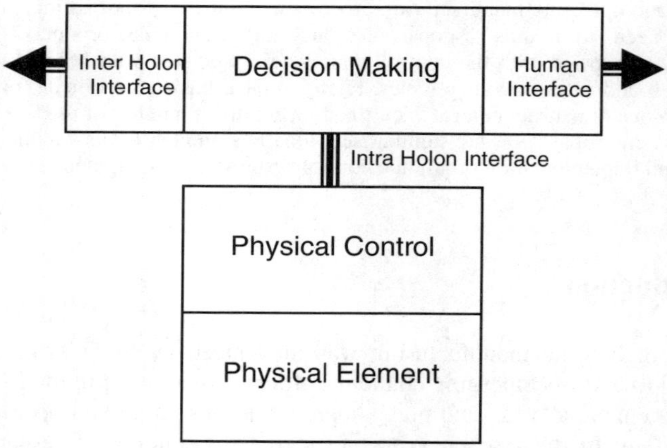

Fig. 13.1. General architecture of a holon

Some key properties of a (holonic) manufacturing system developed from this model are:

- *Autonomy* – the capability of a manufacturing unit to create and control the execution of its own plans and/or strategies (and to maintain its own functions).
- *Cooperation* – the process whereby a set of manufacturing units develop mutually acceptable plans and execute them.
- *Selforganisation* – the ability of manufacturing units to collect and arrange themselves in order to achieve a production goal.
- *Reconfigurability* – the ability of a function of a manufacturing unit to be simply altered in a timely and cost effective manner.

13.1.2 Holonic Manufacturing Systems

We now provide some simple descriptions, definitions and examples of *holons* and *holonic manufacturing systems*. We define a manufacturing holon as "an autonomous and cooperative building block of a manufacturing system for transforming, transporting, storing physical and information objects" [13.17]. It consists of a control part and an optional physical processing part (see Fig. 13.1.) Hence, for example, a suitable combination of a machine tool, an NC controller and an operator interacting via a suitable interface could form a holon which transforms physical objects in a manufacturing environment. Other examples of manufacturing holons could be products and their associated production plans, customer orders and information-processing functions. A holon can itself also consist of other holons which provide the necessary processing, information, and human interfaces to the outside world. A "system of holons which can cooperate to achieve a goal or objective'' is then called a holarchy [13.17]. Holarchies can be created and dissolved dynamically depending on the current needs of the manufacturing process.

Hence, the intention is that a combination of different holons will be responsible for the entire production operation, including not only the production planning and control functions, but also the physical transformation of raw materials into products, the management of the equipment performing the production tasks, and the necessary reporting functions. In this case the set of holons is referred to as a holonic manufacturing system. A holonic-systems view of the manufacturing operation is one of creating a working manufacturing environment from the bottom up. By providing the facilities within holons to both (a) support all production and control functions required to complete production tasks and (b) manage the underlying equipment and systems, a complete production systems is built up like a jigsaw puzzle!

Since 1990 there has been a significant amount of reported research and a wide range of publications produced that refer to control systems in a holonic context:

- *conceptual descriptions* – providing a high-level overview of the way in which holonic control systems might be structured and might function [13.5, 13.6, 13.17, 13.47, 13.49, 13.57, 13.58, 13.62, 13.65, 13.67].
- *specific architectural designs and specifications* – providing detailed descriptions of the different functions of a holon and its interconnection with other holons and the way the holons operate. A range of architectures have been proposed in the literature – some more feasible than others [13.10, 13.12, 13.17, 13.28, 13.29, 13.30, 13.38, 13.45, 13.46, 13.50, 13.56, 13.63, 13.66, 13.67, 13.68].
- *distributed algorithm design* – a number of authors have also developed algorithms, protocols and interaction mechanisms which underpin holonic operating methodologies [13.4, 13.7, 13.8, 13.9, 13.19, 13.21, 13.20, 13.28, 13.33, 13.34, 13.39, 13.43, 13.44, 13.48, 13.53, 13.63, 13.70, 13.71, 13.72].
- *simulated or prototype implementations* – are less prevalent in the literature and have generally been proof of concept level implementations as opposed to industrial implementations [13.1, 13.4, 13.15, 13.14, 13.26, 13.28, 13.29, 13.31, 13.32, 13.36, 13.43, 13.60, 13.61, 13.62, 13.65, 13.66].

Despite the level of intensity applied to this work, there are a significant number of gaps which must be filled before holonic control can become an industrial reality. This is addressed in the latter part of this chapter.

13.1.3 Manufacturing Control in a Holonic Context

Holonic manufacturing represents an alternative approach to designing manufacturing systems to *Computer Integrated Manufacturing (CIM)*, as an integrating methodology for computer control in manufacturing. The difference may be summarised as a contrast between a so called *top down* approach and a so called *bottom up* approach. For the last 15–20 years, CIM has been a blueprint for the design and specification of *hierarchical, centralised* computer-based operations, while holonic manufacturing represents a possible blueprint for *distributed computer-based* manufacturing operations which support local decision making. (See [13.40] for a detailed comparison between CIM and holonic manufacturing.) We conclude this section by summarising some of the primary differences between holonic control solutions and their conventional counterparts. Fig. 13.2 lists the main differences between conventional CIM-based and holonic approaches to production control. A preliminary version of this table was given in [13.47]. Fundamentally, the holonic characteristics are intended to provide for a "plug and play" control environment which is as simple to add to, remove from, or reconfigure as are today's computer networks.

	Conventional control solution	**Holonic control solution**
1	The intended product is closely-considered in the design of the plant and its control system	The holonic resources and their controls are designed independently of any particular products
2	Fixed layered, hierarchical architecture representing the different production control functions	Dynamically changing hierarchy of control functions
3	Command/response mechanism provides the basis for the connection between different production control functions	Interactive interchange / simultaneous solution is possible between different production control functionss
4	Predetermined solution strategy for individual production control functions	Control strategy determined by the nature of the operation required and the different holons involved
5	Typically a centralised solver for each individual production control function	Typically a distributed solver, with co-operative interactions between nodes, and simultaneously addressing more than one control function
6	Control solution time constrained by processing power	Control solution time constrained by communications speed

Fig. 13.2. Characteristics of Conventional and Holonic Control

Like CIM, holonic manufacturing approaches have already exploited and will continue to exploit many existing technologies and methods. For example, the manufacturing control approaches appearing in the holonic literature have many characteristics in common with existing developments in *heterarchical manufacturing control* (see, for example, [13.22, 13.23, 13.24, 13.25, 13.41, 13.42]), *intelligent scheduling* (see [13.52, 13.73] and the references therein) and in *multi agent systems* [13.2, 13.12, 13.35].

13.1.4 A Simple Example of Holonic Manufacturing Operations

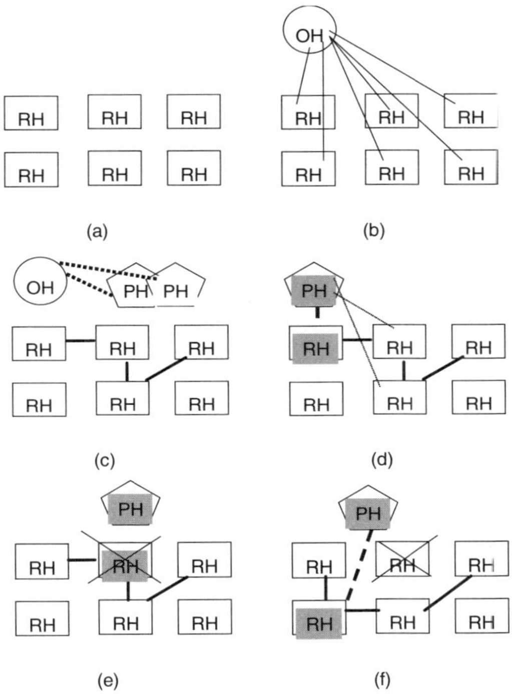

Fig. 13.3. Selforganisation of order processing.

In order to better illustrate the characteristics of holonic control systems described above, we now provide a simple example. This illustration is an extension of an example that appeared in [13.13, 13.47] and is outlined in Fig. 13.3. Initially (referring to Fig. 13.3a), a factory based on the holonic manufacturing principles consists only of a pool of unassigned *resource* holons (RHs). As discussed in Figure 13.2, there is no *a-priori* relationship between the different resources and no pre-determined control code has been written in anticipation of future orders. Hence, upon arrival of an order, an *order* holon (OH) is created which – empowered with customer order requirements and product specifications – begins to negotiate with resource holons regarding the provision of certain manufacturing operations (Fig 13.3b). During the negotiation process, the order holon demands specific properties required from the operation, such as high quality or fast delivery, while the resource holons try to optimise their utilization – for example, to ensure high overall throughput for the factory. At the end of the negotiation, the resource holons form an agreed "production-line" and the order holon initiates the creation of *product or workpiece* holons (PHs) as in Figure 13.3c.

The product holons enter the newly formed "production line" (e.g., from raw-material stock) and immediately bargain for appropriate resources as specified from the order negotiation in order to get processed. Each product holon does so individually and focuses on the next operation(s). Once these operations have been performed at a resource, the product re-initiates the bargaining with the holons representing the remaining (next) operations (Fig 13.3d). The overall organisation of the resource holarchy – initially or subsequently negotiated between order and resource holons – ensures that the product load is efficiently distributed over the available resources in order to achieve the global goals of this holarchy.

In the case of a disturbance (Fig 13.3e), the affected resource holon removes itself from the resource pool and goes to a repair booth. The remaining resource holons re-organise themselves in order to take account of the capacity loss (Fig 13.3f). From the point of view of the product holons, the processing continues as *business as usual*, only with fewer resource holons to negotiate with. After repair, the resource holon rejoins the resource holon pool again.

At the end of the order processing, the order holon is removed and the resource holarchy dissolves into the resource holons, which then try to participate in new order holarchies.

13.2. Manufacturing Requirements Analysis

Manufacturing operations are not an end in themselves, but serve as a means to achieve the business goals of a company. It is therefore essential for an evaluation or comparison of manufacturing concepts to identify the requirements on the manufacturing process against which the concepts should be evaluated. These requirements are derived from the business goals and the given or expected market conditions. Business goals and market conditions, however, may change over time and thus the set of manufacturing requirements may change. A manufacturing approach that has been sufficient until now, may result in a poor performance in the future. Consequently, manufacturing concepts should be evaluated not only against the existing requirements, but also against future (possibly unknown) requirements. This section therefore looks at the current business trends and shows how these will change the manufacturing environment. The new manufacturing requirements are then used to derive requirements on the control of future manufacturing systems. This process is outlined in Fig. 13.4.

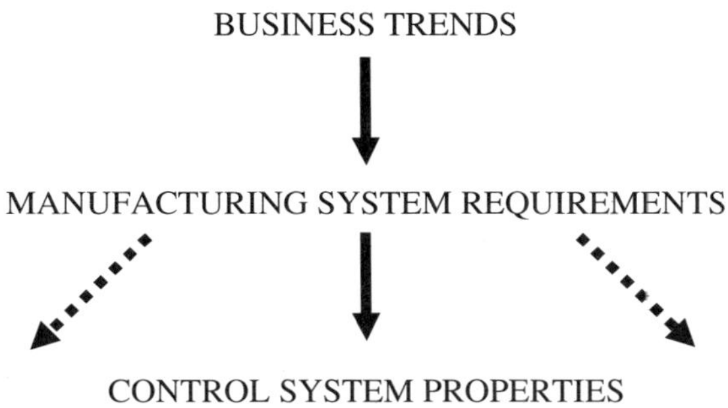

Fig. 13.4. Requirements Break Down Process.

We note that the initial development of these results was made specifically with the automobile industry in mind [13.13]. The results have since been validated however against the requirements of Fast Moving Consumer Goods (FMCG) companies and several other types of manufacturers.

13.2.1 Business Trends

It is difficult to estimate what the business requirements of the 21st century will be. The current requirements of producing goods of a specific quality at low cos will certainly remain in place. But the current market trends suggest that additional requirements will arise which will determine the competitiveness of a company and thus its survival in this century.

Recently, manufacturing industry has been facing a continuous change from a supplier's to a customer's market. The growing surplus of industrial capacity provides the customer with a greater choice, and increases the competition between suppliers. Aware of this power, the customer becomes more demanding and less loyal to a particular product brand. He/she demands constant product innovation, low-cost customisation and better service, and chooses the product which meets his requirements best. In combination with globalisation, these trends will increase in the future.

The consequences for the manufacturing industry are manifold. Companies must shorten product life cycles, reduce time-to-market, increase product variety and instantly satisfy demand, while maintaining quality and reducing investment costs. These consequences imply:

- more complex products (because of more features and more variants),
- faster changing products (because of reduced product life-cycles),
- faster introduction of products (because of reduced time-to-market),
- a volatile output (in total volume and variant mix), and
- reduced investment (per product).

The effects can be summarised as *increasing complexity* and *continual change* with *decreasing costs*.

13.2.2 Manufacturing System Requirements

Most existing requirements placed on a manufacturing operation will still apply in the future. These include guaranteed performance, high reliability of equipment, quality assurance, cost minimisation etc. Given the trends described in the previous section, though, additional requirements will become relevant, if not predominant.

- *Increasing complexity:* A major requirement will be to minimise the complexity of the manufacturing process (despite the likely increases in the variety of products and product ranges). This can be achieved basically by reducing the number of manufacturing system components and by standardising the structure of these components and their interaction. Nevertheless, there is a limit to reduction and standardisation, as a complex product requires a certain set of complex operations. The remaining process complexity must be mastered. This can be achieved on the one hand by creating an intuitive, self-explaining structure of the manufacturing (and control) system, and on the other hand by ensuring a well-defined behaviour upon certain actions and events. Ideally, the control layer of a manufacturing system should be completely transparent to the end-user, and any actions or events should exhibit well-known effects on the overall system performance. In particular, the control layer should not introduce additional complexity and the overall behaviour of a manufacturing system should be well defined under all circumstances.
- *Constant product changes:* Constant product changes require the reuse of existing manufacturing equipment. Buying new equipment either is too costly or takes too much time. Reuse of equipment implies the reuse of the units and the reorganisation of the manufacturing process. Reuse of manufacturing units can be achieved either through flexibility of function or through reconfigurability. A unit is immediately reusable if the new operations required are part of the range and mix of operations of this unit. High functional flexibility thus increases the chances of equipment reuse. Units equipped (up front) with a large range of operations, however, can be very costly. In contrast, the costs of a unit are often reduced considerably if the reuse is provided through manual reconfigurability. For monthly product changes, this is acceptable. Weekly or daily product changes, though, are likely to require instant unit flexibility. An analogous requirement applies to process re-organisation. The manufacturing process must be either flexible or reconfigurable in order to deal with the product changes.
- *Volatile Output:* The volatility of the demand forces the vendors to adapt their output to the market. A product sells only when the market demands it. If a company does not supply the right product at the right time, another company makes the deal. As a consequence, the manufacturing system must be able to vary its production output. This implies scalability of the manufacturing system if the total volume changes, and inter-product flexibility if the product mix changes. Scalability can be achieved either by extending the working time or by adding more resources
- *Reduced Investment and Robustness:* The task of managing change becomes even more difficult if it has to be achieved at decreasing costs. A company might even decide not to provide full flexibility or reconfigurability if the costs are prohibitive. The real challenge is to manage change at low costs. A low investment approach to change management, however, creates a second difficulty, namely that of disturbances. A behaviour which is achieved under scarce resources is vulnerable to (internal and external) disturbances. Future manufacturing operations will therefore

require increasing robustness. Robustness can be achieved either structurally or dynamically. Buffers in terms of material or time slack provide structural robustness. System flexibility allows to adapt the process to failures, for instance by using spare resources or re-routing jobs.

13.2.3 Control System Properties

The requirements on the manufacturing system have also implications for the control of such a system. Many requirements can only be achieved if the control system meets equivalent requirements. Requirements like unit flexibility or reconfigurability are mainly hardware issues, but system responsiveness is certainly impossible without some kind of intelligent control. This subsection therefore looks at the consequences of the new manufacturing requirements for the control, regardless of the actual design and implementation of the control system.

I. The architecture of the control should be decentralised and physically-based.
For even small manufacturing systems, a centralised approach to control is practically impossible. There must be at least some kind of decentralisation. Decentralisation, however, can take many forms. For instance, a system can be functionally or geographically distributed. But in order to allow for maximum flexibility, the decentralisation should be product- and resource-based. In a resource-based architecture, every resource contains all control capabilities necessary to process jobs. In particular, a set of resources is able to allocate jobs to resources without a centralised support. The advantage of the resource-oriented approach is that the system can be changed and scaled up fairly easily. Furthermore, the control corresponds in its structure to the manufacturing system and thus reduces the complexity added by the control system to a minimum. The control activities might even become transparent to the end-user. A similar argument applies to equipping orders and work pieces with the necessary control capabilities to get produced.

II. Control interactions should be abstract, generalised and flexible.
A resource-based control system is certainly easier to change and scale up than a centralised or functionally decentralised system. Maximum changeability, however, is only achieved if dependencies between resources are reduced to a minimum. If one resource is changed, but other resources that rely heavily on exactly this resource and its specific behaviour, then a change of single resource entails a lot of changes at other resources. Consequently, in order to achieve maximum changeability, interaction between resources should be decoupled by ensuring abstract, generalised and flexible interactions.

III. The control should be both reactive and proactive.
In order to respond to short-term changes and disturbances, the control must be reactive. This includes the ability to recognise critical situations, make decisions about the reaction, and perform corresponding actions. In contrast to traditional planning and control approaches, the product- and resource-based architecture also distributes the planning capabilities since they depend strongly on the characteristics of the resources and the product. A resource, for instance, must also participate in the alloca-

tion of jobs or the sequencing of operations. As a result, the control must be reactive and proactive at the same time.

IV. The control should be selforganising.

The need to adapt the manufacturing process in the face of changes or disturbances will affect not only the resources, but also the organisation of the manufacturing process as a whole. Obviously, in a highly responsive manufacturing system, the organisation must be responsive too and this responsiveness should emerge from any (re-)configuration of the resources and rearrangement of the process.

13.2.4 Connecting Business Drivers and Control System Needs

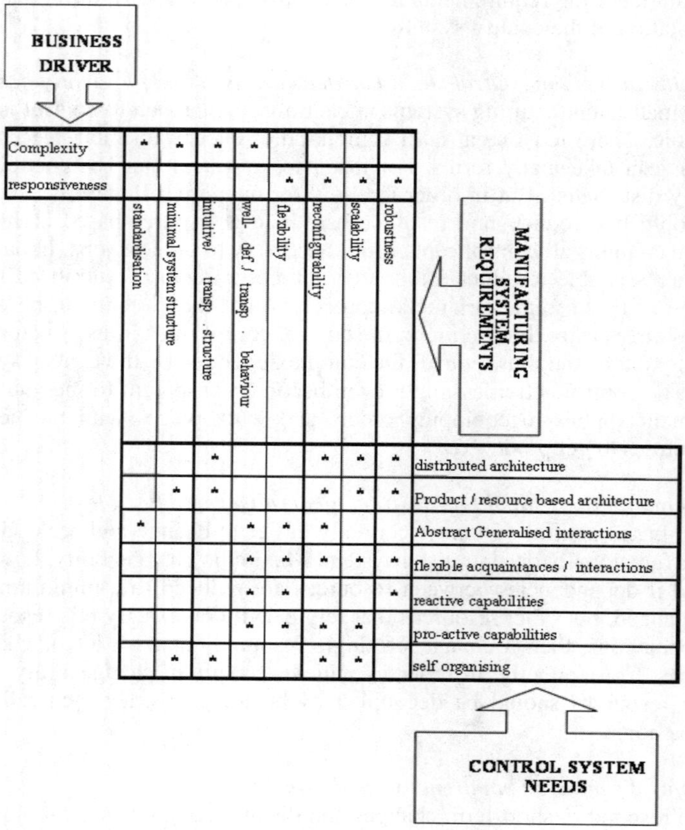

Fig. 13.5. Linking Business Drivers and Control System Needs

The table in Figure 13.5 uses a decomposition approach to connect business drivers to control system needs, by examining the impact of the business drivers on manufacturing system requirements as an intermediary. Where a connection is important between the different levels it is marked with an asterisk.

We note that this requirements deployment is completely focused on establishing control system needs only and is hence somewhat limited as a result. A full requirements deployment would also consider – for example – factors such as human needs in terms of skills, information management, adaptability etc, or equipment requirements such as flexibility of functions, routings, fixtures, reliability of machines, processes etc. We emphasise here that holonic manufacturing is more than simply a control system and we will return to this issue in Section 13.4 in discussing conditions under which a holonic control system implementation is likely to be successful.

13.2.5 Aligning the Holonic Control Approach with Control System Needs

The short description of the holonic vision of manufacturing in Section 13.1.4 indicates that a holonic approach can address many of the requirements (I-IV) identified in Section 13.2.3. The requirements are met because of the basic concepts that underpin the holonic approach:

- Holonic structure – The holonic approach inherently proposes a distributed, product- and resource-based architecture for the manufacturing operations. (Requirement I)
- Autonomy – Each holon has local recognition, decision making, planning, and action taking capabilities, enabling it to behave reactively and pro-actively in a dynamic environment. (Requirements I,III)
- Cooperation – Coordination, negotiation, bargaining, and other cooperation techniques allow holons to flexibly interact with other holons in an abstract form. Because of the dynamic nature of the holarchies, each holon must employ generalised interaction patterns and manage dynamic acquaintances. (Requirement II)
- Selforganisation – Holonic manufacturing systems immediately re-negotiate the organisation of the manufacturing operations whenever the environmental conditions change. (Requirement IV)
- Reconfigurability – Because of the modular approach, holons can be reconfigured locally once the inherent flexibility of the holons has reached its limit. (Requirements II,IV)

To summarise, when fully developed, holonic manufacturing control has the potential to address many of the outstanding problems facing today's industrial control systems. The extent to which holonic control systems have been developed to date is addressed next.

13.3. Holonic Control Systems Development

Architectures for the design of holonic elements and their interconnection and algorithms which define their (inter)operation represent the two key classes of developments in holonic control research to date. The principal focus of this section is to provide a brief review the development of algorithms which support holonic control systems. The reasons for this focus is that:

1. The existence of effective control algorithms is an indicator of the degree of maturity of holonic control research – without them it is not possible to assess the likely performance of a production operation running with holonic systems in place. In contrast, it is expected that numerous architectures for designing and implementing holonic control systems will continue to be proposed (as discussed in Section 13.1) and will also vary as information technology advances.
2. Algorithms supporting holonic control developed thus far can be directly contrasted with those found in conventional production control environments. This enables us to compare and contrast different holonic control developments using a common (conventional) view of production control.
3. Holonic systems architecture developments have been examined elsewhere in this book.

However, to begin, we establish some common ground, so as to be able to uniformly describe the different systems architectures used in the algorithms that follow.

13.3.1 Holonic Control Architectures

In order to simplify the discussions that follow, we will assume a common description of a manufacturing process operating on holonic principles. The control system is assumed to comprise software elements aligned with the different physical entities in the manufacturing environment, namely, *resources, products, customer orders* and *coordinators* of operations. The software element and the physical entity, coupled by a suitable communications network (see Figure 13.1) represent the different manufacturing holons for this process. Each of these holons – once created – is assumed capable of a degree of local reasoning and decision making and an ability to communicate in an interactive manner with other holons. We will discuss the way in which these capabilities support different production planning and control issues in the next section.

For more details on the descriptions and architectures of individual holons and their interconnection systems, the reader is referred to [13.12, 13.17, 13.67, 13.71]. For example, the so called *Product – Resource – Order – Staff Architecture* (PROSA) proposed in [13.68] has been widely adopted and more recently an architecture based only on the nesting of a *Product – Resource* model – the so called *Holonic Component Based Architecture (HCBA)* – has been developed [13.15].

13.3.2 Review of Developments in Holonic Control Algorithms

This section is intended to briefly summarise the current work relating to the algorithms deployed between holons in order to generate production control solutions. For more details the reader is referred to [13.47], in which a comprehensive review of holonic production planning and control is carried out.

We shall describe how the current developments in holonic control research apply to each of the control activities listed in Fig 13.6, (i.e. planning and scheduling, shop floor control and execution and machine & device control).

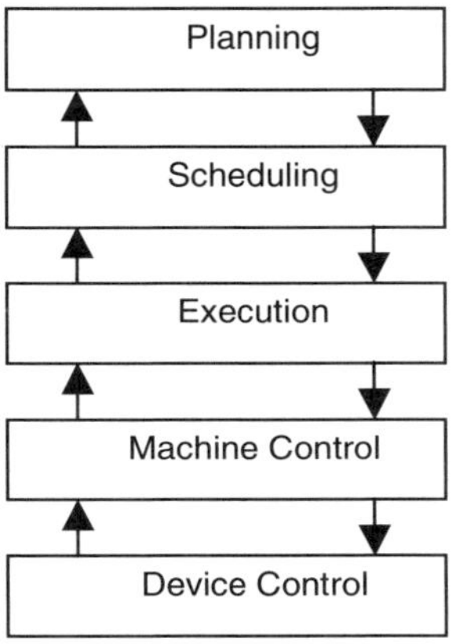

Fig. 13.6. Typical manufacturing control hierarchy

The solutions for holonic control we will be discussing will mainly address only one of the control issues in Fig 13.6 at a time, and can be seen as providing partial holonic solutions of the form in Fig 13.7. That is each, separate control function in the decision making hierarchy in Figure 13.6 can be addressed by a partial holonic solution.

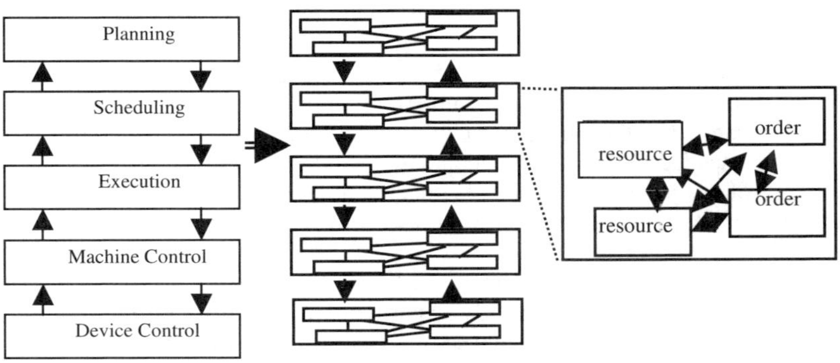

Fig. 13.7. Partial holonic control solution

Planning and Scheduling. We restrict the following discussions on production planning to a) the decomposition of an order into a sequence of production operations and b) the nominal allocation of operations to resource types (but not to specific resources or times). Approaches to holonic planning typically involve a number of the following steps:

1. Each product holon performs a decomposition of the supplied product specification into constituent parts or sub-assemblies.
2. For each product, the manufacturing operations needed are identified (by the product holon).
3. The types of resources needed to provide operations are selected via interaction approach between product and resource holons.
4. An interactive process involving resource holons and product holons is used to determine a suitable set of operations.
5. A full make sequence (assembly plan) is finalised and this normally resides in the product holon.

We note that this assumes – *a priori* – that the products required to fulfil an order have already been identified and also that either the product or the resource is coordinating the planning process. The benefits of a holonic approach compared with more conventional approaches are principally due to the distributed and interactive nature of the planning process, enabling new products and/or production resources to be introduced without major system alterations. The close connection between the individual holons and the physical resources they represent also enables planning to maintain a close alignment with the (dynamically changing) capabilities available on the shop floor. Holonic planning approaches have been reported in [13.4, 13.20, 13.21, 13.28, 13.29, 13.30, 13.31, 13.32, 13.55].

Similarly, we assume that scheduling simply involves (a) the allocation of production operations to specific resources and (b) the specification of the timing (start, duration, completion) for those operations. The key characteristics which typify a holonic scheduling approach are [13.47]:

1. A local decision-making and computational capability associated with each holon.
2. A cooperative interaction strategy which governs the way in which holons exchange information and determine mutually acceptable solutions.
3. An interchange mechanism or protocol which manages the exchange of the message types needed to execute the cooperative strategy.
4. A means of ensuring that the global concerns of the factory are addressed.
5. A degree of central coordination (not present in all solutions).

Work on holonic scheduling to date [13.1, 13.11, 13.28, 13.29, 13.31, 13.32, 13.43, 13.50, 13.53, 13.60] has predominantly experimented with different algorithms and simulated testing although issues such as the desire for an *emergent* schedule vs. a fixed schedule structure, and the relationship between scheduling and execution has been examined also.

Execution and Shop Floor Control. Execution or shop floor control involves the initiation, control, monitoring and termination of tasks and involves actual times and actual production settings. Within a holonic manufacturing system, execution is predominantly concerned with (a) ensuring that the holon is capable of establishing and maintaining autonomous operations and (b) that it undertakes tasks compatible with production requirements even in the face of disruptions. Execution has been addressed in the holonic literature by [13.27, 13.31, 13.32, 13.66], where the autonomous behaviour of the (resource) holons in each case is managed by an internal model of the operations. Such a model is an essential requirement for the holon's self-management. The novel elements of a holonic approach to execution are that a) execution proceeds via a negotiated set of steps rather than a predetermined sequence and that b) the resources (machines) executing the manufacturing operation are also responsible for the decisions made about the timing and nature of the execution. A further important issue is the relationship between execution and scheduling which has been discussed in [13.54, 13.66, 13.69].

Machine and Device Control. In a holonic system, machine control – which involves the initiation, coordination and monitoring of the different machine functions or devices required to support the execution of production tasks by an individual machine – has been largely treated as a conventional control problem coupled to a higher level holonic operation. (See for example [13.3, 13.54, 13.61, 13.62, 13.71].) The focus in these cases has been on achieving effective interfaces. Only in [13.51] is the possibility of a machine itself running on holonic principles truly considered where the interactions of the individual devices which constitute a machine are determined cooperatively. There has been even less work in the device area – that is, the actuation, sensing and feedback control of the physical operations which support a machine – but most of the above comments also apply.

Although developments in both holonic machine control and holonic device control have been limited to date, opportunities for greater flexibility and disturbance handling present themselves in the way in which trajectories and control actions could be negotiated to suit the current operational environment rather than following predetermined paths. One would expect such a system to be more adaptable to changing conditions arising, for example, from wear, damaged parts, faulty components or faulty sensors.

13.3.3 Summary of Developments

The concepts that were described as underpinning the holonic approach in Section 13.2 were the holonic structure and properties of autonomy, cooperation, selforganisation and reconfigurability. Examining the results of this section's review, we see that (a) numerous architectures for holonic manufacturing systems have been proposed, (b) cooperative mechanisms have been explored to a degree within the different production control levels, and (c) requirements for autonomy have been established, particularly with regard to the lower-level control functions in Fig 13.6. However, apart from organisational aspects associated in holonic planning there has been little or no attempt to explicitly address the requirements for selforganisation which underpin the flexible response of a holonic system. (Issues of reconfigurability are primarily associated with architectural rather than algorithmic considerations and therefore can not be assessed directly here.)

In addition, there are more practical issues requiring attention. In the next section we shall summarise a number of outstanding issues that need to be addressed before widespread applications of holonic control systems can be considered.

13.4 Open Issues

In this section we address a number of significant open issues that must be addressed before holonic control solutions can be expected to play any significant part in next generation manufacturing systems. We examine issues in both the design and implementation of holonic control systems and then explore possible application areas beyond discrete manufacturing.

13.4.1 Open Issues in Holonic Control System Design

There are a number of key design issues that must be addressed before holonic control solutions can be expected to play any significant part in next generation manufacturing production systems. These points reflect the fact that to date many of the developments described in Section 13.2 have been of an exploratory nature, and there is a definite need now to move to a more rigorous, scientific assessment of the designs that have been proposed.

- *Availability of proven design methodologies:* Only when definitive methodologies have been produced which provide clear, unambiguous guidelines can it be expected that a level of consistency and reliability will emerge in holonic control system design. To date, there have been no two applications built following the same design methodology and hence it is very difficult to evaluate the potential reusability of any method specified. The lack of effective methodologies also means that researchers in the area have no common base for their own developments which is clearly inefficient.
- *Analysis of the performance of holonic manufacturing systems:* A limitation in the research to date has been the lack of assessment of the relative performance of the control mechanisms that the holonic manufacturing systems support. In particular, holonic manufacturing systems are frequently cited as performing well in the face of disturbances, but there has been little reported evidence of them being shown to do so. Any serious industrial commitment to holonic manufacturing systems in the future will require a demonstrated ability to *improve* performance beyond that of conventional systems. To be fully effective, holonic manufacturing requires a complete reorganisation of production operations, which is a costly undertaking. Therefore, it is very important to show and quantify the benefits as is done for example in [13.14].

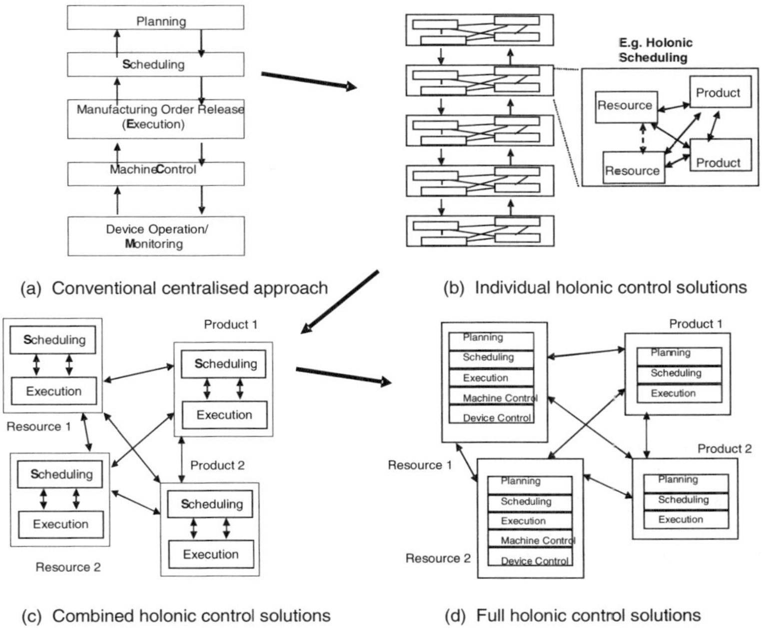

Fig. 13.8. Migration to holonic control algorithms

- *Migration to full holonic manufacturing control algorithms:* The review in Section 13.3 reflects a research activity that has to date aligned itself with the conventional control systems hierarchy in Fig 13.7. That is, distributed, cooperative solutions have been sought for each of the individual problems on this hierarchy. Few authors, however, have truly attempted to question the relatively static *command-response* connections between these layers. These current developments are illustrated in 13.8(b). It is the authors' opinion that a new, more holistic approach is required for the control of manufacturing operations, which seeks to achieve cooperative interaction *across* these layers as well as between elements *within* them. For example, a separate planning and scheduling phase is in fact restrictive, because planning can commit an order to a particular make sequence when in fact more than one may be possible and each option may be more or less desirable depending on the current plant state. Hence combining planning and scheduling may be highly attractive, at least if planning options are not deleted until scheduling is considered. A distributed, interactive approach to *combined* planning and scheduling, to *combined* scheduling and execution or even to *combined* execution and control should present a relatively straightforward migration from the current state of development. (See Fig. 13.8 (c)). From these combined solution approaches, the next migration step is to consider systems which support comprehensive manufacturing holons that can seamlessly integrate all of the five control functions into their operations (Fig. 13.8 (d)). We emphasise that at this stage Fig. 13.8(d) is very much a target, and it has not yet been demonstrated that distributed planning and scheduling functions can coexist with execution and control functions within a single logical unit.

13.4.2 Open Issues in Holonic Control Systems Implementation

Assuming that the a well defined holonic control design process is available with performance guarantees established, it is also important to ensure an effective environment for implementation. The following items identify some of the key issues to be considered in implementing and maintaining an effective holonic control environment.

- *Establishing suitable implementation architectures compatible with existing and future commercial computing systems:* There has been little work done on determining the compatibility of the holonic control with the current or the next generation of industrial control and computing systems. Holonic systems will require a high level of reasoning and computational capability at the shop floor level, coupled with more flexible communication and more dynamic interfaces to human operators and users. Determining how to construct and implement system architectures capable of fully supporting holonic operations while still operating with existing legacy systems will also be a major issue as the capabilities of holonic systems reach industrial strength. In the shorter term, suitable migration approaches for the implementation of intermediate holonic control capabilities are required (see, for example [13.16, 13.27]), and effort is required to ensure systems vendors can access and adopt these approaches.
- *Appropriate environments for implementing holonic control:* For a successful, effective implementation which achieves the target behaviour outlined in Section 13.2, the holonic control system is ideally be complemented by several other features in a production environment:
  - *flexible resources* (e.g. machines, storage and transportation systems) providing redundancies of function, operation and/or routing
  - *flexible networking environments* for both resources and raw materials / products enabling accurate monitoring of highly varied operations (e.g. the latter is the aim of the Auto ID project which Unilever sponsors through its Digital Futures Laboratory)
  - *flexible and empowered manpower base:* HMS enhances the role of the skilled operator in assisting with disturbance handling
  - *flexible business information systems* enabling inter-site interactions between devices, resources and orders to be as flexible as those provided by holonic control systems within an operational site

  Under these conditions the flexible decision making policies afforded by holonic control system create the greatest value. Conversely, the implementation of a holonic control system on a single product, serial, production line with few variations or alternative operating scenarios and little operator intervention possible would be largely pointless.
- *Establishing suitable standards for holonic control systems:* Before any industrial confidence in holonic manufacturing systems can be established, a comprehensive

set of standards is required for the open specification of communications, data formats, systems architectures, algorithms and interfacing of holonic systems. There has to date been no comprehensive study of the requirements for standards in this area.

13.4.3 Other Applications for Holonic Control Systems

Holonic manufacturing has almost exclusively focussed on production applications within the discrete manufacturing domain. We note that there is considerable potential for applying the same approaches within other application domains:

- *Process control systems based on holonic principles:* In [13.18] it is noted that process industries today form a major part of the GDP within the economy of every nation. In general, process industries cover a very large and diverse sector of industries including petrochemicals, polymers, bulk and specialities chemicals and related utilities sectors. Historically, the processes have evolved from small scale, simple units, which were often operated in batch or semi-continuous mode. Energy and primary raw materials were relatively plentiful. Large and attractive profit margins were the basis on which these industries have grown at such a rapid rate. Over the last two decades, however, this sector of manufacturing has also experienced an important change, owing primarily to increasing energy costs and increasingly strict environmental regulations. Growing competitive markets demanding "*mass customisation*" of products and rapid technological innovations are replacing the old style of mass production and *copycat-* type R&D structures. There is also now a growing emphasis on improving efficiency and increasing the profitability of existing plants rather than creating plant expansions. In a similar manner to that described in Section 13.2, a set of rationales has been developed for applying *holonic manufacturing* principles as a technological solution to the growing business concerns in the chemical process industries. The anticipated benefits from holonic approach stem from the use of a distributed control systems architecture that supports flexible operations of units that can dynamically integrate and collaborate with others as and when the production conditions change.
- *Holonic manufacturing applications in the supply chain:* The restrictions to date of holonic control applications to production applications alone is rather artificial and a consequence of the origins of this movement from within the production control community. Holonic control concepts are applicable in any operation where there is benefit to be extracted from an increased level of autonomous, distributed decision making. As noted above, this is closely aligned to the flexibility and reconfigurability of the physical resources that will execute the actions resulting from these decisions. We note that a number of applications in supply chain logistics clearly fit this description. In fact, noting that holonic control concepts are best suited to environments in which a high degree customisation and of response to disturbances is required, it may in fact be more appropriate to prioritise holonic control applications for those activities in the supply chain *closest* to the end-customer. This is depicted in Figure 13.9. Stated simply, the closer the operation is to the end customer, the more the need for rapid customer response and the greater the need to introduce late customisation in order to address a specific customer need. In the case of detergent powder manufacture, for example, while the basic

chemicals are produced in bulk and even the detergent manufacturing is quite generic, the packaging of the product and its subsequent distribution processes are very highly geared towards changing customer needs. Hence the operations need to be able to respond quickly to a new order, a packaging variation or an increased volume requirement.

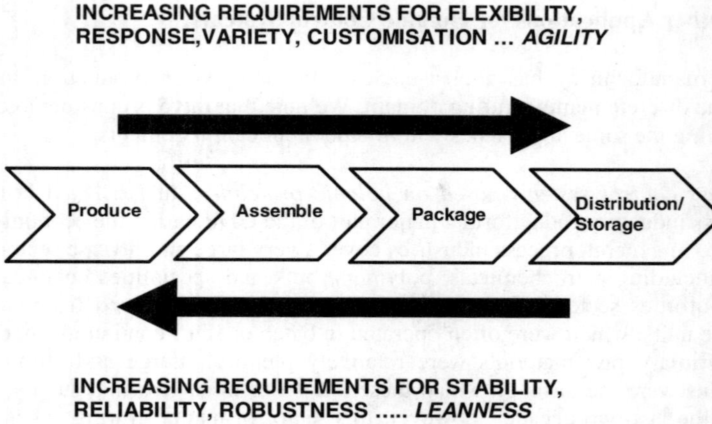

Fig. 13.9. Requirements variation along the supply chain

We conclude this section by noting that even restricting applications of holonic control system to the manufacturing supply chain may also be unnecessarily constraining. If the principles and their implementation can be well proven and demonstrated to be repeatable, then countless other opportunities present themselves.

13.5 Acknowledgements

The authors are grateful for the support and help of Hartwig Baumgärtel and Sven Broekner at Daimler Chrysler; Jeremy Matson, Jin Lung Chirn, Martin Kollingbaum, Alan Thorne, C Y Wong and Nirav Chokshi at the University of Cambridge; Dennis Jarvis, Jacquie Jarvis, Andrew Lucas and Martyn Fletcher at Agent Oriented Software; Roy Johnson at OMRON; and Andrew McDonald at Unilever.

References

[13.1] Agre, J., Elsley, G., McFarlane, D., Cheng, J., and Gunn, B., 1994, Holonic Control Of Cooling Control System, *Proceedings Of Rensslaers Manufacturing Conference* (New York).

[13.2] Baker, A., 1998, A Survey Of Factory Control Algorithms Which Can Be Implemented In A Multi-Agent Heterarchy: Dispatching, Scheduling, And Pull. *Journal of Manufacturing Systems*, 17,(4),297–320.

[13.3] Bengoa, A., 1996, An Approach To Holonic Components In Control Of Machine Tools. *Annals of CIRP*, 45,(1).

[13.4] Biswas, G., Sugato, B., And Saad, A., 1995, Holonic Planning And Scheduling For Assembly Tasks, Internal Report *CIS-95-01*, Centre For Intelligent Systems, Vanderbilt University, Nashville, TN.

[13.5] Bongaerts, L., Monostori, L., McFarlane, D., And Kadar, B., 1998a, Hierarchy in Distributed Shop-Floor Control, *Proceedings of IMS'98 – ESPRIT Workshop in Intelligent Manufacturing Systems* (Lausanne, Switzerland).

[13.6] Bongaerts, L., Monostori, L., McFarlane, D., And Kadar, B., 2000, Hierarchy in Distributed Shop-Floor Control. *Computers in Industry*, v43, n2, pp123–137.

[13.7] Bongaerts, L., Van Brussel, H., And Valckenaers, P., 1998, Schedule Execution Using Perturbation Analysis, *Proceedings of IEEE International Conference on Robotics And Automation* (Leuven, Belgium), 2747–2752.

[13.8] Bongaerts, L. B., Van Brussel H. Valckenaers, P. Peeters, P, 1997, Reactive Scheduling in Holonic Manufacturing Systems: Architecture, Dynamic Model And Cooperation Strategy, *Proceedings of Advanced Summer Institute of The Network of Excellence on Intelligent Control And Integrated Manufacturing Systems* (Budapest).

[13.9] Bongaerts, L. V., P. Brussel, Van, H. Peeters, P., 1997, Schedule Execution in Holonic Manufacturing Systems, *Proceedings of 29th CIRP International Seminar on Manufacturing Systems* (Osaka).

[13.10] Brennan, R., Balasubmramanian, S., And Norrie, D., 1997, Dynamic Control Architecture For Advanced Manufacturing Systems, *Proceedings of International Conference on Intelligent Systems For Advanced Manufacturing* (Pittsburgh, PA).

[13.11] Brown, J., And Mccarragher, B., 1998, Maintenance Resource Allocation Using Decentralised Co-Operative Control, Internal Report, Australian National University, (Canberra, Australia).

[13.12] Bussmann, S., 1998, An Agent-Oriented Architecture For Holonic Manufacturing Control, *First Open Workshop, Proceedings of IMS98 – ESPRIT Workshop in Intelligent Manufacturing Systems* (Lausanne, Switzerland).

[13.13] Bussmann, S., And McFarlane, D., 1999, Rationales For Holonic Control Systems, *Proceedings of IMS99* (Leuven, Belgium).

[13.14] Bussmann, S., Sieverding, J, 2001, Holonic Control of an Engine Assembly Plant – An Industrial Evaluation, Proceedings of IEEE SMC Conference, Tucson.

[13.15] Chirn, J.-L., And McFarlane, D. C., 1999, A Holonic Component-Based Architecture For Manufacturing Control Systems, *Proceedings of MAS99* (Vienna).

[13.16] Chirn, J.-L., And McFarlane, D. C., 2000, Application of A Holonic Component-Based Architecture For The Control of A Robot Assembly Cell, *in Proceedings of IEEE Robotics And Automation, 2000* (San Francisco).

[13.17] Christensen, J., 1994, Holonic Manufacturing Systems – Initial Architecture And Standards Directions, First European Conference on Holonic Manufacturing Systems, (Hannover, Germany).

[13.18] Chokshi, N, McFarlane, D, 2001, Rationales for Holonic Manufacturing Systems in the Chemical Process Industries, Proceedings of HOLOMAS 2001, France.

[13.19] Czerwinski, C., And Luh, P., 1994, Scheduling Products With Bills of Materials Using An Improved Lagrangian Relaxation Technique. *IEEE Transactions on Robotics And Automation*, 10, (2).

[13.20] Deen, S. M., 1993, Co-Operation Issues in Holonic Manufacturing Systems. H. Yoshikawa, Goossenaerts, J, Ed., *Information Infrastructure Systems For Manufacturing*, (Elsevier Science B V), 401–412.

[13.21] Deen, S. M., 1994, A Co-Operation Framework For Holonic Interactions in Manufacturing. *CKBS94: Proceedings of The Second International Working Conference on Cooperating Knowledge Based Systems*, 103–124.

[13.22] Dilts, D. M., Boyd, N. P., And Whorms, H. H., 1991, The Evolution of Control Architectures For Automated Manufacturing Systems. *Journal of Manufacturing Systems*, 10, (1), 79–93.

[13.23] Duffie, N., Chitturi, R., Mou, J., 1988, Fault Tolerant Heterarchical Control of Heterogeneous Manufacturing Entities. *Journal of Manufacturing Systems*, 7, (4).

[13.24] Duffie, N., And Piper, R., 1987, Non Hierarchical Control of A Flexible Manufacturing Cell. *Robotics And Computer Integrated Manufacturing*, 3,(2).

[13.25] Duffie, N., And Prabdu, V., 1994, Real Time Distributed Scheduling of Heterarchical Manufacturing Systems. *Journal of Manufacturing Systems*, 13, (2).

[13.26] Durna, M., Erkmen, I., Erkmen, A., 1998, Holonic Grasping, *IEEE International Conference on Intelligent Robots And Systems* (Victoria, Canada), 140–145.

[13.27] Gayed, N., Jarvis, D., And Jarvis, J., 1998, A Strategy For The Migration of Existing Manufacturing Systems To Holonic Systems, *IEEE International Conference on Systems, Man And Cybernetics* (San Diego), 319–324.

[13.28] Gou, L., Hasegawa, T., Luh, P., Tamura, S., And Oblak, J., 1994, Holonic Planning And Scheduling For A Robotic Assembly Testbed, *Proceedings of The 4th Rensselaer International Conference on Computer Integrated Manufacturing and Automation Technology*. (Rensselaer, NY).

[13.29] Gou, L., Luh, P., And Kyoya, Y., 1998, Holonic Manufacturing Scheduling: Architecture, Cooperation, Mechamism, And Implementation. *Computers in Industry*, 37, (3), 213–231.

[13.30] Hasegawa, T., Gou, L., Tamura, S., Luh, P., And Oblak, J., 1994, Holonic Planning And Scheduling Architecture For Manufacturing, *Proceedings of The International Working Conference on Cooperative Knowledge Based Systems* (Keele, UK) ,141–157.

[13.31] Heikkila, T., Jarviluoma, M., And Hasemann, J., 1995, *Holonic Control of a Manufacturing Robot Cell*, Internal Report, VTT Automation (Oulu, Finland).

[13.32] Heikkila, T., Jarviluoma, M., And Juntunen, T., 1997, Holonic Control For Manufacturing Systems: Design of A Manufacturing Robot Cell. *Integrated Computer Aided Engineering*, 4,,202–218.

[13.33] Hoitomt, D., Luh, P., And Pattipati, K., 1993, A Practical Approach To Job Shop Scheduling Problems. *IEEE Transactions on Robotics And Automation*, 9,(1).

[13.34] Honma, N., Abe, K., Sato, M., And Takeda, H., 1998, Adaptive Evolution of Holon Networks By An Autonomous Decentralized Method. *Applied Mathematics And Computation*, 91, (1), 43–91.

[13.35] Jennings, N. R., 1996, Coordination Techniques For Distributed Artificial Intelligence. G. M. P. Ohare And N. R. Jennings, Eds., *Foundations of Distributed Artificial Intelligence*, (Wiley), 429–447.

[13.36] Kawage, M. et al, 1994, Holonic Control of An Assembly Robot, *Annual Conference of Robotics Society of Japan* (Japan).

[13.37] Koestler, A., 1967, *The Ghost in The Machine* (Hutchinson).

[13.38] Kruth, J. P. D., J. Prianggada,I.T. Ginderachter, T Van. Wyns, J., 1996, An NC Holon Architecture, *Proceedings of CNMU96 Machine Tool National Conference* (Bucharest).

[13.39] Kusumi, N., Hirasawa, K., And Obayashi, M., 1998, Holonic Control System Based on A Universal Learning Network. *Electrical Engineering in Japan*, 124,(4),64–72.

[13.40] Langer, G.: *A Methodology And Architecture For Holonic Multi-Cell Control Systems*, PhD Thesis, Technical University of Denmark, (Lyngby, Denmark, 1999).

[13.41] Lin, G., And Solberg, J., 1992, Integrated Shop Floor Control Using Autonomous Agents. *IIE Transactions*, 24,(3),169–206.

[13.42] Lin, G., And Solberg, J., 1994, Autonomous Control For Open Manufacturing Systems. S. Joshi, Smith, J, Ed., *Computer Control of Flexible Manufacturing Systems*, (Chapman-Hall), London, 169–206.

[13.43] Marcus, A., Kis Vancza, T., And Monostori, L., 1996, A Market Approach To Holonic Manufacturing. *Annals of CIRP*, 45,(1),433–436.

[13.44] Marik, V, Pechoucek, M, (eds.), 2000, Proceedings of Holomas 2000 – First International Workshop on Industrial Applications of Holonic and Multi-Agent Systems, London.

[13.45] Matthews, J., 1995, Organisational Foundations of Intelligent Manufacturing Systems – The Holonic Viewpoint. *Computer Integrated Manufacturing Systems*, 8,(4),237–243.

[13.46] McFarlane, D., Marett, B., Elsley, G., And Jarvis, D., 1995, Application of Holonic Methodologies To Problem Diagnosis in A Steel Rod Mill, *Proceedings of IEEE Conference on Systems, Man And Cybernetics* (Vancouver).

[13.47] McFarlane, D., and Bussmann, S, 2000, State of the Art of Holonic Systems in Production Planning and Control, International Journal of Production Planning and Control, v11, n6, pp522–536.

[13.48] McFarlane, D. C., 1995, Holonic Manufacturing Systems in Continuous Processing: Concepts And Control Requirements, *Proceedings of ASI 95* (Portugal).

[13.49] MESA International, 1995, “Controls Definition & MES To Controls Data Flow Possibilities”, White Paper, www.mesa.org.

[13.50] Ng, A., Yeung, R., And Cheung, E., 1996, HSCS – The Design of A Holonic Shopfloor Control System, *Proceedings of IEEE Conference on Emerging Technologies And Factory Automation*,179–185.

[13.51] Overmars, A., And Toncich, D., 1996, Hybrid FMS Control Architectures Based on Holonic Principles. *International Journal of Flexible Manufacturing Systems*, 8,,263–278.

[13.52] Prosser, P., And Buchanan, I., 1994, Intelligent Scheduling: Past, Present And Future. *Intelligent Systems Engineering* (Summer Issue).

[13.53] Ramos, C., 1996, A Holonic Approach For Task Scheduling in Manufacturing Systems, *Proceedings of IEEE Conference on Robotics And Automation* (Minneapolis) , 2511–2516.

[13.54] Rannanjarvi, L, And Heikkila, T, 1998, Software Development For Holonic Manufacturing Systems. *Computers in Industry*, 37,(3),233–253.

[13.55] Saad, A., Kawamura, K., And Biswas, G., 1997, Performance Evaluation of Contract Net Based Hierarchical Scheduling For Flexible Manufacturing Systems, Internal Report, Center For Intelligent Systems, Vanderbilt University, (Nashville).

[13.56] Smith, R. G., 1980, The Contract Net Protocol: High-Level Communication And Control in A Distributed Problem Solver. *IEEE Transactions on Computers*, C-29,(12).

[13.57] Sousa, P., Ramos,C, 1998, Dynamic Scheduling Holon For Manufacturing Orders. *Journal of Intelligent Manufacturing*, 9,(2),107–112.

[13.58] Suda, 1989, Future Factory System Formulated in Japan. *Techno Japan*, 22,(10),15–25.

[13.59] Suda, 1990, Future Factory System Formulated in Japan (2). *Techno Japan*, 23,(3),51–61.

[13.60] Sugimura, N., Hiroi, M., Moriwaki, T., And Hozumi, K., 1996, A Study on Holonic Scheduling For Manufacturing System of Composite Parts, *Japan/USA Symposium on Flexible Manufacturing*, 1407–1410.

[13.61] Tanaya, P., Detand, J., And Kruth, J.-P., 1997, Holonic Machine Controller: A Study And Implementation of Holonic Behaviour To Current NC Controller. *Computers in Industry*, 33,,325–333.

[13.62] Tanaya, P., Detand, J., Kruth, J.-P., And Valkenaers, P., 1998, Object-Oriented Execution Model For A Machine Controller Holon. *European Journal of Control*, 4,(4),345–361.

[13.63] Tharumarajah, A., And Wells, A., 1996, Behaviour-Based Approach To Scheduling in Distributed Manufacturing Systems. *Integrated Computer-Aided Engineering* (Special Issue on Intelligent Manufacturing Systems).

[13.64] Thompson, D., And Hughes, D., 1996, Holonic Approach To CIM Architecture And Systems Design, *Proceedings of IEE Colloquium on Systems Approach To Manufacturing*, England.

[13.65] Valckenaers, P., Bonneville, F., Van Brussel, H., Wyns, J., 1994, Results of The Holonic Control System Benchmark At K U Leuven, *Proceedings of Computer Integrated Manufacturing and Automation Technology* (Troy, NY) ,128–133.

[13.66] Valckenaers, P., Van Brussel, H., Bongaerts, L., And Bonneville, F., 1995, Programming, Scheduling And Control of Flexible Assembly Systems. *Computers in Industry*, 26,,209–218.

[13.67] Van Brussel, H. V., P. Bongaerts, L. Wyns, J., 1995, Architectural And System Design Issues in Holonic Manufacturing Systems, *Proceedings of 3rd IFAC Workshop on Intelligent Manufacturing Systems* (Bucharest).

[13.68] Van Brussel, H. W., J. Valckenaers, P Bongaerts, L. Peeters P., 1998, Reference Architecture For Holonic Manufacturing Systems. *Computers in Industry*, 37,(3),255–274.

[13.69] Wyns, J., Ginderachter, T., Valckenaers, P., Van Brussel, H., 1997, Integration of Resource Allocation And Process Control in Holonic Manufacturing Systems, *29th CIRP International Seminar on Manufacturing Systems* (Osaka), p57–62.

[13.70] Wyns, J. B., Van. H. Valckenaers, P. Bongaerts, L., 1996, Workstation Architecture in Holonic Manufacturings Systems, *28th CIRP International Seminar in Manufacturing Systems* (Johannesburg).

[13.71] Zhang, X., And Norrie, D., 1999a, Holonic Control At The Production And Controller Levels, *Proceedings of IMS99* (Leuven, Belgium).

[13.72] Zhang, X., And Norrie, D. H., 1999b, Dynamic Reconfiguration of Holonic Lower Level Control, *Proceedings of IPMM '99* (Honolulu).

[13.73] Zweben, M., And Fox, M., 1994, ''Intelligent Scheduling.'' (Morgan Kaufman).

Printing: Mercedes-Druck, Berlin
Binding: Stein+Lehmann, Berlin